Synergic Influence of Gaseous, Particulate, and Biological Pollutants on Human Health

Synergic Influence of Gaseous, Particulate, and Biological Pollutants on Human Health

Editor

Jozef Stefan Pastuszka

Associate Professor of Environmental Engineering
Head of the Chair of Air Protection
Silesian University of Technology
Gliwice
Poland

CRC Press is an imprint of the
Taylor & Francis Group, an **informa** business
A SCIENCE PUBLISHERS BOOK

Cover illustrations reproduced by kind courtesy of Dr. Józef Pastuszka (editor of the book) and Prof. Ewa Talik (University of Silesia in Katowice, Poland).

CRC Press
Taylor & Francis Group
6000 Broken Sound Parkway NW, Suite 300
Boca Raton, FL 33487-2742

First issued in paperback 2019

ISBN-13: 978-1-4987-1511-9 (hbk)
ISBN-13: 978-0-367-37717-5 (pbk)

Library of Congress Cataloging-in-Publication Data

Synergic influence of gaseous, particulate, and biological pollutants on human health / editor, Jozef S. Pastuszka.
p. ; cm.
Includes bibliographical references and index.
ISBN 978-1-4987-1511-9 (hardcover : alk. paper)
I. Pastuszka, Jozef S. (Jozef Stefan), 1953- , editor.
[DNLM: 1. Air Pollutants--adverse effects. 2. Air Pollution--adverse effects. WA 754]

RA576
363.739'2--dc23 2015028950

Visit the Taylor & Francis Web site at
http://www.taylorandfrancis.com

and the CRC Press Web site at
http://www.crcpress.com

Preface

Airborne substances are of different physical forms and include gases, vapors and aerosols including bioaerosols. Their inhalation may cause adverse effects on human health. This book gives brief information about the important properties of these air pollutants, their environmental distributions, and examines the nature of these health hazards, as well as the attendant problems for science and society. The different roles and attributes of population-level and individual level studies are discussed critically. Many of the recent research methods, especially in such matters as exposure assessment and the handling of interactive effects, is described in detail.

Jozef Stefan Pastuszka

Contents

Introduction

The beginning of the 20th century was the time of rapidly growing manufactures and new industrial plants. They emitted gases and dust particles into the atmosphere. The first observations indicated that these gases and dust particles could probably have a negative influence on human health. Since that time our knowledge on the subject has evolved enormously and therefore should be conveyed in this book.

First, it was important to find which chemical substances were released into the air, and to estimate the amounts of these emitted pollutants. The next task was to study the chemical and physical properties of these pollutants and their transport in the environment. Finally, the adverse health effects caused by the exposure of human and animal organisms to these pollutants should be investigated. It took a relatively long time to discover that the adverse health effects depend on both: the toxicity of the pollutant closely connected with its chemical properties, and on the absorbed dose of the airborne substance. Since that moment, the general method of the prediction of the adverse health effects for various pollutants has quickly been developed.

This method can be divided into the qualitative and quantitative parts. The first part uses the available toxicological information of the pollutant while the second is based on the exposure assessment and the dose calculation. By the application of this method, the precise prognosis of the adverse health effects for a fast growing number of airborne pollutants could be prepared. Unfortunately, new environmental, epidemiological, and toxicological studies on air pollutants complicated this method again. The discovery of the new classes of air pollutants seems to be especially important: carcinogens and allergens cause quite different adverse health effects than "classical" toxins. For this reason, and also because of the studies on other, specific pollutants, such as bioaerosols, fibrous aerosols, and ultrafine particles, there still isn't one, homogenous method of prognosis of the adverse health effects for a population exposed to polluted air. In this book, we try to show the actual knowledge on this subject. It should be noted, however, that recently a critical approach to the analysis of the dose-effect relationship of only one, specific pollutant is becoming very popular. Inhalation of only one kind of pollutant is rather rare. Typically, various pollutants are inhaled together at the same time. Therefore, according to the new paradigm, the interactions between these pollutants should also be added into the total prognosis of the adverse health effects resulting from the exposure to a huge number of air pollutants. These problems still need future studies but certainly even now we should change our perspective and our understanding of air pollution. We try to face these problems in this book. We will be satisfied if we are able to encourage the readers' reflections, and maybe also their own studies in this field.

PART 1

Air Pollutants and Qualitative Analysis of their Adverse Health Effects

1

Environmental Characteristics of Gaseous Pollutants and Related Adverse Health Effects

Lucyna Falkowska

Introduction to air pollution

The atmosphere is a space surrounding the Earth, filled with gases, aerosols, water vapor, and tiny ice crystals, all of which participate in processes both far from and close to the surfaces of contact with land or oceans.

The concentration of chemical components in the atmosphere depends on the distance from the Earth surface and on physical conditions such as temperature, pressure, turbulent mixing and the quantity of solar radiation energy which reaches each layer of the atmosphere. They all perform important functions in the "Earth's life", but for humans and all plant and animal species, the most essential layer is the troposphere, the closest to the planet's surface (Sainfeld and Pandis 1998; Finlayson-Pitts and Pitts 2000; Falkowska and Lewandowska 2009). It is here that the highest concentration of gases and aerosols can be found, as well as almost all the humidity accumulated in clouds of various types. Here the turbulent mixing processes, caused by rapid changes of pressure and temperature, are very intense.

Despite its potentially stable nature, the atmosphere is undergoing gradual evolution. It is a dynamic system which is exposed to constant interactions with the hydrosphere, lithosphere, and biosphere. The processes of mass exchange which take place among them are decisive factors in the formation of principal components: gases and aerosols, which take part in the cycles of biochemical and geochemical elements.

University of Gdansk, Institute of Oceanography, Department of Marine Chemistry and Environmental Protection, 46 Aleja Piłsudskiego, 81-378 Gdynia, Poland.
E-mail: l.falkowska@ug.edu.pl; lucynafalkowska@gmail.com

Gases penetrate into the atmosphere as a result of biological and microbiological activity, volcanic eruptions, and radioactive disintegration where they undergo chemical transformations. Solar radiation is the initiator of the main chemical reactions, which are called photochemical reactions. As a result of these reactions, gas particles undergo dissociation, excitation, and ionization, and the energy is transferred to other molecules. Chemical reactions in the atmosphere take place during molecular or turbulent diffusion when particles collide. At such times new gases are formed, with different characteristics and different properties. As a consequence of such processes, enormous changes occur in gas particle concentrations in the atmosphere.

The time which particles spend in the atmosphere may vary, ranging from fractions of a second to millions of years. This depends both on the properties of gases and the kinetics of chemical and photochemical reactions. The removal of gases from the atmosphere is governed by chemical reactions and physical processes, which result in: molecules being converted into aerosol particles, washed out or assimilated and respired by land and sea organisms.

The main gas components of the air (nitrogen, oxygen, argon and water vapor) have been well-studied and described in detail in various handbooks (Graedel 1980; Monahan 1983; Holland 1984; Falkowska and Korzeniewski 1998; Seinfeld and Pandis 1998; Finlayson-Pitts and Pitts 2000; Falkowska and Lewandowska 2009). Their concentrations are stable and any possible fluctuations are minor. However, major trace gases (CO_2, CH_4, CO, NO, NO_2, N_2O, NH_3), minor trace gases (VOC) and trace metals (Hg) are characterized by great changeability in time and space. The present-day chemical composition of the atmosphere is quite different from the natural, which existed before the Industrial Revolution of the 18th century. Over urbanized areas, substances which are commonly referred as pollutants have been introduced into the air through man's industrial activity. They are to be found in trace amounts, which are nonetheless higher than the natural background.

Air pollution can be defined as follows: *air pollutants are solid, liquid, and gaseous substances which though present in the atmosphere, are alien to its natural composition; or are themselves natural substances occurring in excessive amounts, posing health threats to humans, harmful to plants and animals, and adversely affecting the climate and the utilization of certain elements of the environment.*

Gaseous pollutants of the atmosphere can be primary or secondary. The former are directly emitted from biological, geogenic, and anthropogenic sources, and include:

- carbon compounds—CO, CO_2, CH_4, VOC;
- nitrogen compounds—NO, N_2O, NH_3;
- sulphur compounds—H_2S, SO_2, $(CH_3)_2S$;
- mercury compounds—Hg^0, CH_3Hg.

Primary gaseous pollutants can have an adverse effect on the fauna and flora as well as on people. For instance, when leaves are exposed to 10 ppm NO, the process of photosynthesis is impaired. SO_2 damages both plants' respiratory pores and the protective layer of wax which covers needles (Monahan 1983). Sulphur compounds are responsible for the creation of reductive smog, which can be lethal to humans.

Respiration is the main way of introducing noxious gases into the system and the inhalation of CO reduces the amount of oxygen in the bloodstream. High concentrations can lead to headaches, dizziness, unconsciousness, and death.

Secondary pollutants are not emitted directly from land and ocean sources, but are formed in the atmosphere out of primary pollutants—their precursors. Major secondary pollutants include:

- ozone—created during photochemical reactions involving NO_x, CO, VOC,
- NO_2 and HNO_3—formed from NO,
- Both inorganic and organic aerosols, which are formed as a result of gas-to-particle reactions, and which contain primary gaseous pollutants and water vapor.

It follows that air pollutants are substances originating from both natural and anthropogenic sources (CO_2, CH_4, N_2O), as well as those emitted during fuel and biomass burning, which are harmful to all life forms. The former kind are classed as greenhouse gases and, despite the increase in their global concentrations, they only influence humans and the ecosystem by triggering changes in living conditions (temperature, precipitation, and weather). Other gases, such as nitrogen and sulphur oxides, VOC and ozone, have local or regional influence. They contribute to the formation of reductive (wintertime) smog or photochemical (summertime) smog, having an adverse effect on anthropogenic or natural non-living structures, and causing reduction in the air's visibility. It should be noted that short-term harmful effects may accumulate in the atmosphere to create a long-term harmful effect.

Carbon oxide

Carbon monoxide is one of the main reactive trace gases in the earth's atmosphere. Carbon oxide has both natural and anthropogenic sources on land and over water (Table 1-1). These include: fuel and biomass combustion and oxidation of methane and non-methane hydrocarbons (NMHC). Despite considerable uncertainty concerning the total volume of CO emissions from all sources, it is reckoned that about two thirds of the carbon oxide in the atmosphere results from anthropogenic activity, methane oxidation included (IPCC 1995). The Northern Hemisphere contains about twice as much CO as the Southern Hemisphere. Because of its relatively short lifetime and distinct emission patterns, CO has large gradients in the atmosphere, and its global burden of about 360 Tg is more uncertain than those of CH_4 or N_2O (IPCC 2007). Some natural CO sources are chlorophyll degradation, which takes place in autumn, and the decomposition of other organic non-vegetal matter. The gas is produced during fuel combustion, forest and savanna fires as well as during methane and NMHC oxidation.

Concentration of CO in the troposphere ranges from 40 to 200 ppb (45–229 $\mu g\ m^{-3}$), and its atmospheric residence time is between 30 and 90 days. As CO emission is closely related to incomplete combustion of fuels, its high concentrations—up to 50–100 ppm (572–1145 $\mu g\ m^{-3}$)—can occur in specific situations such as during traffic rush hours.

Table 1-1. Estimation of carbon oxide emission sources and sinks typical for the last decade of the 20th century (IPCC 1995).

Sources	CO [Tg yr^{-1}] (Tg = 10^{12} g)
Technological	300–550
Biomass combustion	300–700
Biogenic	60–160
Oceans	20–200
Methane oxidation	400–1000
NMHC oxidation	200–600
Total	1800–2700
Sink	CO [Tg yr^{-1}]
Reactions with OH·	1400–2600
Fixation in the soil	250–640
Diffusion into the stratosphere	~ 100
Total	2100–3000

The main processes of carbon monoxide removal are those of oxidation to carbon dioxide with radicals, oxygen, and ozone. However, the most effective are hydroxyl and hydroperoxyl radicals:

$$HO^{\cdot} + CO \rightarrow CO_2 + H \tag{1-1}$$

$$H + O_2 + M \rightarrow HO_2^{\cdot} + M \tag{1-2}$$

$$HO_2^{\cdot} + CO \rightarrow HO^{\cdot} + CO_2 \tag{1-3}$$

$$HO_2^{\cdot} + NO \rightarrow HO^{\cdot} + NO_2 \tag{1-4}$$

$$HO_2^{\cdot} + HO_2^{\cdot} \rightarrow H_2O_2 + O_2 \tag{1-5}$$

$$H_2O_2 + h\upsilon \rightarrow 2\ HO^{\cdot} \tag{1-6}$$

Beside radicals, CO removal reactions also involve methane and the formation of methyl radicals.

The fixation of carbon oxide in the soil involves the participation of microorganisms. It has been estimated that globally the process can result in $4.1 \cdot 10^{14}$ g CO per year.

A similar amount of CO is produced during fuel combustion ($6.4 \cdot 10^{14}$ g year^{-1}), and through methane oxidation by hydroxyl radicals ($4.0 \cdot 10^{14}$ g year^{-1}) (Bartholomew and Alexander 1981).

The 8-hour limit of CO concentration levels (10 mg m^{-3}) has been exceeded at 6 test stations located within the EU: four traffic stations, one urban background station and one industrial station, located in Italy, Bulgaria, and Bosnia & Herzegovina respectively (Mol et al. 2011).

In the USA, exceeding of CO concentration limits in the air haven't been reported for 20 years (9 ppm = 10.3 mg m^{-3}) but, both in the USA and the EU, a decreasing CO concentration trend can be observed (Fig. 1-1a,b).

Carbon oxide (CO) is a toxic gas which can cause fatal asphyxiation (http://cfpub.epa.gov/ncer_abstracts/ index.cfm). The presence of COHb in the blood is dependent on CO concentration in the air, duration of exposure, and type of physical human

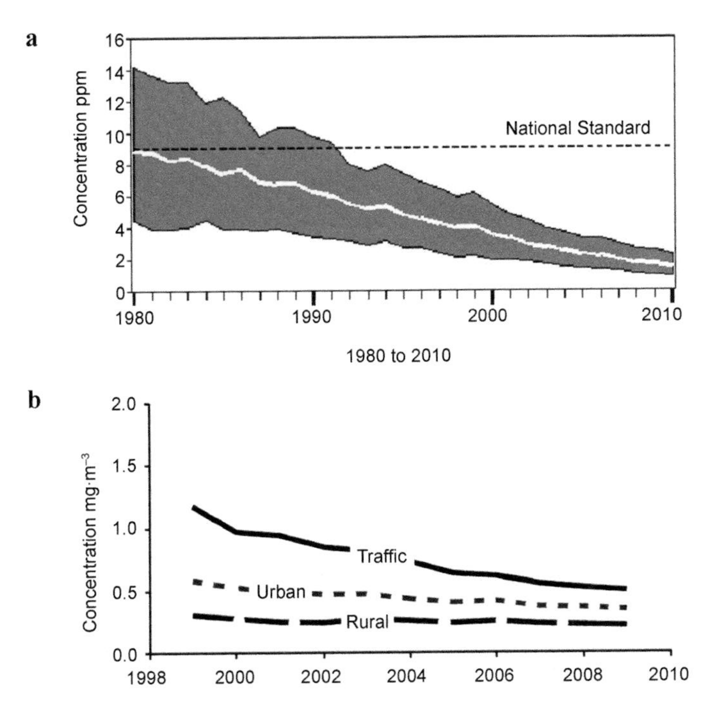

Fig. 1-1. The trends in the annual average CO concentration (a) 8-hour average based on 104 sites in US (http://www. epa.gov/air trends/carbon.html) (b) annual mean in EU at each station type (Mol et al. 2011).

activity. It was found that a healthy and relaxed adult, after 10 hours of exposure to the concentration of CO ~ 250 ppmv, has 30% COHb in the blood, and becomes unconscious while the hard-working adult attains this level three time faster (WHO 1979; Badr and Probert 1995). However, our knowledge of the effects of exposure to moderate doses of CO on health and human behavior is incomplete.

Nitrogen compounds

Nitrogen oxides come from natural emission sources (biochemical and photochemical transformations) as well as from anthropogenic activity. Nitrogen oxide is emitted into the atmosphere in much larger quantities than NO_2. Reactions which take place naturally do not lead to the formation of nitrogen oxides from molecular nitrogen but this can happen at high temperatures generated by, for example, solar radiation or electrostatic discharges. When exposed to high voltage, nitrogen changes into its allotropic form, which is very active chemically and reacts with oxygen to form NO. It is this form of nitrogen which is probably predominant during electric storms.

High temperatures generated by human activity (industrial furnaces and car engines) can lead to N_2 combining with O_2 to form nitrogen oxides (Falkowska and Korzeniewski 1998; Seinfeld and Pandis 1998; Finlayson-Pitts and Pitts 2000). The mechanism of their creation, however, is very complicated.

First, oxygen and nitrogen atoms are formed at a high temperature:

$$O_2 + M \rightarrow O + O + M \quad (1\text{-}7)$$

$$N_2 + M \rightarrow N + N + M \quad (1\text{-}8)$$

Breaking the O-O bonding requires 118 kcal mol^{-1} and in the case of nitrogen (N-N), as much as 225 kcal mol^{-1}. Thus the energy required for these reactions needs to be high in order to enable the formation of atomic oxygen and nitrogen, and to trigger the reaction chain which leads to NO formation:

$$N_2 + O \rightarrow NO + N \quad (1\text{-}9)$$

$$N + O_2 \rightarrow NO + O \quad (1\text{-}10)$$

Summary $N_2 + O_2 \rightarrow 2\ NO$

Reactions 1-9 and 1-10 initiate further transformations and lead to the formation of nitrogen dioxide (NO_2):

$$2\ NO + O_2 \rightarrow 2\ NO_2 \quad (1\text{-}11)$$

The range of processes which nitrogen oxides undergo in the atmosphere under the influence of sunlight and in dark phases of the day is presented in Fig. 1-2. Nitrogen dioxide (NO_2) is a very active gas which absorbs most of the UV light spectrum and the visible spectrum. Reaction 1-12, describing the photo-dissociation of nitrogen dioxide, occurs at a wavelength of 398 nm and takes place more rapidly as the light's angle of incidence increases. Over 430 nm, photo-dissociation does not occur, and the only products are excited molecules (Falkowska and Korzeniewski 1998; Seinfeld and Pandis 1998; Finlayson-Pitts and Pitts 2000).

$$NO_2 + h\nu \rightarrow O + NO \quad (1\text{-}12)$$

The photo-dissociation reaction takes place continually during the day and leads to a considerable increase in nitrogen oxide concentrations. As a result of photo-dissociation, other reactions are initiated, possibly involving other nitrogen compounds, oxygen, ozone, and VOC. This is because solar radiation provides enough energy to break the bonds in both nitrogen oxides and volatile hydrocarbons.

Nitrogen dioxide is removed from the atmosphere through oxidation reactions, leading to the formation of nitrogen salts, organic nitrogen or nitric acid fumes. Droplets and fumes of nitric acid are removed from the atmosphere with rain, snowfall, or aerosol dry rain.

The biomass burning also enhances the biogenic emissions of nitric oxide and nitrous oxide from soil. It is believed that these emissions are related to increased concentrations of ammonium found in soil following burning. Ammonium, a major nitrogen component of the burn ash, is the substrate in nitrification, which is the microbial process responsible for the production of nitric oxide and nitrous oxide. The enhanced biogenic soil emissions of nitric oxide and nitrous oxide may be comparable to or even surpass the instantaneous production of these gases during biomass burning. Nitrogen oxides of natural origin are distributed fairly evenly over the globe, but the anthropogenic emission sources contribute considerably more and have a great effect both regionally and locally. Human activity is responsible for releasing into the

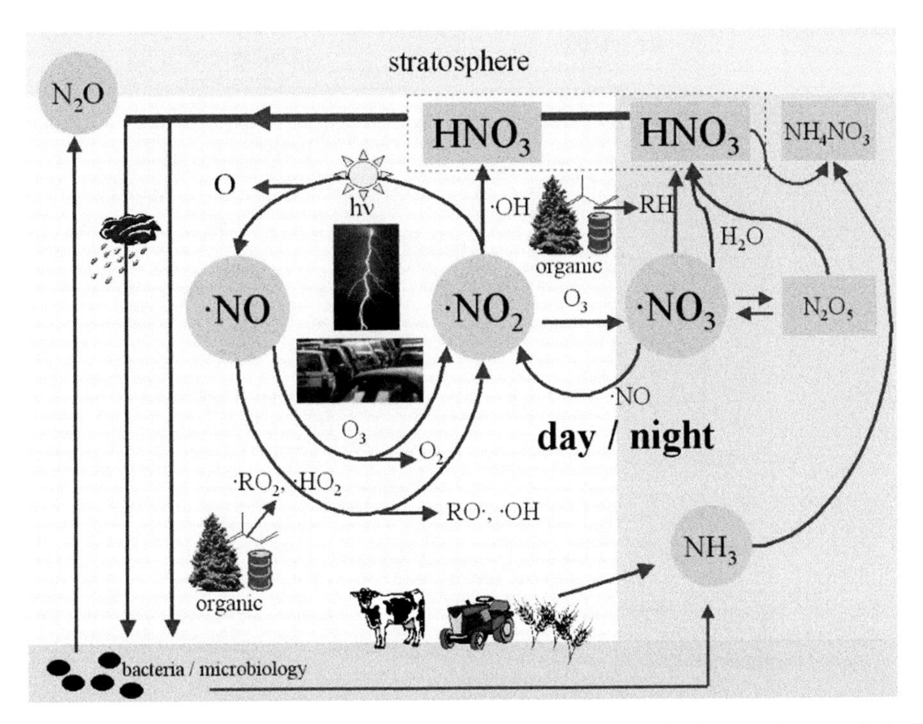

Fig. 1-2. Transformation range of nitrogen oxides originating from natural and anthropogenic sources both in the light and the dark phase of the day (Elmar Uherek, www.atmosphere.mpg.de).

atmosphere approximately $50 \cdot 10^7$ tones of nitrogen per year. Among all sources of NO_x ($NO_x = NO + NO_2$) in the atmosphere, arable land (5.6 TgN year^{-1}), electrostatic discharges (5 TgN year^{-1}), and shipping (3 TgN year^{-1}), are together responsible for only about half the amount that is produced through fuel combustion (30 TgN year^{-1}) (Evans 2006). Nitrogen oxides of anthropogenic origin accumulate particularly in urbanized and industrial areas, where their concentrations can be 10–100 times higher than elsewhere. Nitrogen oxide emission is also associated with density of population, road traffic, energy production, and waste removal.

Until recently it was believed that combustion resulted mostly in nitrogen oxide (Fig. 1-3) and that in urbanized areas the result of the reaction of NO with O_3 was a form of NO_2 (Fig. 1-2). However, on the basis of long-term observations carried out as part of an atmosphere monitoring scheme (Jenkin 2004a,b), it turned out that direct NO_2 emission is also significant (the proportion of NO_x in exhaust fumes can be up to 20%) and its significance increases considerably as a result of fitting catalytic converters in diesel engines. This is noted mainly in the atmosphere of large polluted cities, where a large number of personal and heavy goods vehicles are equipped with catalytic converters (Carslaw 2005).

In addition to NO_2, HONO (nitrous acid) is also directly emitted with car exhaust fumes to the amount of approximately 0.5–1% NO_x (Kurtenbach et al. 2001; Gutzwiller et al. 2002). Taking into consideration the rapid photolysis of HONO, this can lead locally to the oxidation of volatile organic compounds (VOC) and constitute

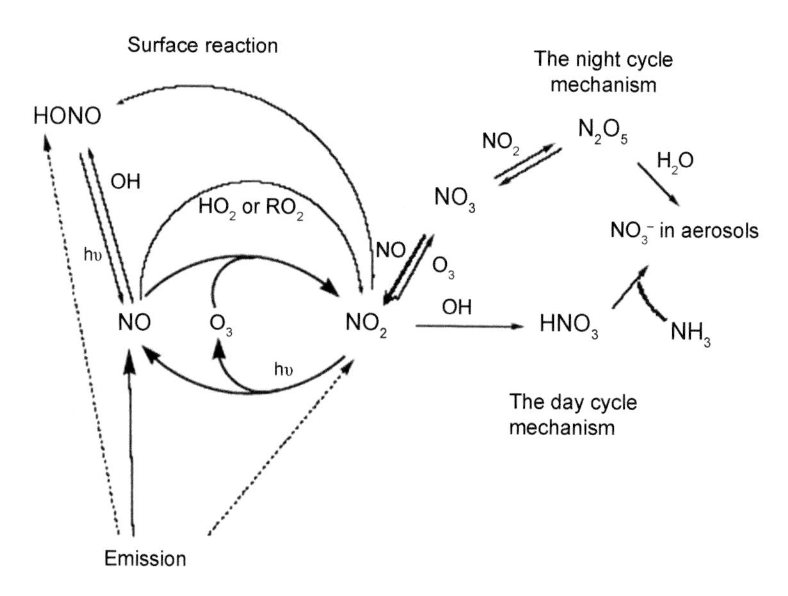

Fig. 1-3. Chemical and photochemical reactions of NO_y in an urbanized atmosphere. The graph shows transformations of NO into NO_2 and subsequent NO_2 oxidation into inorganic nitrogen forms in aerosols (adapted from Jenkin 2006).

an additional source of NO, which then undergoes oxidation to NO_2. Considerable amounts of NO_2 are formed from HONO as a result of photosensitive reactions on the surface of aerosols which have organic compounds in their composition (George et al. 2005).

As well as diurnal variation, related to photochemical transformations, nitrogen oxides exhibit seasonal variation. This is most evident in the temperate zone, where higher demand for energy during the winter season leads to increased NO_x emission from fuel combustion. Significant seasonal variations in NO_2 concentration, with winter maximum and summer minimum, may be observed in the Baltic countries. In winter, the atmospheric residence time is longer due to low photochemical activity and reduced vertical mixing (Bartnicki et al. 2008).

Nitrogen oxides originating from traffic have a significant influence on the quality of air in cities. The importance of mobile sources has been noted both in Poland and in many other European countries. The volume of nitrogen oxide emissions from vehicles is almost equal to that of large, constant point sources such as energy plants and industrial furnaces (Fig. 1-4).

In large European cities, annual mean concentrations of nitrogen oxides can fall within a range from 20–30 mg m^{-3} or else exceed 40 mg m^{-3} (Fig. 1-5). Most large Polish cities (Warsaw, Tricity, Poznan, Wroclaw, and the Silesian Agglomeration), as well the capital cities of Germany, Portugal, and Denmark, are less affected by the problem of nitrogen oxide pollution than other urban centers in Europe. This is mostly due to a smaller number of vehicles, as well as to their climatic and geographic conditions. The air of London, Rome, and Madrid, cities inhabited by millions of people, is characterized by particularly high NO_x concentrations.

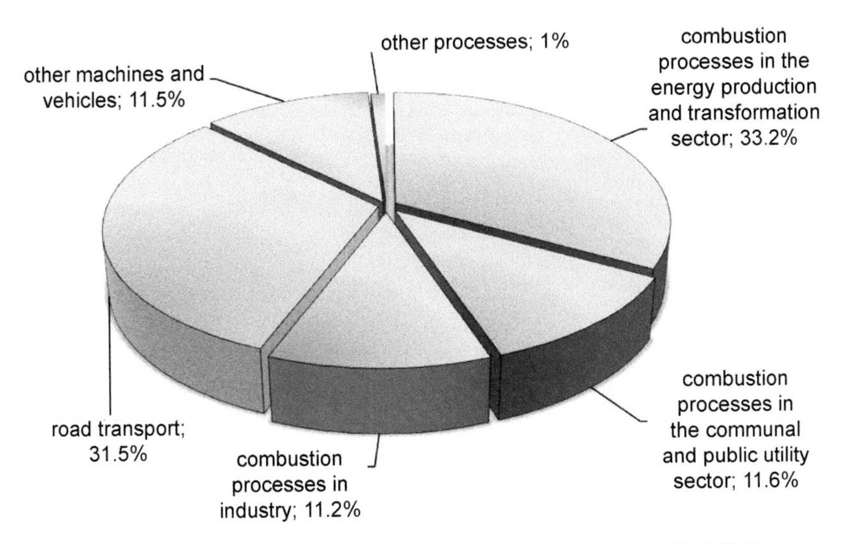

Fig. 1-4. Emission of NO_2 in Poland in 2009–2010, by the main sectors (EPI 2012).

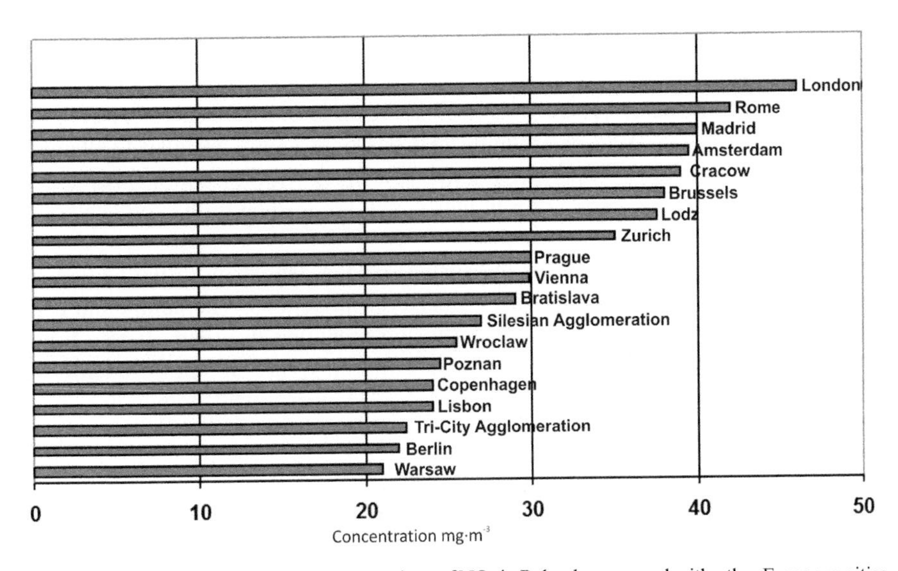

Fig. 1-5. Variations in annual mean concentrations of NO_x in Poland, compared with other European cities (EPI 2003).

In the United States of America, NO_2 concentrations in the air are below the National Standard (0.1 mg m^{-3}) and show a constant downward trend (Fig. 1-6). Between 1980 and 2010, a 52% drop was observed in the annual mean concentration of nitrogen dioxide.

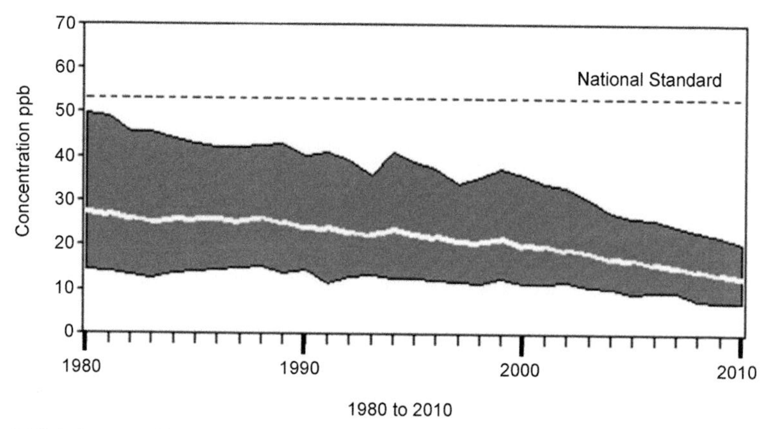

Fig. 1-6. Variations trend in the annual arithmetic average of nitrogen oxides in the USA, determined on the basis of measurements carried out in 1980–2010 at 81 test stations (http://www.epa.gov/airtrends/nitrogen.html).

Nitrogen oxides are characterized by various levels of toxicity. For example, NO_2 (brown, with a strong, acrid smell) is four times more toxic than the colorless, odorless NO. These compounds are collectively marked as NO_x due to the fact that one is easily transformed into another. They are responsible for acid rains and are the main cause of photochemical smog as well as the formation of highly toxic secondary pollutants, such as ozone or aromatic hydrocarbons. They are also to blame for corroding stone buildings and metal structures. Like NO, they are the precursors of carcinogenic and mutagenic nitrosamines which are formed in the soil (Falkowska and Korzeniewski 1998).

The toxicity of NO_2 consists in causing oxygen deficiency in the body, which in turn results in reduced immunity to bacterial infections. The compound causes irritation of the eyes and the airways, causing breathing disorders. It also triggers allergies, including asthma (especially in children living in smog-ridden cities). Prolonged exposure to nitrogen dioxide may lead to biochemical, immunological, and morphological changes in animals (EEA 1997). Owing to its adverse effect on human health, nitrogen dioxide is classed as a dangerous compound and concentration limits have been determined for it in the air (Table 1-2). It has also been taken into account in the air quality index (EEA 1997).

Table 1-2. Limit and threshold values for NO_2 and NO_X as set out in the 2008 Air Quality Directive (EEA 2011).

Objective	Averaging period	Limit or threshold value [μg m^3]	Number of permissible exceeds
Human health	One hour	200	18 hours per year
Human health	Calendar year	40	
Alert (a)	One hour	400	
Vegetation (b)	Calendar year	30	

Symbols: (a) To be measured over three consecutive hours at locations representative of air quality over at least 100 km^2 or an entire zone or agglomeration, whichever is smaller.
(b) As oxides of nitrogen

Ozone

Most chemical processes occurring in the atmosphere are oxidation processes. Among trace gases, ozone, ubiquitous in large cities and industrial areas, is a very strong oxidant. It is often called "bad ozone" as it is to be found close to the Earth's surface and causes severe irritation of the respiratory system. In healthy individuals, even a small quantity of ozone close to the Earth's surface can trigger soreness of the throat, chest pain, coughing, nausea, and obstructions in blood circulation. People who are ill or sensitive can suffer much worse consequences of breathing air with ozone such as bronchitis, emphysema, or heart attack (Bascom et al. 1996). Tropospheric ozone is also harmful, even toxic, to plants. It causes the decay of the assimilation apparatus, damages leaves, and diminishes harvests of cultivable plants and ornamental flowers, as well as biomass production (Lefohn 1997). In the USA, it is estimated that 90% reduction in farm produce is caused by air pollution. In addition to the aforementioned consequences, the presence of ozone in the troposphere has at least two more serious effects.

First of all, ozone is a greenhouse gas and absorbs UV radiation, particularly in the upper part of the troposphere where positive radiative forcing amounts to between 0.2 and 0.35 W m^{-2} (Houghton et al. 1996). This can account for 10–25% of radiative forcing caused by carbon dioxide.

Secondly, tropospheric ozone participates in the production of hydroxyl radicals. Since ozone is a potent oxidant, it contributes to the formation of another extremely powerful oxidant, which is predominant in reactions with methane and other gases (Warneck 1988). Conducive conditions for this can be found in the tropical troposphere. Strong solar radiation and very high air humidity are favorable to the formation of radicals from ozone which undergoes photolysis (1-13, 1-14).

$$O_3 + hv\ (\lambda \leq 1180\ \text{nm}) \rightarrow O + O_2 \qquad (1\text{-}13)$$

$$O_3 + hv\ (\lambda \leq 320\ \text{nm}) \rightarrow O^* + O_2 \qquad (1\text{-}14)$$

At short light wavelengths, the lysis of an ozone molecule leads to the creation of very reactive excited forms of oxygen. Their atmospheric residence time is very short, as they soon return to their basic state as a result of colliding with other molecules, but a small proportion of them react with water vapor:

$$O^* + H_2O \rightarrow 2HO^{\bullet} \qquad (1\text{-}15)$$

In effect, free hydroxyl radicals are formed, whose function in the troposphere is that of detergents aiding atmospheric self-purification by removing volatile organic compounds. In this way, ozone indirectly contributes to the eradication of hydro chlorofluorocarbons (HCFCs), which are substitutes of CFCs in the destruction of stratospheric ozone. This purifying effect reduces the volume of potentially harmful gases reaching the stratosphere, at other times preventing it altogether. Figure 1-7 shows a simplified cycle of tropospheric ozone. The only external source of ozone in the atmosphere is its import from the stratosphere in medium and high latitudes, which is related to polar jet stream currents. Other sources are located on land and are connected to human activity.

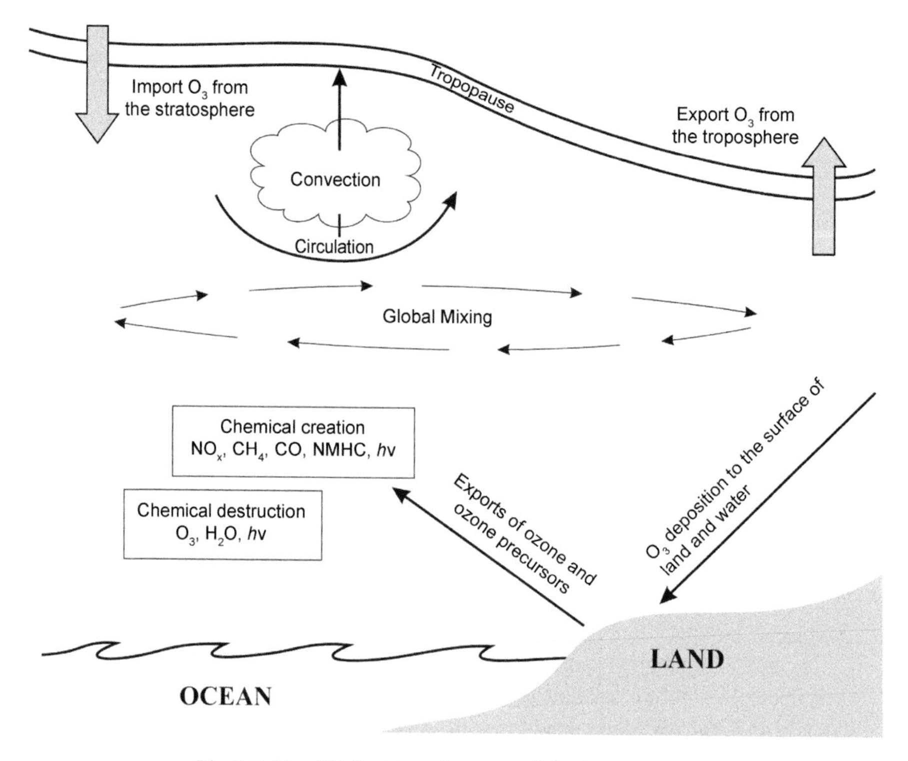

Fig. 1-7. Simplified pattern of ozone cycle in the troposphere.

The main precursors of ozone in the troposphere are nitrogen oxides, carbon oxide, methane, and other non-methane hydrocarbons (NMHCs). They are all products of fuel and biomass combustion, but each of them also has an important source in the biosphere (Crutzen and Andreae 1990; Jacob et al. 1996). The anthropogenic contribution to the supply of tropospheric ozone results from additional nitrogen oxides formed during the combustion of various fuels. In the intertropical area, biomass combustion plays a significant role in the production of nitrogen oxides, especially in the dry season, as huge areas of savannas and tropical forests are consumed by fire, often through human activity (http://rapidfire.sci.gsfc.nasa.gov/firemaps). These sources of NO_x (NO + NO_2) have become predominant, including combustion products released from aircraft engines in the upper part of the troposphere. NO and NO_2 undergo multiple reactions, leading to the creation of nitric acid fumes and nitrate aerosols which are then washed out with precipitation. Another gas closely related to ozone production is carbon oxide.

The creation of ozone as a result of photochemical oxidation of carbon oxide takes place with nitrogen oxides present at concentrations of > 10 pptv. Under these conditions, a sequence of reactions takes place:

$$CO + HO^{\bullet} \rightarrow CO_2 + H \quad (1\text{-}16)$$

$$H + O_2^{\bullet} + M \rightarrow HOO^{\bullet} + M \quad (1\text{-}17)$$

$HOO^{\bullet} + NO \rightarrow HO^{\bullet} + NO_2$ (1-18)

$NO_2 + h\upsilon \rightarrow NO + O\ (\lambda < 400\ nm)$ (1-19)

$O + O_2 + M \rightarrow O_3 + M$ (1-20)

Summary $CO + 2O_2 \rightarrow CO_2 + O_3$

Transformations taking place between the hydroxyl radical and the peroxy radical happen within a matter of seconds, and are controlled by reaction 1-18. This cycle of changes leads to the formation of atomic oxygen and, subsequently, ozone. It follows from the above equations that out of 1 particle of CO, 1 molecule of ozone is formed.

When nitrogen oxide concentrations are below the critical value of 10 pptv, the reactions of carbon oxide with radicals do not result in the formation of ozone, but instead to its destruction:

$CO + HO^{\bullet} \rightarrow CO_2 + H$ (1-21)

$H + O_2 + M \rightarrow HOO^{\bullet} + M$ (1-22)

$HOO^{\bullet} + O_3 \rightarrow HO^{\bullet} + 2O_2$ (1-23)

Summary $CO + O_3 \rightarrow CO_2 + O_2$

It follows from the sequence of reactions 1-21–1-23 that oxidation of one molecule of CO requires one molecule of ozone. Scientists have no doubts that the rise in ozone concentration results from increased NO_x emission. According to forecasts, in 2020 the emission of NO_x may have grown by 350% since 1990. Therefore, unless some serious action is undertaken to control the release of pollutants into the atmosphere, one has to assume that the levels of tropospheric ozone will also increase (van Aardenne et al. 1999).

Ozone concentration levels in Europe and the USA are similar, the only difference being that in Europe, ozone concentration does not show a statistically significant downward trend (www.eea.europa.eu). In the USA, a decreasing tendency was observed (a 28% decrease in National Average) between 1980 and 2010 (Fig. 1-8). Ozone pollution has to be considered a serious problem when assessing air quality

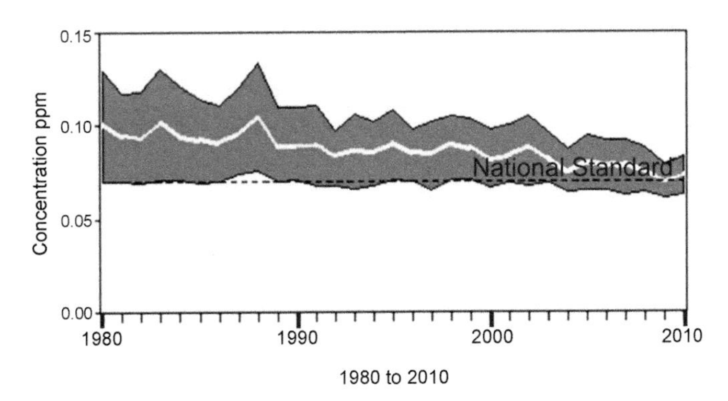

Fig. 1-8. Variations in the annual 4th maximum 8-hour average in ozone concentration observed in the USA between 1980 and 2010 at 247 stations (www.epa.gov/airtrends/ozone.html).

and, while the WHO has indicated 0.08 ppm as the air quality standard, five out of 1560 test stations have noted exceeds of the value of 180 $\mu g\ m^{-3}$ (0.09 ppm) in which the excessive values persisted for about one hour (www. eea.europa.eu).

Sulphur dioxide

Sulphur compounds in the atmosphere come from natural on-land and sea sources as well from anthropogenic sources. On land, volcanic eruptions and organic matter decomposition processes constitute a very rich source of the gaseous forms of this element in the atmosphere (Andreae and Crutzen 1997). After the eruption of the Tambora volcano in Indonesia, 50 Tg of S was released into the atmosphere and for the following two years (1815 and 1816) England, USA and Canada had no summer. Huge amounts of sulphur in aerosols are supplied into the atmosphere by seas and oceans and, in addition to this, gaseous sulphur compounds are released following a biochemical transformation in the sea water. One of these is dimethyl sulphide $(CH_3)_2S$ which, after several stages of oxidation in the atmosphere, first to SO_2, then to SO_4^{2-}, becomes a precursor of new particles.

Sulphur dioxide is an air pollutant regarded as being a key indicator of overall air quality. In chemical transformations in the atmosphere, all reduced forms of sulphur of both natural and anthropogenic origin undergo oxidation to one gaseous compound—sulphur dioxide. The continental background for sulphur dioxide concentration remains at 20 ppt to 1 ppb, but in the marine boundary layer, SO_2 is found at 20 to 50 ppt. In a polluted atmosphere, sulphur dioxide stays on the level of several hundred parts per billion (Berresheim et al. 1995).

Sulphur dioxide reactions in the atmosphere take place under the influence of physical factors, the most important of which are temperature, humidity, sunlight, wind speed, and the composition and direction of incoming air masses. Solar radiation triggers photochemical reactions with the participation of SO_2. In the lower part of the atmosphere, reached only by a small fraction of UV radiation, ~ 384 nm light in the UV-A region is capable of exciting the triplet state in a particle:

$$SO_2 + h\upsilon \rightarrow {}^3SO_2 \qquad (1\text{-}24)$$

Solar radiation of higher energy in the UV-B range (~ 294 nm) is capable of provoking the singlet state:

$$SO_2 + h\upsilon \rightarrow {}^1SO_2 \qquad (1\text{-}25)$$

During the dark phase of the day, clean air contains only a few particles of sulphur dioxide per million, and oxidation reactions of SO_2 to SO_3 and the consequent formation of sulphuric acid fumes takes place very slowly. In sunlight, on the other hand, within the concentration range of 5–30 ppm and at relative humidity of 32–91%, the formation of H_2SO_4 removes SO_2 at the pace of about 0.1% per hour (Finlayson-Pitts and Pitts 2000). The summary reaction can be presented as follows:

$$SO_2 + \tfrac{1}{2} O_2 + H_2O \rightarrow H_2SO_4 \qquad (1\text{-}26)$$

The reaction is in fact more complicated and takes place in many stages with the presence of other molecules. In the troposphere, sulphur dioxide oxidation takes place faster in the presence of hydrocarbons and nitrogen oxides, gaseous components which are fixed elements of a polluted atmosphere and together take part in the creation of photochemical smog (see the Section "Smog" of this Chapter). Large amounts of oxidants which accompany smog accelerate sulphur dioxide oxidation by up to even 10% per hour.

Some of the most important reactions taking place with the participation of SO_2 in the atmosphere are reactions in the clouds, accompanied by liquefied aerosols. They take place much faster there than in dry and clean air. Only about 20% of SO_2 in the atmosphere is oxidized during a homorganic reaction in the gaseous phase; the remaining 80% undergoes oxidation during heterogenic reactions in the liquid phase (Möller 1980). It is possible to distinguish two main ways through which S(IV) is converted to S(VI): the first one involves a direct inclusion of SO_2 in liquefied aerosols and cloud droplets, which results in the formation of sulphuric acid; the second one involves oxidation in the gaseous phase through reactions with hydroxyl radicals and the creation of sulphuric acid fumes, which condense on surrounding aerosols. If aerosol concentration is very low, a reaction with water vapor and ammonium can take place, leading to the creation of a new aerosol particle.

The inclusion of sulphur in liquefied aerosols in the presence of H_2O_2 is the most effective process of S(IV) oxidation to S(VI), which is proven in the studies by Hegg (1985) and by Langner and Rodhe (1991). In these cases, the authors showed that a rapid SO_2 reaction with H_2O_2, taking place in droplets, is globally responsible for about 80% of sulphates formed in clouds. When hydrogen peroxide runs out or when pH rises in cloud droplets, the role of the SO_2 oxidant in the liquid phase can be taken over by ozone, whereas in the marine boundary layer (MBL), the role of an oxidant is performed by sea aerosols.

The oxidation of sulphur dioxide in the liquid phase takes place relatively fast in the presence of ammonium:

$$NH_3 + SO_2 + H_2O \rightarrow NH_4^+ + HSO_3^- \quad (1\text{-}27)$$

$$NH_3 + HSO_3^- \rightarrow NH_4^+ + SO_3^{2-} \quad (1\text{-}28)$$

The products of reactions 1-27 and 1-28 are hydrogen sulphite and sulphite ions. Ammonium is the catalyst of both reactions.

SO_2 oxidation reactions in water solutions play an important role in the troposphere, where nearly all of the water vapor resides, whereas outside the troposphere oxidation takes place mostly in the gaseous form.

Some of the most active oxidants are hydroxyl radicals, which can participate in the following reactions:

$$HO^\bullet + SO_2 + M \rightarrow HOSO_2^\bullet + M \quad (1\text{-}29)$$

$$HOSO_2^\bullet + O_2 \rightarrow HOSO_2O_2^\bullet \quad (1\text{-}30a)$$

$$\rightarrow HO_2^\bullet + SO_3 \quad (1\text{-}30b)$$

The free radical can subsequently react with nitrogen oxide, oxidizing it to nitrogen dioxide:

$$HOSO_2O_2^{\cdot} + NO \rightarrow HOSO_2O^{\cdot} + NO_2 \quad (1\text{-}31)$$

or, the newly-formed sulphur trioxide, in the presence of water and the M-particle, is quickly converted to sulphuric acid:

$$SO_3 + H_2O + M \rightarrow H_2SO_4 + M \quad (1\text{-}32)$$

The presence of high SO_2 concentrations in the atmosphere has a negative effect on plants and humans. Its' direct influence manifests itself in damaged leaves and needles, while the protective wax layer on needles also becomes impaired and membranes and respiratory pores are destroyed, disrupting respiration and water balance. An indirect influence of SO_2 on plants results from increased acidity of rain and soil. Various species of trees, bushes, and forest undergrowth have different levels of sensitivity to chemical pollutants in the air but, generally speaking, deciduous plants are less sensitive than coniferous plants. This is mainly because in moderate climate zones, plants shed leaves every year, meaning that they are not exposed to the harmful influence of SO_2 for as long as needles. Constant exposure of leaves to SO_2 causes *Chlorosis*, the yellowing or even whitening of normally green parts of leaves or stems. Plants are most vulnerable to the effect of sulphur dioxide during the day, when the gas is able to penetrate into the tissue through open pores.

The presence of sulphur dioxide in the air is also harmful to humans. It causes breathing difficulties, which particularly affect individuals suffering from chronic diseases of the respiratory system, such as asthma. Death occurs at a SO_2 concentration of 500 ppm, but laboratory tests on animals showed no adverse effects whatsoever at 5 ppm SO_2 (Monahan 1983).

Sulphur dioxide was at least partly responsible for several serious incidents related to air pollution. One of them took place in December 1930 in Belgium, in the Meuse river valley, when a thermal inversion trapped fumes from industrial sources in a small area. Sulphur dioxide reached 38 ppm. About 60 people and large numbers of cattle died in the incident. In October 1948, a similar situation made 40% of the population of Donor, Pennsylvania ill, resulting in the demise of 20 people. SO_2 concentration was at 2 ppm. In December 1952 in London, during a period of thermal inversion and fog lasting five days, the average death rate was exceeded by about 3500–4000 deaths. People had difficulty breathing at a SO_2 concentration of 1.3 ppm (Monahan 1983).

Many countries have concentration norms for substances considered to be pollutants (Table 1-3) and the WHO quotes 0.13 ppm (212 $\mu g\ m^{-3}$) as the air quality standard. Poland's norms are being adapted to those of the EU, although in many respects it falls into the category of countries whose air pollution norms are very low.

At the end of the 20th century, most European regions experienced a marked decrease in SO_2 emissions. The total reduction reached nearly 70%, but there is wide diversity between the many countries and regions of Europe. In USA, in the last 30 years, the mean average concentration of sulphur dioxide has dropped by 83% and is about ten times lower than the national standard (Fig. 1-9). While Europe and North America were the 20th century "leaders" in terms of sulphur emission into the atmosphere, nowadays it is Asia. This is largely due to the rapid development

Table 1-3. Sulphur dioxide concentration standards [μg m^{-3}] in different measurement periods (Ordinance of the Minister of Environmental Protection, Natural Resources and Forestry of 18.09.1998, http://www.ciop.pl/3589.htmlhttp:/ /www.ellaz.pl/polska/ksia-powietrze.htm#powietrze7).

Compound	Period of Measurements	Poland	EU	Germany	US
SO_2	30 min	500	350 (since 2005)	400 (3 h)	650 (1 h)
	24 h	150 125 (since 2005)	250–350 125 (since 2005)	140	365
	Year	40 30 (since 2005)	8–20 (2 years after the enforcement of the Directive - 20)	60	80

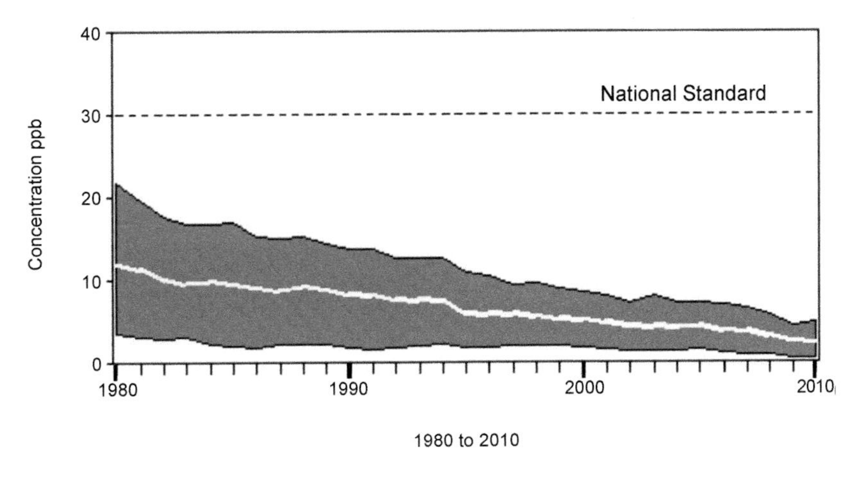

1980 to 2010

Fig. 1-9. Variations in the annual arithmetic average of sulphur dioxide concentrations, indicating an 83% decrease during the period 1980–2010, on the basis of measurements carried out at 12 stations in the USA (available http://www.epa.gov/airtrends/sulfur.html).

of Chinese economy, in which increased utilization of rich hard coal resources as a source of energy has resulted in very high SO_2 emissions. According to Akimoto et al. (1994), in the South-West Pacific region, at the end of the 1980s, 85% of sulphur was introduced into the atmosphere by China (nearly 10 Tg(S) $year^{-1}$).

ROS—Hydrogen peroxide (H_2O_2)

Once an oxygen particle, in itself harmless and indispensable for most living organisms, enters into metabolic reactions, it becomes the main precursor of reactive oxygen species—ROS. These include not only free oxygen radicals and its non-radical derivatives, but also other non-radicals, which are oxidants and/or are easily converted into radicals (HOCl, HOBr, O_3, $ONOO_2$, $^1O_2$, and H_2O_2) (Halliwell and Gutteridge 2006).

A key role in the formation of ROS in the environment is played by photochemical processes (Fig. 1-10). ROS are created in the atmosphere and surface sea waters under the influence of UV radiation.

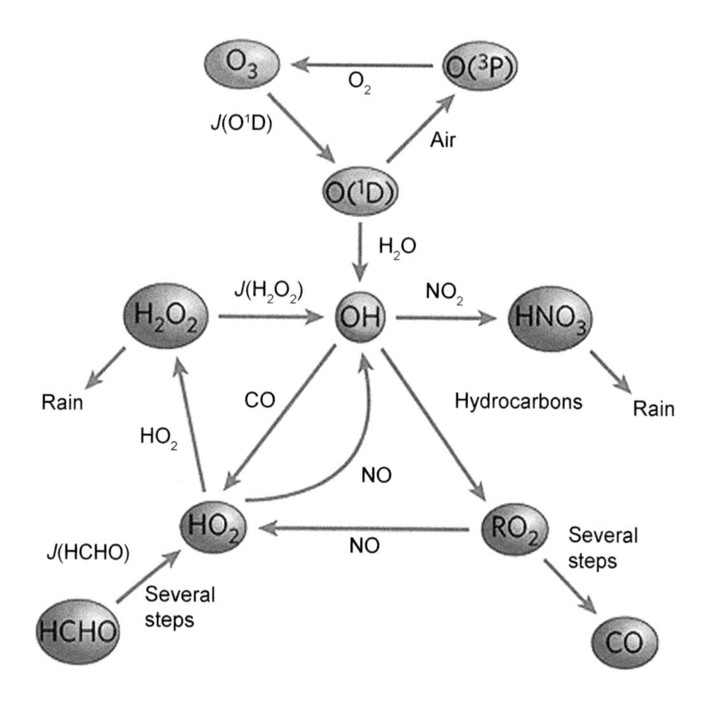

Fig. 1-10. The formation and chemistry of atmospheric hydroxyl radicals (Wennberg 2006) (available http://www.nature.com/nature/journal/v442/n7099/fig_tab/442145a_F1.html).

The main route for hydroxyl-radical (OH) generation is shown in the green ovals (Fig. 1-10). Ozone is broken up by ultraviolet light to form atomic oxygen in the ^1D excited electronic states; this occurs with a frequency denoted by $J(O^1D)$. The excited oxygen atoms react with water vapor to form OH radicals. Alternatively, the excited atoms may be recycled back to ozone, via the ^3P electronic ground state. The concentration of OH is usually less than 1 part per trillion per volume of gas. This reflects the rate of OH formation as described above; the rate of OH losses in reactions with, for example, nitrogen dioxide (NO_2), carbon monoxide (CO), and hydrocarbons such as methane; and the efficiency with which OH is recycled from its photochemical siblings HO_2 and RO_2 (where R is any hydrocarbon chain) through reactions with the common pollutant nitric oxide (NO). Recycling with NO is the mechanism responsible for the formation of ozone in the lower atmosphere. $J(H_2O_2)$ and J(HCHO) are the frequencies with which the respective molecules are fragmented by ultraviolet radiation (Wennberg 2006).

The products of oxygen excitation in the triplet state are O, O_3, $^1O_2$, and the products of its reduction are neutral compounds (H_2O_2) and radicals ($O_2^{\cdot}$, $HO^{\cdot}$) (Aguirre et al. 2005; Wennberg 2006).

ROS vary in terms of chemical properties: life span, diffusion ability, and chemical reactivity. Strong oxidants in photochemical reactions, such as O_3, H_2O_2, and $HO^{\cdot}$, are an important component of air pollution because they participate directly in the sequence of oxidation reactions of S, N, C, and halide compounds. The production of ROS is determined by meteorological and climatic conditions in a given area and, because of this, ROS are believed to potentially have a toxic effect on human health, plants, and crops at much lower concentrations than their precursors. H_2O_2 and organic peroxides (ROOH) are key components of the photo oxidation of volatile organic compounds (VOC). According to scientists, the devastation and reduction of forest areas in North America, Central Europe, and some higher parts of European mountains caused by peroxides are more significant and more threatening than the ones caused by acid rain deposition (Martin et al. 1997; Deng and Zuo 1999). However, the potential harmful effect of atmospheric H_2O_2 or other ROS has not been demonstrated yet, and it remains to be a debated subject (Lee et al. 2000).

One of the key roles of hydrogen peroxide in the atmosphere is sulphur dioxide (IV) oxidation to sulphuric acid (VI), a compound which determines the acidity of clouds, fog, and wet precipitation. It is believed that among ROS, H_2O_2 is the most efficient reagent in this reaction (Brandt and Eldik 1995; Deng and Zuo 1999). Because of this, H_2O_2 can be considered a purity indicator for air masses (Martin et al. 1997). Kelly and coworkers (1985) discovered that SO_2 conversion to H_2SO_4 with H_2O_2 participation takes place very rapidly in a highly acidic environment ($pH < 4.5$), while at $pH > 5$ the role of the main SO_2 oxidant in the atmosphere is taken over by O_3. The oxidation of SO_2 by other oxidants, such as O_3 and O_2 in the presence of transition metals ions (Fe and Mn) as catalysts is decidedly slower (Brandt and Eldik 1995). However, the accessibility of H_2O_2 in the atmosphere is determined by the relative proportions between the emission of sulphur dioxide of the anthropogenic origin and sulphuric acid deposition.

Each of the processes which affect the speed of H_2O_2 production or disintegration in the gaseous or liquid phase, changes its concentration in the atmosphere (Deng and Zuo 1999). The study performed by these authors showed that higher solar radiation, temperature, the presence of volatile organic compounds (VOC), and water vapor in the air are favorable conditions for the formation of H_2O_2 and may result in higher concentrations of this compound.

The most important peroxide of the gaseous phase, H_2O_2, is a product of a radical recombination reaction (Peña et al. 2001):

$$HO_2^{\cdot} + HO_2^{\cdot} \rightarrow H_2O_2 + O_2 \quad (1\text{-}33)$$

Nitrogen oxide influences the efficiency of the above reaction. Even though the reaction speed between NO_x and H_2O_2 is very slow, NO_x can act as an intermediary in H_2O_2 production and disintegration processes, entering into reactions with the radical $HO_2^{\cdot}$ and forming nitric acid (V).

$$HO_2^{\cdot} + NO \rightarrow HO^{\cdot} + NO_2 \quad (1\text{-}34)$$

$$NO_2 + HO^{\cdot} \rightarrow HNO_3 \quad (1\text{-}35)$$

It has been discovered that H_2O_2 concentration drops as the concentration of NO_3^-, SO_4^{2-}, and H^+ ions increases (Deng and Zuo 1999). The acids HNO_3 and H_2SO_4 are identified as the first predominant source of H^+ in acid precipitation. In addition, H^+ promotes the dissolution of iron suspended in atmospheric aerosols. Dissolved iron (as well as Cu and Mn) is an active catalyst in the reaction of H_2O_2 with dissolved SO_2 (Zuo and Hoigné 1992; Deng and Zuo 1999).

Hydrogen peroxide, apart from taking part in SO_2 oxidation, is also an efficient source of ROS ($^{\bullet}OH$, $HO_2^{\bullet}$). Several H_2O_2 disintegration mechanisms can be distinguished in the atmosphere (Zuo and Hoigné 1992):

$$H_2O_2 + h\nu\ (\lambda \leq 360\ nm) \rightarrow 2\ HO^{\bullet} \qquad (1\text{-}36)$$

$$H_2O_2 + HO^{\bullet} \rightarrow H_2O + HO_2^{\bullet} \qquad (1\text{-}37)$$

Due to the high solubility of gaseous H_2O_2 in water, it is believed that heterogenic processes (washing out with rain, dry deposition, absorption on the surface of aerosols) are the predominant method of removing atmospheric H_2O_2. The removal of peroxides and hydro peroxides from the atmosphere by wet and dry precipitation reduces the atmospheric oxidation capacity (Lee et al. 2000).

H_2O_2 is one of the most plentiful oxidants found in the natural snow cap of the polar areas and research done in these areas was used to evaluate the variations which have occurred in its concentration since the Industrial Revolution. Measurements taken in ice cores in central Greenland showed that hydrogen dioxide concentrations increased in the last decade of the 20th century from 125 ng g^{-1} to 255 ng g^{-1} (Anklin and Bales 1997). These results are consistent with the foreseen increase in methane, nitrogen oxides, or carbon oxide emissions. Hydrogen peroxide concentration deviated from the natural background as early as the mid-19th century, pior to the onset of any significant pollution by nitrogen oxides. In the last few years, however, the upward trend of hydrogen peroxide concentration has been even more noticeable, with defined amplitudes between the winter minimum and the summer maximum, which have tripled since 1970.

Volatile Organic Compounds (VOC)

The term volatile organic compounds (VOC) refers to the organic compounds which reside in the atmosphere in a gaseous form, excluding carbon oxide and carbon dioxide. They are also referred to as non-methane volatile organic compounds (NMVOC) or non-methane hydrocarbons (NMHC).

A natural source which releases volatile organic compounds into the atmosphere may be found in the biochemical transformations in plants. In 1960, Went (1960) was the first to state that the natural emission of VOC from foliage can have a considerable influence on the chemistry of the Earth's atmosphere. Since then, the statement has been confirmed by a large number of studies and several compounds have been identified, among others isoprene (C_5H_8), which is widely emitted from both deciduous and coniferous trees; and α-pinen and β-pinen—terpenes present in the ethereal oils of many plant species, mostly conifers (Rasmussen and Went 1965).

Isoprene is a unique biogenic hydrocarbon emitted from plants during their photosynthetic activity. An increase in temperature and in solar radiation contributes to a higher emission of this compound from plants exposed to changeable external conditions (Fig. 1-11). Not only does the emission of isoprene and terpenes vary considerably, but the biochemical and biophysical processes which control the emission are also characterized by a great level of variability. Isoprene is probably an additional product of photosynthesis or photorespiration or both of these processes at the same time. As a result, isoprene emission depends on light and temperature. When there is no light, emission drops to zero.

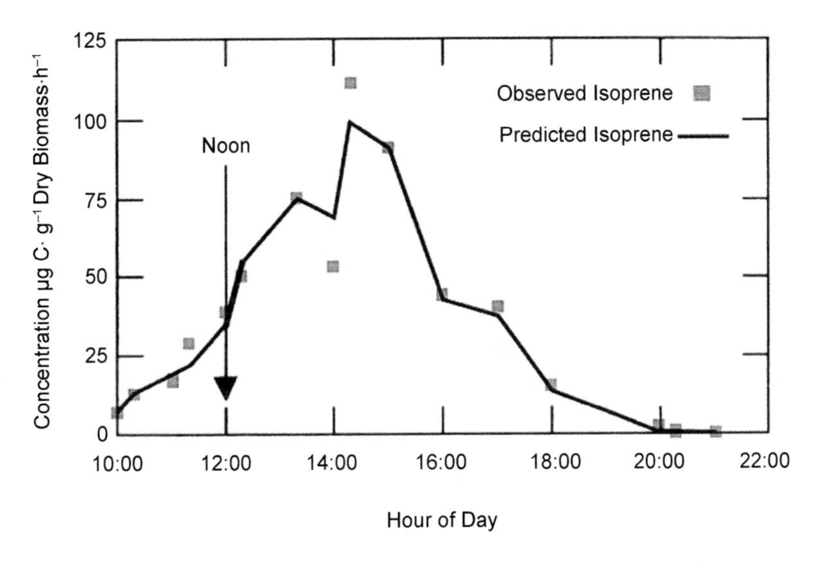

Fig. 1-11. Diurnal variability of isoprene emission from the red oak in Alabama (USA). The squares signify the observed results; the line represents the results of simulation taking into account solar radiation and temperature (adapted from Ebisch 1992).

In contrast to isoprene, the emission of terpene appears to be triggered by biophysical processes related to the pressure of terpenoid fumes and their volume in the ethereal oils of resins and saponins. As a result, the emission of terpene does not depend much on sunlight, but more on the temperature of the surroundings. Thus, the natural emission of isoprene and terpene ought to be associated mostly with the vegetative season. An increase in temperature from 25 to 35°C can quadruple the speed of biogenic isoprene emission from coniferous trees, and cause a 1.5-fold increase in terpene emission from deciduous trees (Lamb et al. 1987). Because of this, the highest emission levels of biogenic hydrocarbons can be expected in the tropics, where isoprene is a predominant component. This is the effect of not only high temperatures, but also the huge volume of plant biomass to be found there. A global assessment of biogenic volatile organic compounds (VOC) emissions (Table 1-4) indicates that plant biomass has a far greater input than anthropogenic processes.

Table 1-4. Assessment of global emission of biogenic VOC (Tg yr^{-1}) from basic sources (Guenther et al. 1995).

Source	Isoprene	Monoterpene	Other more reactive organics[a]	Other less reactive organics[b]	Total VOC
Woods	372	95	177	177	821
Crops	24	6	45	45	120
Shrubs	103	25	33	33	194
Ocean	0	0	2,5	2.5	5
Other	4	1	2	2	9
All	503	127	260	260	1150

[a)]—defined as having lifespan < 1 day under typical tropospheric conditions.
[b)]—defined as having lifespan of > 1 day.

The natural emission of non-methane hydrocarbons (ethylene, propylene, isoprene, propane, benzene, and toluene) has been identified all over the world (Fig. 1-12). NMHC streams originating from the biogenic sources of the Amazon, burning forests and savannas of the tropical regions, and termite heaps suggest mutual conjugation between oxidants ($HO^{\bullet}$, O_3) and organic substances in the atmosphere. Moreover,

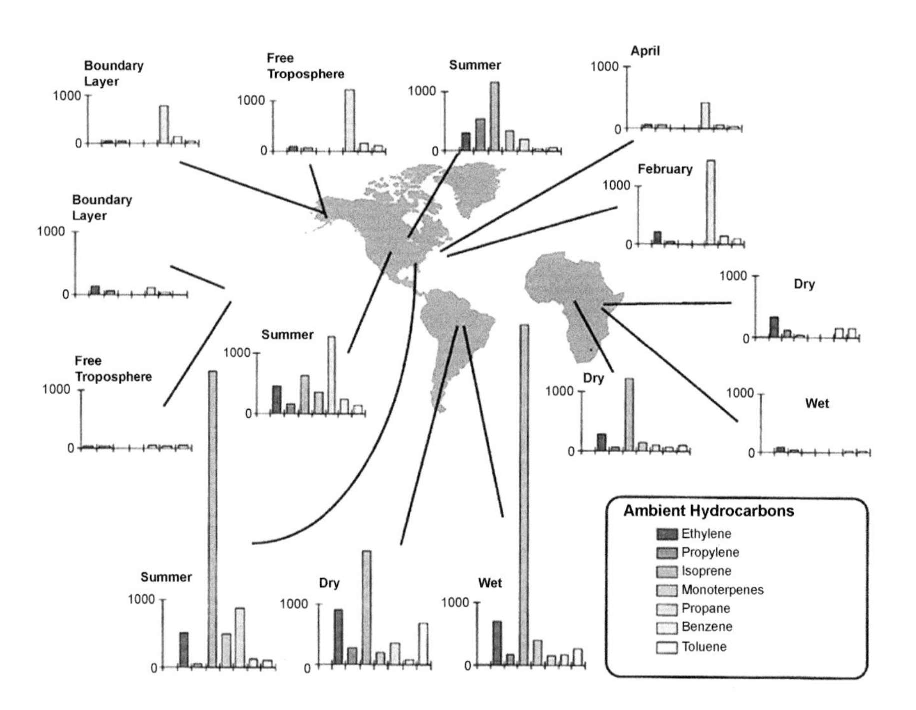

Fig. 1-12. Natural and anthropogenic sources and depositories of non-methane hydrocarbons at different times and in different regions. The observed concentrations [pptv] show strong seasonal variability in the emission of NMHC into the air in rural areas and fields distant from urban centers, as well as clear differentiation between the planetary boundary layer and upper troposphere (adapted from Ebisch 1992).

there is proof of the existence of synergistic reactions, which take place between both biogenic and anthropogenic non-methane hydrocarbons and the production of tropospheric ozone.

Most non-methane hydrocarbons of natural and anthropogenic origins presented in Fig. 1-12 have a clear seasonal cycle of variability, although the depicted changes are different for specific compounds. VOC emitted on-land during fuel combustion and industrial activity occur at their highest concentrations mostly in winter, while low levels of concentration are observed mainly in summer and autumn (Wahner et al. 1994). A similar seasonal variability is exhibited by most non-methane hydrocarbons of an anthropogenic origin.

In the tropical region of the globe, a very large plume of trace gases and aerosols is carried to the atmosphere, a direct result of biomass fires. Burning is a widely used method of cleansing the land from forests or grass in order to use it as arable fields or pastures. About 16 million km^2 consist of tropical or subtropical savannas which burn every five years on an average and in many regions even more often than that (Table 1-5). It could be said that fire is the most important tool for farmers in the tropics.

Table 1-5. Global estimates of annual amounts of biomass burning and of consequent release of carbon into the atmosphere (Andreae 1993).

	Biomass burned (Tg dry mass·yr^{-1})		Carbon released	
	Tropical	Extra tropical	Total	Tg yr^{-1}
Forest	1260	1150	2410	1080
Savanna	3690	-	3690	1660
Biomass fuel	1720	220	1940	880
Charcoal	20	1	21	82
Agricultural Waste in Fields	420	420	850	380
World total	7110	1790	8910	4080

Direct measurements taken during a savanna fire in Kruger National Park detected the presence of a range of volatile carbon compounds such as: aromatic hydrocarbons, aldehydes, alkanes, alkenes, alcohols, acids, carbonyl compounds, organic acids, chlorohydrocarbons, carbon monoxide and carbon dioxide, methane, methyl chloride, methyl bromide, methyl iodide, carbonyl oxide and carbon isotopes in aerosols, and elemental carbon (Andreae et al. 1994). It follows that smoke carries a very complex mixture, containing particles of aerosol, hydrocarbons, partly-oxidized hydrocarbons and carbon oxide, all from a single source. The quantity of volatile carbon emitted during a fire and its qualitative characteristics depend not only on the type of flora, but also on temperature, humidity content, wind direction, and speed. The two latter parameters denote the spatial scale of a fire (Fig. 1-13).

The influence of hydrocarbons emitted during biomass burning is extremely significant on the chemistry of the atmosphere as well on climate. The large scale of fires which sweep tropical Africa, Latin America, and South-East Asia, leads in the dry season to an increase in tropospheric ozone approaching values typical of highly urbanized and industrialized areas (Andreae et al. 1992).

Fig. 1-13. Smog plume from vegetation fire at Indonesia in 1997. Image courtesy NASA GSFC Scientific Visualization Studio, based on data from TOMS (www.atmosphere.mpg.de).

Beside natural processes (Table 1-4), the sources of VOC in the atmosphere are biomass burning (Table 1-5) as well as anthropogenic processes (Table 1-6).

Table 1-6. Assessment of the global anthropogenic emission of non-methane volatile organic compounds into the atmosphere (Middleton 1995).

Activity	Emission [Tg yr^{-1}] ($Tg = 10^{12}$ g)
Production and distribution of fuels	
petroleum	8
natural gas	2
oil refinery	5
petrol distribution	2.5
Fuel consumption	
Coal	3.5
Wood	25.0
Harvest remnants + waste	14.5
Charcoal	2.5
Manure	3.0
Road transport	36.0
Chemical industry	2.0
Use of solvents	20.0
Uncontrolled waste burning	8.0
Other	10.0
TOTAL	142.0

The highest emission of volatile organic compounds into the atmosphere originates from combustion engines in road transport, during incomplete combustion of fuel and its evaporation. It is the main source of alkanes and aromatic hydrocarbons. Car exhaust fumes include unburnt or only partly burnt fuels, gear oils and gases released from petrol. The use of solvents and wood combustion are two different categories of sources which have a considerable contribution to the global VOC emission. The former relates to such sectors of the industry where production involves the use of various organic solvents (the chemical industry, production of paints and varnishes, cosmetics, footwear, cars, paper, etc.), while the latter refers to both intentional and accidental burning of wood.

About 600 various compounds have been identified, each of which affects air quality in its own unique way. Owing to an increase in urbanization in recent years, a new and hitherto overlooked source of VOC has been revealed in disposal sites of many different kinds of waste. Monitoring of VOC from organic waste material detected the presence of 155 volatile compounds. Among them, benzene, toluene, ethyl benzene, and naphthalene have been identified, all of which are classed as major pollutants due to their harmful influence on human health (Statheropoulos et al. 2005). Benzene in particular possesses carcinogenic qualities. Its concentration was determined to be at 0.8–12.7 $\mu g\ m^{-3}$, while only as much as 1 $\mu g\ m^{-3}$ creates the risk of leukemia over a prolonged period of time, according to the WHO (1987). Other pollutants harmful to humans also occurred in high concentrations: toluene 8.1 $\mu g\ m^{-3}$; ethyl benzene 12.7 $\mu g \cdot m^{-3}$; naphthalene 1.4–3.4 $\mu g\ m^{-3}$. Additionally, the presence of the organic-sulfur compound DMS has been detected at disposal sites. Its value of 16.9 $\mu g\ m^{-3}$ has a substantially harmful impact on the receiving population. Organic-sulfides can be generated from human feaces and food wastes (Statheropoulos et al. 2005). According to the authors of the study, research results prove the existence of threats generated by disposal sites not only for their staff, but also for the population in cities.

Smog

Photochemical smog first appeared in the 1940s in Los Angeles, but the term smog had been used before, at the beginning of the 20th century, to describe the mixture of smoke and fog containing sulphur dioxide, which prevailed over London in winter, when the most common fuel was heavily sulphurized coal. The smog in the air over Los Angeles is formed on the basis of oxidants such as ozone, hydron peroxide, organic peroxides of the ROOR—type, organic hydro peroxides (ROOH) and peroxyacetyl nitrate (PAN). This is why it is called oxidating or photochemical smog. It tends to appear in the summer and the mixture of chemical components in the air is as unpleasant as with reductive smog, exhibiting similar qualities. At the relative air humidity of 60%, the visibility drops below 300 meters, people experience irritation of the skin and eyes, and have difficulty breathing.

The first stage of the creation of photochemical smog is the formation of oxidants in the air, especially ozone. When the growths speed of oxidant concentration reaches 0.15 ppm per hour or faster, smog starts to become more aggravating (Kinney and Özkaynak 1991).

Three main components of photochemical smog have been identified:

- ultraviolet radiation,
- nitrogen oxides (see the Section "Nitrogen compounds" of this Chapter),
- VOC (see the Section VOC of this Chapter).

The two latter components are products of fuel combustion, mostly in cities where numerous emission sources are responsible for air quality. The means of road transport constitute linear emission sources. Energy plants and large factories are point sources, while low dispersed emission constitutes a surface source.

The products of oxidative substances, transformations are radicals considered to be substrates in the formative reactions of hydrogen peroxide or ozone, so the presence of VOC in the atmosphere initiates the creation of numerous new compounds participating in the formation of photochemical smog. It is assumed that benzene is one of the typical components of exhaust fumes and is emitted only by vehicles in use. Using it as a solvent is forbidden in the EU countries (Directive 89/677/EEC). Toluene and xylene are also to be found in car exhaust fumes, but can also be emitted from factories producing or using organic solvents.

Nitrogen oxides are formed during the burning of fossil fuels in furnaces and combustion engines, when a sufficiently high temperature makes it possible for the bonds in oxygen and nitrogen particles to break. Next, it enables atomic oxygen to combine with atomic nitrogen and form a NO particle. The mean global average concentration of NO_x remains at the level of 3 ppb, but in the centers of large cities where traffic is very intense, exhaust fumes are such an efficient source that nitrogen oxide concentration may increase by up to threefold.

When photochemical smog is forming, nitrogen oxide concentrations undergo significant changes (Fig. 1-14). It is characteristic that in early hours of the morning, NO concentrations considerably drop, while NO_2 increases.

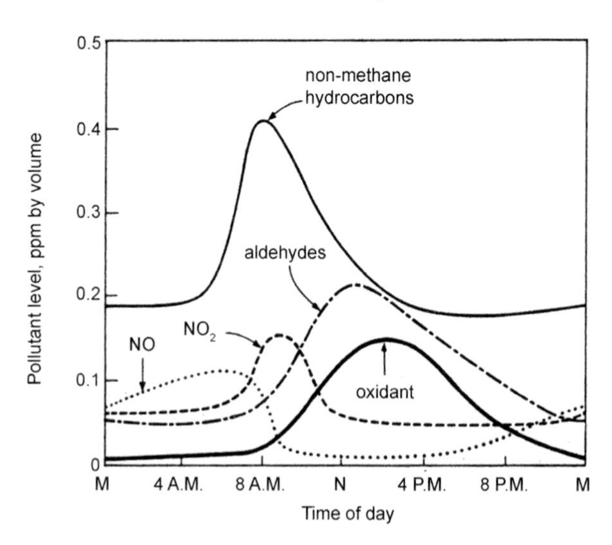

Fig. 1-14. Typical concentration changes of basic components of photochemical smog, occurring during the day with the participation of solar radiation (Monahan 1983).

During the day low NO concentration persists, while the concentration level of oxidants increases. Hydrocarbons also occur at maximum concentrations in the morning hours, after which they experience a drop in value. Both nitrogen oxides and hydrocarbons behave differently during the formation of smog than under normal circumstances. The most significant difference is a rapid jump in NO_2 concentration accompanied by a simultaneous drop in NO concentration when photo-dissociation occurs (see the reaction 1-12). What is more, the decrease in hydrocarbons concentration is so rapid that sluggish reactions with ozone and atomic oxygen cannot serve as an explanation for it (Monahan 1983).

One of the most commonly found oxidants in a polluted atmosphere (containing olefins and NO_X) is peroxyacetyl nitrate (PAN). Its presence in the air accelerates the formation of photochemical smog and in the absence of light a series of reactions take place with the participation of PAN. The decomposition of peroxyacetyl nitrate leads to the release of a range of organic and inorganic radicals which participate in the formation of smog and accelerate chemical reactions. Additionally, NO, released at night, undergoes conversion to a photochemically active form through reactions with radicals. The presence of NO_2 initiates the whole process. Another transformation product is nitrous acid (HONO), formed with the participation of hydroxyl radicals, whose presence depend on solar radiation.

A significant visibility reduction in smog results from the formation of aerosols. They are created as a result of the conversion of gases, such as nitrogen and sulphur oxides, fumes of nitrous and nitric acids and hydrocarbons. The presence of suspended particles is conducive to the sorption of gases and radicals on their surface.

The mechanism of photochemical smog formation has been explained by numerous researchers over the years (Pitts and Finlayson 1975; Falls and Seinfeld 1978; Graedel 1980; Whitten et al. 1980; Monahan 1983; Finlayson-Pitts and Pitts 2000).

The intensification of photochemical smog is closely connected to car traffic, and consequent volume of exhaust fumes in the air. An external factor which aids the formation of photochemical smog and its duration is thermal inversion. It is characterized by a positive temperature gradient with height and a high degree of static balance stability. The inversion layer creates a barrier and does not allow for mixing and transport of the air beyond itself. Hence, the longer the thermal inversion remains over a smog-ridden area, the higher the concentration of combustion products and secondary aerosols in the air close to the Earth's surface. In favorable geographic conditions, smog lingering time can extend from a few to a dozen or so days (Falkowska and Korzeniewski 1998).

The consequences of photochemical smog are very serious. In young plants, trees or shrubs, contact with ozone and peroxyacetyl nitrate leads to the build-up of a glossy coating which irritates external tissue and causes foliage to brown. A several hours long exposure to PAN at a concentration of 0.02–0.05 ppm diminishes crops. Damage to needles and their browning occurs at ozone concentration of 0.05 ppm, and PAN concentration of 0.01 ppm. Strong oxidants are thought to be responsible for about 75% of damage among plants. The harmful consequences of smog are also borne by people and the most vulnerable group are asthma sufferers, who start to experience the aggravating effects of smog at very low levels of oxidant concentrations (Seltzer et al. 1986). An analysis of the effect of the Los Angeles smog on the human body showed a close relation between the concentration of oxidants combined with air temperature and mortality (Kinney and Özkaynak 1991). Gaseous pollutants acting in synergy with dusts and aerosols especially advanced the death of people with respiratory and circulatory conditions. There are limits of permissible and even emergency concentration levels for the gaseous components of smog (Fig. 1-15). If these values are exceeded, the local authorities are obliged to inform the public about it and to undertake counter-measures which could lead to lowering the concentrations of aggravating and dangerous substances. This can be done through temporary restrictions on car traffic in the city, or by limiting industrial activity. Children and sensitive individuals should remain at home while the smog persists.

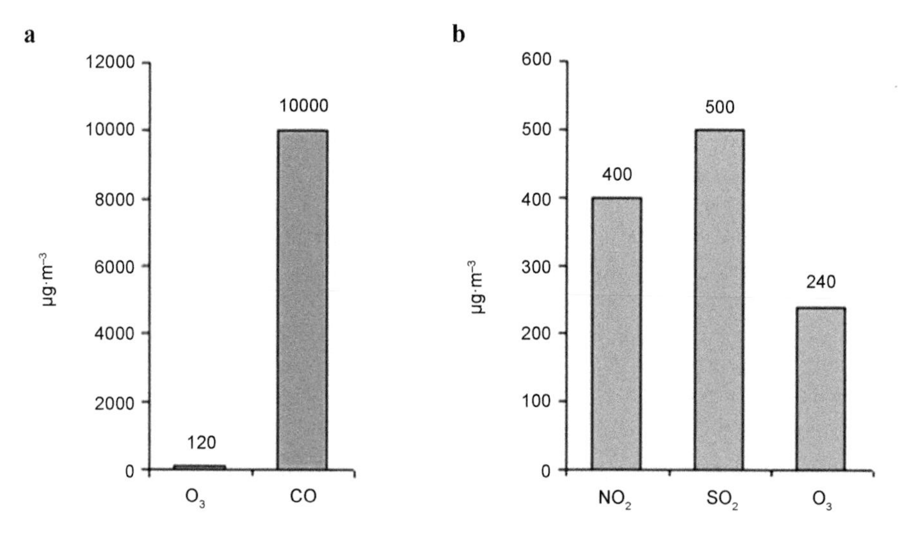

Fig. 1-15. (a) Permissible values for ozone and carbon oxide concentration levels as 8-hour averages and (b) emergency values for nitrogen dioxide, sulphur dioxide, and ozone concentrations as 1-hour averages in EU countries (Paweł Jezioro, www.atmosphere.mpg.de).

Elemental mercury

Mercury is a mobile element, as it is the only metal which in normal conditions occurs in the gaseous form. This unusual quality of mercury enables it to travel long distances in the atmosphere, giving its influence a global character.

Mercury is emitted into the air from natural and anthropogenic sources. The main source of mercury is the process of weathering, which causes the emergence of Hg from rocks and minerals. It can also be released during degasification from the Earth's crust (Gustin 2012) and in processes related to volcanic activity. It is estimated that the volume of yearly mercury emission from volcanic eruptions amounts to 80–4,000 tones (Pyle and Mather 2003). Other natural sources of mercury are geothermic processes, biomass burning, as well as reemission from sea waters. The yearly transportation of mercury from land and water to the atmosphere by way of reemission is about 2,000 tones Hg·year^{-1}, while biomass fires alone release 590–930 tones of mercury every year (Wang et al. 2004).

In the last 500 years, man has introduced about 923,000 tones of mercury, extracted mostly from the stable and hydrophobic form which is cinnabar (HgS). Today, contemporary anthropogenic sources are industrial processes, fossil fuel combustion, metal industry, cement production, mining, agriculture, waste deposition sites, and incineration plants and crematoria. The total mercury emission in Europe in 2005 was 226.5 tones (Pacyna et al. 2006b), of which Poland was contributing about 9% (Hławiczka 2008).

Gaseous elemental mercury (GEM) is the predominant form in the atmosphere, constituting about 99% of total mercury (Pacyna et al. 2006a,b). Owing to its low reactivity and solubility in water, its residence time in the air is long, and amounts to 0.5 to 2 years, according to Weiss-Penzias et al. (2003). Holmes et al. (2006), based on the latest model tests, proposed the time to be from 0.5 to 1.7 years. GEM in the ambient air lingers long enough to be dispersed and transported across the globe and deposited in places that are distant from the emission source. The mean concentration of total gaseous mercury in the ambient air has been estimated at about 1.5 ng m^{-3} (Wänberg et al. 2001), and the reactive gaseous mercury (RGM) and total mercury in aerosol particles (TPM) at between 1.0 and 100 pg m^{-3} (Stratton and Lindberg 1995; Wänberg et al. 2001).

The concentrations of particular forms of mercury in the air of clean unpolluted areas vary considerably (Table 1-7). An increase in air pollution occurs in the vicinity of factories and mines where Hg concentration, close to the ground, can reach 200 ng m^{-3} (Pacyna et al. 2006b).

In the atmosphere, mercury undergoes chemical transformations in the gaseous and liquid forms and in heterogenic aerosols (Seigneur et al. 1994). The pace and direction of these transformations are influenced by, among others, the oxidative-reductive conditions of the environment. The most important Hg^0 oxidants in the liquid phase are O_3 and HClO, and HSO_3^-, and in the gaseous phase—O_3, Cl_2, and

Table 1-7. Concentrations of different forms of mercury in the ambient air of different regions.

Regions	TGM [ng m^{-3}]	TPM [pg m^{-3}]	RGM [pg m^{-3}]	MeHg [pg m^{-3}]	Literature
Europe					Schroeder and Munthe 1998
- non-polluted area	0.66–6.20	0.5–200			
- polluted area	1.96–33.8	10–20,000	1.0–50	1.0–20	
Georgia, US					Jansen and Edgerton 2005
- non-polluted area	1.59		4.5		
- polluted area	1.67		6.1		
Arctic	1.54–1.6		0.5–3.5		Steffen et al. 2003
Korea					Kim et al. 2005
- polluted area	5.06				
China					Shang et al. 2003
- polluted area	7.09		37.5		

H_2O_2, while SO_2 and CO are the most significant reductors of Hg^{2+} (Schroder et al. 1991; Morel et al. 1998).

Gaseous mercury oxidation through ozone to the reactive form takes place in the gaseous and liquid phases. It is conducted with the highest efficiency during the day and the kinetics of this process increase as the pH of cloud droplets is reduced:

$$Hg(0)_g + O_3 \rightarrow HgO_{(g,s)} + O_2 \tag{1-38}$$

$$Hg(0)_g + O_3 + H^+ \rightarrow Hg^{2+}_{(aq)} + OH^- + O_2 \tag{1-39}$$

During the day, there also occur reactions of mercury oxidation with hydrogen peroxide, while at night the role of an oxidant is performed by $NO_3\cdot$. Both of the reactions take place only in the gaseous phase:

$$Hg(0)_g + H_2O_2 \rightarrow Hg(OH)_{2(g,s)} \tag{1-40}$$

$$Hg(0)_g + NO_3\cdot \rightarrow HgO_{(g,\,s)} + NO_2 \tag{1-41}$$

Mercury undergoes oxidation under the influence of the hydroxyl radical. The process takes place in the gaseous phase during the day as a two-stage reaction:

stage I $$Hg(0)_g + HO^{\bullet} \rightarrow Hg^+_{(aq)} + OH^- \tag{1-42}$$

stage II $$Hg^+_{(aq)} + HO^{\bullet} \rightarrow Hg^{2+}_{(aq)} + OH^- \tag{1-43}$$

Chlorine compounds are very important oxidants:

$$Cl_2 + Hg(0)_g \rightarrow HgCl_{2(g,\,aq)} \tag{1-44}$$

$$2Cl^- + Hg(0)_g \rightarrow HgCl_{2(g,\,aq)} \tag{1-45}$$

$$HOCl_{(aq)} + Hg(0)_g \rightarrow Hg^{2+}_{(aq)} + Cl^-_{(aq)} + OH^-_{(aq)} \tag{1-46}$$

$$OCl^-_{(aq)} + Hg(0)_g + H^+ \rightarrow Hg^{2+}_{(aq)} + Cl^-_{(aq)} + OH^-_{(aq)} \tag{1-47}$$

Gaseous mercury created in the above reactions has a relatively short atmospheric residence time (from a few hours to several days) and soon undergoes further changes. It forms complex compounds with water droplets or is associated to aerosol particles. These reactions lead to mercury being removed from the atmosphere through deposition (Weiss-Penzias et al. 2003). Reactive gaseous mercury is removed from the air mostly as a result of wet deposition, and associated to particles owing to dry gravitational sedimentation (Wang et al. 2004).

All forms of mercury have an adverse effect on the functioning of living organisms. The level, exposure time, the routes of exposure, and the differences in the transportation and the metabolism of inorganic (Hg^{2+}) and organic (RHgX and R_1HgR_2) forms of Hg are responsible for the differences in their distribution in various tissues and organs, and for their biological effects and their toxicity (Zalups and Lash 1994). The difference in the bimolecular way of inorganic and organic mercury compounds' activity results from their varied chemical structure and different levels of reactivity. Owing to mercury's considerable affinity with thiol groups (SH), mercury compounds deprive albumins, glutathione (GSH), cysteine, homocysteine, and N-acethylcysteine of SH groups (Zalups and Lash 1994; Milaeva 2006).

The bonding of mercury with endogenous ligands, such as glutathione metallothioneins, which is responsible, for example, for removing reactive forms of oxygen (ROS) and xenobiotics, leads to increased ROS concentration and oxidative stress (Zalups 2009; Monteiro et al. 2010).

The main route of elemental mercury penetration into the human body is through the respiratory system. Trace amounts are absorbed through the skin, and less than 10^{-3}% undergoes absorption via the digestive system (Affelska–Jercha 1999; Risher et al. 2003).

Exposure to elemental mercury is mostly related to occupational or housing exposure or religious or cultural factors. Occupational exposure results from mishandling mercury, accidental leaks, and inadequate ventilation in the workplace. Housing exposure can occur as a result of damaging measurement instruments (Stier et al. 1998; Ozuah 2000; Risher et al. 2003).

To this day, metallic mercury is used by some religions in ritual practices. Examples of such religions are Santeria (cult of African and catholic deities on Cuba), Voodoo (a Haitian set of beliefs and rituals), and Palo Mayombe (a covert form of ancestor worship practiced mostly in the Caribbean) (Risher et al. 2003). The cultural use of elemental mercury consists in attributing it with spiritual quality and magical power. Rituals involve applying mercury on the body, swallowing the metallic form, and boiling and spilling it around the house (Ozuah 2000).

Despite its high changeability in the atmosphere, indoors Hg^0 can continue to pose a chronic threat to the present and future inhabitants for about a year (Risher et al. 2003; Wang and Pehkonen 2004). This type of exposal particularly affects children. Mercury reaches higher concentrations in the body in children as compared to adults, after the same exposure time. Besides, mercury fumes are heavier than the air, so they possess the highest concentrations close to the ground. Consequently, the closer to the floor a person's nostrils are, the more fumes are inhaled and absorbed through the lungs. Children inhale more air per body mass unit, so they breathe in more air than adults (Ozuah 2000). Inhaled mercury fumes reach the alveoli, from where, thanks to their mono-atomic form, they penetrate into the bloodstream. Some of them become oxidized to Hg^{2+} in erythrocytes, but a large proportion of mercury, thanks to its good solubility in fats (5–50 mg dm^{-3}), is able to penetrate the blood—brain and blood—placenta barriers. It is accumulated in the brain tissue, where it gets oxidized to Hg^{2+} with the participation of catalyses and hydrogen peroxide, and in this form it is retained for years (Stier et al. 1998; Affelska–Jercha 1999; Rooney 2007).

The medical consequences of exposal to metallic mercury fumes depend on the scale and period of exposal as well as the age and the health condition of the exposed individuals (Affelska–Jercha 1999; Risher et al. 2003). Accumulation in the brain tissue leads to the occurrence of fits, emotional instability, insomnia, memory loss, fatigue, headaches, polyneuropathy, cognitive dysfunction, psychomotoric disorders, hearing difficulties, impaired vision, slower conduction of nerve impulses, concentration disorders, false electromyography (EMG) results, and paraesthesia. Motoric disorders are usually reversible after terminating the exposal, but cognitive dysfunction (mostly memory deficits), can last longer or be permanent in case of chronic occupational exposures (Lech 1998; Risher et al. 2003). Cutaneous lesions, which often appear on the palms and the soles of the feet, are combined with redness, itchy rash and skin shedding. Acrodynia is a syndrome of symptoms, occurring mostly in children and is characterized by pain in the skin covering the limbs, the redness of limbs and the nose, excessive perspiration, and sometimes gastro-intestinal symptoms. It can be combined with anorexia and increased sensitivity to light. Mercury fumes inhaled in high concentration over a few days lead to the inflammation of the oral cavity, the intestines, the bronchial tubes, the bronchiole, and the lungs.

Kidneys are particularly sensitive to the influence of metallic mercury. Prolonged respiratory exposure can result in severe kidney insufficiency and damage with necrotic degeneration in the proximal tubules (Lech 1998; Affelska–Jercha 1999; Risher et al. 2003). The total period of metallic mercury's biological half-life in the body is about two months (from about 30 to 90 days), in the blood—about three days. Smaller animals with less body mass eliminate mercury faster than large ones and cold-blooded ones (Affelska–Jercha 1999; Risher et al. 2003). No carcinogenic properties have been found for elemental mercury, regardless of the type of exposure or the species of animals, including humans (Risher et al. 2003).

References

Aardenne, J.A. van, G.R. Carmichael, H. Levy II, D. Streets and L. Hordijk. 1999. Anthropogenic NO_x emissions in Asia in period 1990–2020. Atmos. Environ. 33: 633–646.

Affelska–Jercha, A. 1999. Toxic effects of mercury in a professional and environmental exposure. Medycyna Pracy L 4: 305–314.

Akimoto, H., D.D. Davis and S.C. Liu. 1994. Atmospheric chemistry of the East Asian Northwest Pacific region. pp. 71–82. *In*: R.G. Prinn [ed.]. Global Atmospheric-Biospheric Chemistry. Plenum Press, New York, London, USA, U.K.

Anklin, M. and R.C. Bales. 1997. Recent increase in H_2O_2 concentration at Summit, Greenland. J. Geophys. Res. 102: 1909919104.

Andreae, M.O. 1993. The influence of tropical biomass burning on climate and the atmospheric environment. pp. 113–150. *In*: R.S. Oremland [ed.]. Biogeochemistry of Global Change: Radiatively Active Trace Gases. Chapman & Hall, New York, USA.

Andreae, M.O., A. Chapuis, B. Cros, J. Fontan, G. Helas, C. Justice, Y.J. Kaufman, A. Minga and D. Nganga. 1992. Ozone and Aitken nuclei over equatorial Africa: airborne observations during DECAFE 88. J. Geophys. Res. 97: 6137–6148.

Andreae, M.O. and P.J. Crutzen. 1997. Atmospheric aerosols: Biogeochemical sources and role in atmospheric chemistry. Science 276: 1052–1058.

Andreae, M.O., J. Fishman, M. Garstang, J.G. Goldammer, C.O. Justice, J.S. Levine, R.J. Scholes, B.J. Stocks, A.M. Thompson, B. van Wilgen and STARE/TRACE-A SAFARI-92 Science Team. 1994. Biomass burning in the global environment: First results from the IGAC/BIBEX field Campaign STARE/TRACE-A/SAFARI-92. pp. 83–101. *In*: R.G. Prinn [ed.]. Global Atmospheric-Biospheric Chemistry. Plenum Press, New York, London, USA, U.K.

Aquirre, J., M. Rios-Momberg, D. Hewitt and W. Hansberg. 2005. Reactive oxygen species and development in microbial eukaryotes. Trends Microbial. 13(3): 111–118.

Badr, O. and S.D. Prober. 1995. Sinks and environmental impacts for atmospheric carbon monoxide. Applied Energy 50(4): 339–372.

Bartholomew, G.W. and M. Alexsander. 1981. Soil as a sink for atmospheric carbon monoxide. Science 212: 1389–1391.

Bartnicki, J., A. Gusev, W. Aas, H. Fagerli and S. Valiyaveetil. 2008. Atmospheric supply of nitrogen, lead, cadmium, mercury and dioxins/furans to the Baltic Sea in 2006. EMEP report to HELCOM Available at: http://www.helcom.fi/envoronment2/hazsubs/EMEP/en_GB/emep2006/.

Bascom, R., P.A. Bromberg, D.A. Costa, R. Devlin, D.W. Dockery, M.W. Frampton, W. Lambert, J.M. Samet, F.E. Speizer and M. Utell. 1996. Health effects of outdoor air pollution. Am. J. Respir. Crit. Care Med. 153: 3–50.

Berresheim, H., P.H. Wine and D.D. Davis. 1995. Sulfur in the atmosphere. pp. 251–307. *In*: H.B. Singh [ed.]. Composition, Chemistry, and Climate of the Atmosphere. Van Nostrand-Reinhold, New York, USA.
Brandt, C. and R. van Eldik. 1995. Transition metal-catalysed oxidation of sulfur (IV) oxides. Atmospheric-relevant processes and mechanisms. Chem. Rev. 95: 119–190.
Carslaw, D.C. 2005. Evidence for an increasing NO_2/NO_X emission ratio from road traffic emissions. Atmos. Environ. 39: 4793–4802.
Crutzen, P.J. and M.O. Andreae. 1990. Biomass burning in the tropics: Impact on atmospheric chemistry and biogeochemical cycle. Science 250: 1669–1678.
Deng, Y. and Y. Zuo. 1999. Factors affecting the levels of hydrogen peroxide in rainwater. Atmos. Environ. 33: 1469–1478.
Ebisch, R. 1992. Atmospheric chemistry, global change, and NCAR. Challenges and opportunities at the National Center for Atmospheric Research. Boulder, Colorado, 1–44.
[EEA] European Environment Agency. 1997. Monograph, No.5, Assessment and Management of Urban Air Quality in Europe. Copenhagen.
[EEA] Air quality in Europe. 2011. EEA Technical Report No 12/2011, Copenhagen, Denmark.
[EPI] Environmental Protection Institute (IOŚ), Warszawa, 2003. Raport. Stan środowiska w Polsce w latach 1996–2001. Warszawa.
[EPI] Environmental Protection Institute (IOŚ), Warszawa, 2012, Raport krajowy bilans emisji SO_2, NO_X, CO, NMLZO, NH_3, pyłów, metali ciężkich i TZO za lata 2009-2010 w układzie klasyfikacji SNAP i NFR. (in polish). (available at: www.kobize.pl/materialy/...krajowe/2012/Raport_LRTAP_2010.pdf).
Evans, M. 2006. Global Atmospheric Cycle for Fixed Nitrogen. ACCENT- Access to Laboratory Data, Understanding and Quantifying the Atmospheric Nitrogen Cycle, An ACCENT Barnsdale Expert Meeting, -189.
Falkowska, L. and K. Korzeniewski. 1998. Atmospheric Chemistry (in Polish). Gdańsk University Press, Gdańsk, Poland.
Falkowska, L. and A. Lewandowska. 2009. Aerosols and Gases in Earth Atmosphere—Global Change (in Polish). Gdańsk University Press, Gdańsk, Poland.
Falls, A.H. and J.H. Seinfeld. 1978. Continued development of a kinetic mechanism for photochemical smog. Environ. Sci. and Technol. 12: 1398–1406.
Finlayson-Pitts, B.J. and J.N. Pitts. 1986. Atmospheric Chemistry: Fundamentals and Experimental Techniques. John Wiley & Son, New York, N.Y., USA.
Finlayson-Pitts, B.J. and J.N. Pitts. 2000. Chemistry of the Upper and Lower Atmosphere. Academic Press, San Diego, San Francisco, New York, Boston, London, Sydney, Tokyo.
George, C., R.S. Strekowski, J. Kleffmann, K. Stemmler and M. Ammann. 2005. Photoenhanced uptake of gaseous NO_2 on solid organic compounds: a photochemical source of HONO? Faraday Discuss 130: 195–210.
Graedel, T.E. 1980. Atmospheric photochemistry. pp. 107–143. *In*: O. Hutzinger [ed.]. The Handbook of Environmental Chemistry, Vol. 2, Part A. Springer Verlag, Berlin, Germany.
Guenther, A. 1995. A global model of natural volatile organic compound emissions. J. Geophys. Res. 95: 3599–3617.
Gustin, M.S. 2012. Exchange of mercury between the atmosphere and terrestrial ecosystems. pp. 423–454. *In*: G. Liu, Y. Cai and N. O'Driscoll [eds.]. Environmental Chemistry and Toxicology of Mercury. John Wiley & Sons, Inc., New York, N.Y., USA.
Gutzwiller, L., F. Arens, U. Baltensperger, H.W. Gäggeler and M. Amman. 2002. Significance of semivolatile diesel exhaust organics for secondary HONO formation. Environ. Sci. Technol. 36: 677–682.
Halliwell, B. and J.M.C. Gutteridge. 1984. Free Radicals in Biology and Medicine, 2nd ed. Oxford University Press, Oxford, UK.
Hegg, D.A. 1985. The importance of liquid-phase oxidation of SO_2 in the troposphere. J. Geophys. Res. 90: 3773–3779.
Hławiczka, S. 2008. Mercury in the Atmospheric Environment (in Polish). Works & Studies No. 73. IEE PAS, Zabrze, Poland.
Holland, H.D. 1984. The Chemistry of the Atmosphere and Oceans. J. Wiley, Chichester, UK.
Holmes, C.D., D.J. Jacob and X. Yang. 2006. Global lifetime of elemental mercury against oxidation by atomic bromine in the free troposphere. Geophys. Res. Lett. 33, L20808, 1-5. Doi:10.1029/2006GL027176.
Houghton, J.J., L.G. MeiroFilho, B.A. Callander, N. Harris, A. Kattenberg and K. Maskell. 1996. Climate Change 1995: The Science of Climate Change, Contribution of Working Group I to the Second

Assessment Report of the Intergovernmental Panel on Climate Change. Cambridge University Press, Cambridge, UK.

[IPPC] Intergovernmental Panel on Climate Change. 1995. World Meteorological Office, United Nations Environmental Programme, Radiative Forcing of Climate Change. The 1994 Report of the Scientific Assessment Working Group of IPCC. Summary of Policymakers 1–89.

[IPPC] Intergovernmental Panel on Climate Change. 2007. Working Group III to the Intergovernmental Panel on Climate Change. Fourth Assessment Report Climate Change 2007: Mitigation of Climate Change. Summary of Policymakers 1–36. (Available at: http://www.ipcc.ch/publications_and_data/ar4/wg1/en/ch2.html).

Jacob, D.J., B.G. Heikes, S.M. Fan, J.A. Logan, D.L. Mauzerall, J.D. Bradshaw, H.D. Singh, G.L. Gregory, R.W. Talbot, D.R. Blake and G.W. Sachse. 1996. Origin of ozone and NO_x in the tropical troposphere. A photochemical analysis of aircraft observations over the South Atlantic Basin, J. Geophys. Res. 101: 24235–24350.

Jansen, J. and E. Edgerton. 2005. An Aries, Search, and Mercury Update. Presentation in the Atmospheric Research and Analysis, Inc. (Available at /http://www.atmosphericresearch.com/publications/archive/search.htm).

Jenkin, M.E. 2004a. Analysis of the sources and partitioning of oxidant in the UK. Part 1: The NO_X dependence of annual mean concentrations of nitrogen dioxide and ozone. Atmos. Environ. 38: 5117–5129.

Jenkin, M.E. 2004b. Analysis of the sources and partitioning of oxidant in the UK. Part 2: Contributions of nitrogen dioxide emissions and background ozone at a kerbside location in London. Atmos. Environ. 38: 5131–5138.

Jenkin, M.E. 2006. Atmospheric Chemical Processing of Nitrogen Species, ACCENT- Access to Laboratory Data, Understanding and Quantifying the Atmospheric Nitrogen Cycle, An ACCENT Barnsdale Expert Meeting. -189.

Kelly, T.J., P.H. Daum and S.E. Schwartz. 1985. Measurements of peroxides in cloudwater and rain. J. Geoph. Res. 90: 7861–7871.

Kim, K.H., R. Ebinghaus, W.H. Schroeder, P. Blanchard, H.H. Kock, A. Steffen, F.A. Froude, M.Y. Kim, S. Hong and J.H. Kim. 2005. Atmospheric mercury concentrations from several observatories in the Northern Hemisphere. J. Atmos. Chem. 50: 1–24.

Kinney, P.L. and H. Özkaynak. 1991. Associations of daily mortality and air pollution in Los Angeles Country. Environ. Res. 54: 99–120.

Kurtenbach, R., K.H. Becker, J.A.G. Gomes, J. Kleffmann, J.C. Lorzer, M. Spittler, P. Wiesen, R. Ackermann, A. Geyer and U. Platt. 2001. Investigations of emissions and heterogeneous formation of HONO in a road traffic tunnel. Atmos. Environ. 35: 3385–3394.

Lamb, B., A. Guenther, D.F. Gay and H. Westberg. 1987. A national inventory of biogenic hydrocarbon emissions. Atmos. Environ. 21: 1695–1705.

Langner, J. and H. Rodhe. 1991. A global three-dimensional model of the tropospheric sulfur cycle. J. Atmos. Chem. 13: 255–263.

Lech, T. 1998. Effect of mercury and its compounds in the environment of modern man (in Polish). Wiadomości Chemiczne 52: 87–100.

Lee, M., B.G. Heikes and D.W. O'Sullivan. 2000. Hydrogen peroxide and organic hydroperoxide in the troposphere: a review. Atmos. Environ. 34: 3475–3494.

Lefohn, A.S. 1997. Science, Uncertainty, and EPA's new ozone standards. Environ. Sci. Technol. 31(6): 280A–284A.

Martin, D., M. Tsiwou, B. Bonsang, C. Abonnel, T. Carey, M. Springer-Young. A. Pszenny and K. Suhre. 1997. Hydrogen peroxide in the marine atmospheric boundary layer during the Atlantic Stratocumulus Transition Experiment/Marine Aerosol and Gas Exchange experiment in the eastern subtropical North Atlantic. J. Geoph. Res. 102, (D5): 6003–6015.

Metz, N. 2005. Wide Emission Trend 1950 to 2050 Road Transport and all Sources. E U R O C H A M P Workshop in Andechs Chemistry, Transport and Impacts of Atmospheric Pollutants With Focus on Fine Particulates, 10–12 October 2005 (Available at: http://www.eurochamp.org/datapool/page/28/Metz.pdf).

Middleton, P. 1995. Sources of air pollutants. pp. 88–119. *In*: H.B. Singh [ed.]. Composition, Chemistry, and Climate of the Atmosphere. Van Nostrand Reinhold, New York.

Milaeva, E.R. 2006. The role of radical reactions in organomercurials impact on lipid peroxidation. J. Inorg. Biochem. 100: 905–915.

Milne, P.J., D.D. Riemer, R.G. Zika and L.E. Brand. 1995. Measurement of vertical distribution of isoprene in surface seawater, its chemical fate, and its emission from several phytoplankton monocultures. Mar. Chemistry 48: 237–244.

Möller, D. 1980. Kinetic model of atmospheric SO_2 oxidation based on published data. Atmos. Environ. 14: 1129–1144.

Mol, W., P. van Hooydonk and F. de Leeuw. 2011. The state of the air quality in 2009 and the European exchange of monitoring information in 2010, ETC/ACM Technical Paper 1/2011.

Monahan, S.E. 1983. Environmental Chemistry. Brooks/Cole Publishing Company Monterey, California, USA.

Monteiro, D.A., F.T. Rantin and A.L. Kalinin. 2010. Inorganic mercury exposure: toxicological effects, oxidative stress biomarkers and bioaccumulation in the tropical freshwater fish matrinxa, Brycon amazonicus (Spix and Agassiz, 1829), Ecotoxicology 19: 105–123. DOI 10.1007/s10646-009-0395-1.

Morel, F.M.M., A.M.L. Kracpicl and M. Amyot. 1998. The chemical cycle and bioaccumulation of mercury. Ann. Rev. Ecol. Syst. 29: 543–566.

Ozuah, P.O. 2000. Mercury poisoning. Curr. Probl. Pediatr. 30: 91–99. Doi:10.1067/ mps.2000. 104054.

Pacyna, E.G., J.M. Pacyna, J. Fudala, E. Strzelecka-Jastrzab, S. Hlawiczka and D. Panasiuk. 2006a. Mercury emissions to the atmosphere from anthropogenic sources in Europe in 2000 and their scenarios until 2020. Sci. Total Environ. 370: 147–156.

Pacyna, E.G., J.M. Pacyna, F. Steenhuisen and S. Wilson. 2006b. Global anthropogenic mercury emission inventory for 2000. Atmos. Environ. 40: 4048–4063.

Peña, R.M., S. Garcia, C. Herrero and T. Lucas. 2001. Measurement and analysis of hydrogen peroxide rainwater levels in a Northwest region of Spain. Atmos. Environ. 35: 209–219.

Pitts, J.N., Jr. and B.J. Finlayson. 1975. Mechanism of photochemical air pollution. Angewandte Chemie (International Edition in English). 14: 1–15.

Pyle, D.M. and T.A. Mather. 2003. The importance of volcanic emissions for the global atmospheric mercury cycle. Atmos. Environ. 37: 5115–5124.

Rasmussen, R.A. and F. Went. 1965. Volatile organic material of planet origin in the atmosphere. Proc. Natl. Acad. Sci. 53: 215–220.

Risher, J.F., R.A. Nickle and S.N. Amler. 2003. Elemental mercury poisoning in occupational and residential settings. Int. J. Hyg. Environ. Health 206: 371–379.

Rooney, J.P.K. 2007. The role of thiols, dithiols, nutritional factors and interacting ligands in the toxicology of mercury. Toxicology 234:145–156.

Schroeder, W.H. and J. Munthe. 1998. Atmospheric mercury—an overview. Atmos. Environ. 29: 809–822.

Schroder, W.H., G. Yarwood and H. Niki. 1991. Transformation processes involving Hg species in atmosphere—results from a literature survey. Water, Air Soil. Pollut. 56: 653–666.

Seigneur, C., J. Wrobel and E.A. Constantine. 1994. A chemical kinetic mechanism for atmospheric inorganic mercury. Environ. Sci. Technol. 28: 1589–1597.

Seinfeld, J.H. and S.N. Pandis. 1998. Atmospheric Chemistry and Physics, from Air Pollution to Climate Change. J. Wiley & Sons Inc., New York, Chichester, Weinheim, Brisbane, Singapore, Toronto, 1326 pp.

Seltzer, J., B.G. Bigby and M. Stulbarg. 1986. O_3—induced change in bronchial reactivity to methacholine and airway inflammation in humans. J. Appl. Physiol. 60: 1321–1326.

Shang, L., X. Feng, W. Zheng and H. Yan. 2003. Preliminary study of the distribution of gaseous mercury species in the air of Guiyang city, China. J. Phys. IV. 107: 1219–1222.

Statheropoulos, M., A. Agapiou and G. Pallis. 2005. A study of volatile organic compounds evolved in urban waste disposal bins. Atmos. Environ. 39: 4639–4645.

Steffen, A., W. Schroeder, L. Poissant and R. Macdonald. 2003. Mercury in the arctic atmosphere. pp. 124–142. *In*: T. Bidleman [ed.]. The Canadian Arctic Contaminants Assessment Report II: Sources, Occurrence, Trends and Pathways in the Physical Environment. Department of Indian Affairs and Northern Development, Ottawa, Canada.

Stier, P.A. and R.A. Gordon. 1998. Psychiatric aspects of mercury poisoning. Medical Update for Psychiatrists 3(4): 144–147.

Stratton, W.J. and S.E. Lindberg. 1995. Use of a refluxing mist chamber for measurement of gas—phase mercury (II) species in the atmosphere. Water, Air, Soil Pollut. 80: 1269–1278.

Wängberg, I., J. Munthe, N. Pirrone, A. Inverfeldt, E. Bahlman, P. Costa, R. Ebinghaus, X. Feng, R. Ferrara, K. Gardfelt, H. Kock, E. Lanzillotta, Y. Mamane, F. Mas, E. Melamed, Y. Osnat, E. Prestbo, J.

Sommar, S. Schmolke, G. Dpain, F. Spovieri and G. Tuncel. 2001. Atmospheric mercury distribution in Northern Europe and in Mediterranean region, Atmos. Environ. 35: 3019–3025.

Wahner, A., F. Rohrer, D.H. Ehhalt, E. Atlas and B. Ridley. 1994. Global measurements of photochemically active compounds. pp. 205–222. *In*: R.G. Prinn [ed.]. Global Atmospheric-Biospheric Chemistry. Plenum Press, New York, London.

Wang, Q., D. Kim, D.D. Dionysiou, G.A. Sorial and D. Timberlake. 2004. Sources and remediation for mercury contamination in aquatic systems the literature review. Environ. Pollut. 131: 323–336.

Wang, Z. and S.O. Pehkonen. 2004. Oxidation of elemental mercury by aqueous bromine: atmospheric implications. Atmos. Environ. 38: 3675–3688.

Warneck, M.L. 1988. The temporal and spatial distribution of tropospheric nitrous oxides. J. Geophys. Res. 86: 7185–7195.

Weiss-Penzias, P., D.A. Jaffe, A. Mc Clintick, E.M. Prestbo and M.S. Landis. 2003. Gaseous elemental mercury in the marine boundary layer: evidence for rapid removal in anthropogenic pollution. Environ. Sci. Technol. 37(17): 3755–3763.

Wennberg, P.O. 2006. Atmospheric chemistry: radicals follow the Sun. Nature. 442: 145–146 (doi:10.1038/442145a) http://www.nature.com/nature/journal/v442/n7099/fig_tab/442145a_F1.html.

Wennberg, P.O., R.C. Cohen, R.M. Stimpfle, J.P. Koplow, J.G. Anderson, R.J. Salawitch, D.W. Fahey, E.L. Woodbridge, E.R. Keim, R.S. Gao, C.R. Webster, R.D. May, D.W. Toohey, L.M. Avallone, M.H. Proffitt, M. Loewenstein, J.R. Podolske, K.R. Chan and S.C. Wofsy. 1994. Removal of stratospheric O_3 by radicals: *in situ* measurements of OH, HO_2, NO, NO_2, ClO, and BrO. Science 26: 389–404.

Went, F.W. 1960. Organic matter in the atmosphere and its possible relation to petroleum formation. Proc. Natl. Acad. Sci. (USA) 46: 212–221.

Whitten, G.Z., H. Hoho and J.P. Killus. 1980. The carbon-bond mechanism: a condensed kinetic mechanism for photochemical smog. Environ. Sci. and Technol. 14: 690–700.

[WHO] World Health Organization. 1979. Environmental Health Criteria 13. Carbon Monoxide, Geneva, Switzerland.

[WHO] World Health Organization. 1987. Air Quality Guidelines for Europe. WHO Regional Office for Europe, Copenhagen. Denmark.

Wiener, J.G., D.P. Krabbenhoft, G.H. Heinz and A.M. Scheuhammer. 2003. Ecotoxicology of mercury. pp. 409–463. *In*: D.J. Hoffman, B.A. Rattner, G.A. Burton, and J. Cairns [eds.]. Handbook of Ecotoxicology. Lewis Publ., Boca Raton, FL, USA.

Zalups, R.K. 2009. Molecular interactions with mercury in the kidney. Pharmacol. Rev. 52(1): 114–140.

Zalups, R.K. and L.H. Lash. 1994. Advances in understanding the renal transport and toxicity of mercury. J. Toxicol. Environ. Health 42: 1–44.

Zuo, Y. and J. Hoigné. 1992. Formation and hydrogen peroxide and depletion of oxalic acid in atmospheric water by photolysis of iron (III)-oxalato complexes. Environ. Sci. Tech. 26: 1014–1022.

Links:

www.atmosphere.mpg.de
www.epa.gov/airtrends/nitrogen.html
www.epa.gov/airtrends/sulfur.html
www.epa.gov/airtrends/ozone.html
www. eea.europa.eu
www.ciop.pl/3589.htmlhttp
www.ellaz.pl/polska/ksia-powietrze.htm#powietrze7
www.nature.com/nature/journal
www.rapidfire.sci.gsfc.nasa.gov/firemaps

2

Atmospheric Aerosol Particles

Kikuo Okada

Introduction

The term "atmospheric aerosol particles" means the particles suspended in the air, generally about 0.001 to 100 μm in size (Willeke and Baron 1993).

Since there are many aerosol sources, atmospheric aerosol particles composed of various materials are present in a wide size range. Thus the atmospheric aerosols form a size spectrum that extends over about four orders of magnitude. After the generation, the growth of particles occurs by coagulation and gas-to-particle conversion processes in the atmosphere. The growth corresponds to the modification of aerosol particles and would make aerosol particles of mixed composition.

Aerosol particles influence the atmospheric processes such as cloud formation and atmospheric radiation processes and then modify the earth's climate. Moreover, the effects of aerosol particles on environment and human health are of importance for people living mainly in urban areas.

In this chapter, features and behavior of aerosol particles in the atmosphere near the surface, where people mainly live, will be discussed.

Annual emission rates of various aerosol particles

Aerosol particles are emitted from the various surface sources as a primary product (particles are emitted directly to the atmosphere). The typical particles of natural origins are wind-blown mineral particles and sea-salt particles that are generated by the mechanical disintegration of the earth's surface. Primary bioaerosol particles such as pollen, spore, and bacteria are emitted from the biogenic sources. Volcanic mineral particles were injected into the atmosphere by the volcanic eruption. The mass

Meteorological Research Institute, 1-1 Nagamine, Tsukuba, Ibaraki 305-0052, Japan.
E-mail: kokada@mri-jma.go.jp; kokadasan@kni.biglobe.ne.jp

concentrations of these primary particles (natural origin) are generally dominant in the size range of > 2 μm diameter.

In addition to primary aerosol particles, aerosol particles are formed in the atmosphere from gaseous precursors through gas-to-particle conversion (secondary particles).

Table 2-1 shows the estimated annual fluxes of the major mass sources of atmospheric aerosol particles to the atmosphere in the 1980s (Kiehl and Rodhe 1995). The fluxes were modified from Andreae (1985). Emission of mineral particles from soil dust has been classified as a natural source, although a certain amount of anthropogenic influence is likely to exist. Particles emitted by biomass burning are not purely of anthropogenic origin.

The estimated values of each aerosol type are in a wide range. If we look the best values, the estimated flux of aerosols originated from anthropogenic sources is present

Table 2-1. Estimated annual fluxes of the major mass sources of atmospheric aerosol particles to the atmosphere in the 1980s (Kiehl and Rodhe 1995).

Source	Flux unit [Tg yr^{-1}] (dry mass) Estimated flux			Particle size* category	Optical depth
	low	high	best		
Natural					
Primary					
Soil dust (mineral aerosol)	1000	3000	1500	mainly coarse	0.023
Sea salt	1000	10,000	1300	coarse	0.003
Volcanic dust	4	10,000	30	coarse	0.001
Biological debris	26	80	50	coarse	0.002
Secondary					
Sulfates from biogenic gases	80	150	130	fine	0.021
Sulfates from volcanic SO_2	5	60	20	fine	0.003
Organic matter from biogenic VOC	40	200	60	fine	0.017
Nitrates from NOx	15	50	30	fine and coarse	0.001
Total: Natural	2200	23,500	3100		0.071
Anthropogenic					
Primary					
Industrial dust, etc. (except soot)	40	130	100	fine and coarse	0.004
Soot	5	20	10	mainly fine	0.006
Secondary					
Sulfates from SO_2	170	250	190	fine	0.032
Biomass burning	60	150	90	fine	0.027
Nitrates from NOx	25	65	50	mainly coarse	0.002
Organics from anthropogenic VOC	5	25	10	fine	0.003
Total: Anthropogenic	300	650	450		0.075
TOTAL	2500	24,000	3600		0.149

*Coarse and fine size categories refer to mean particle diameter above and below 1 μm, respectively.

in about 13% of the total flux. However, secondary anthropogenic aerosols account for 60% of total secondary aerosols.

Aerosol concentration and size distribution

Coulier (1875a,b) in France first discovered that condensation of water vapor occurs more readily in unfiltered air that in filtered air when air is expanded adiabatically. It means that particles are needed for the condensation of water vapor. Aitken (1880) in England also obtained similar findings without the information on Coulier's experiments, and thoroughly studied the number concentrations of total particles for the first time using his developed dust counter (Aitken 1923). The number concentrations of total particles were measured with the Aitken dust counter at various locations by many researchers (Landsberg 1938) (Table 2-2).

Table 2-2. Number concentrations of total particles at various locations (Landsberg 1938).

Locality	No. of places	No. of observations [cm^{-3}]	Average concentrations [cm^{-3}]	Average maximum [cm^{-3}]	Average minimum [cm^{-3}]
City	28	2,500	147,000	379,000	49,000
Town	15	4,700	34,300	114,000	5,900
Country inland	25	3,500	9,500	66,500	1,560
Country seashore	21	2,700	9,500	33,400	1,560
Mountain:					
500–100 m	13	870	6,000	36,000	1,390
1000–2000 m	16	1,000	2,300	9,830	450
2000 m	25	190	950	5,300	160
Islands	7	480	9,200	43,600	460

Historical review of atmospheric condensation nuclei was found in Spurny (2000) and McMurry (2000).

The comprehensive size distribution of aerosol particles has not been obtained for a long period. Based on the size distribution measurements over different parts of the world, Junge and co-workers in the 1950s obtained the number-size distributions of aerosol particles in the radius range of 0.004–20 μm and found that the size distributions follow a power law function over the radius range of 0.1–20 μm (Junge 1950, 1963). This power-law form of the size distribution was termed the Junge distribution. The radius range was subdivided into three groups. The particles smaller than 0.1 μm are called Aitken particles. The larger particles are classified into large (0.1–1 μm radius) and giant particles (> 1 μm radius). The methods of aerosol measurement before the 1960s were reviewed by Spurny (1999).

In the mid 1960s, a new instrument (electrical mobility analyzer) was developed for the continuous measurement of aerosol number concentrations in the size range of 0.008–0.5 μm (Whitby and Clark 1966; Clark and Whitby 1967). In the 1969 Pasadena Study, number-size distributions of aerosols were measured with the electrical mobility analyzer and optical counter. A total of 342 aerosol size distributions have been obtained in Pasadena during August and September 1969, over the diameter range of 0.008–6.8 μm (Whitby et al. 1972). The volume-size distributions of the smog aerosols universally showed bimodal with the saddle point in the 1- to 2-μm size range. Most of the mass of particles smaller than 1 μm diameter was contributed by photochemical reactions during smog period, whereas most of the mass of particles larger than a few microns was originated from other sources. It was also found that the volume concentration of aerosols smaller than 1.05 μm was correlated well with aerosol light scattering coefficient.

It can be seen from Fig. 2-1 that the complete number-size distribution of aerosols in the urban atmosphere is composed of three modes (Whitby 1978); the nucleus (or nucleation) mode that has peaks at diameter D around 0.01 μm, the accumulation

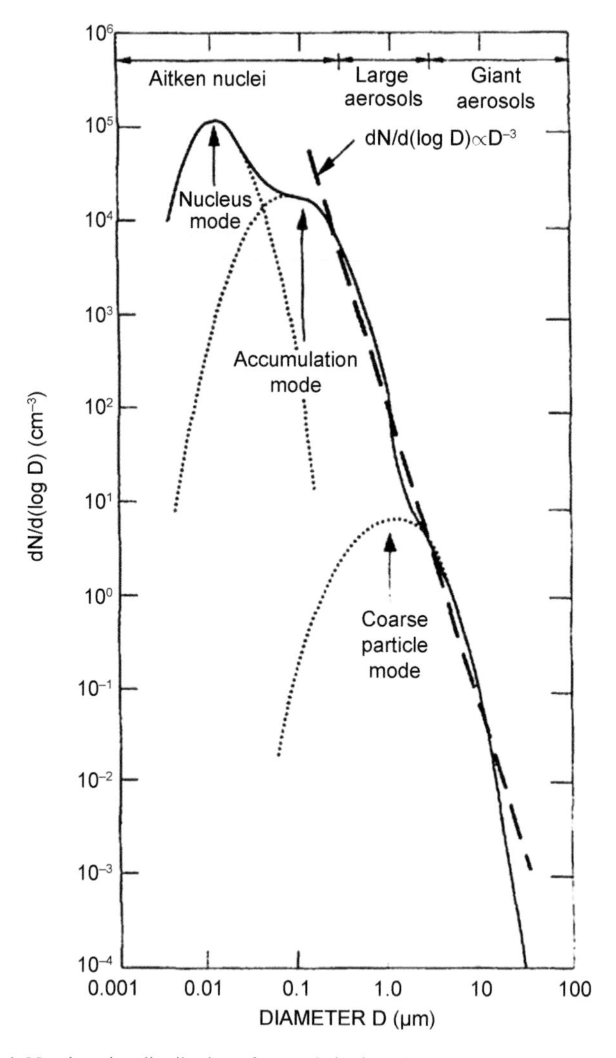

Fig. 2-1. Number-size distribution of aerosols in the urban atmosphere (Whitby 1978).

mode that has peaks at D around 0.1 μm, and the coarse particle mode that has peaks at D around 1 μm. Particles with $D \leq 0.2$ μm dominate the total number concentration; that is, the concentration measured by the Aitken nucleus counter. Hence, these small particles are termed Aitken nuclei (Aitken particles).

Particles with $D < 2$ μm and $D > 2$ μm are referred to as "fine particles" and "coarse particles", respectively. The distinction between fine particles and coarse particles is a fundamental matter. The fine and coarse particles are usually chemically different. The physicochemical processes such as condensation and coagulation produce fine particles while mechanical processes produce mostly coarse particles.

Jaenicke (1993) showed the number size distributions of six aerosol types; that is, polar, background, maritime, remote continental, desert dust storm, rural, and urban (Fig. 2-2). The number-size distribution of each aerosol type was described by the sum of three lognormal functions. The main differences among the distributions are for the smallest and the largest particles. This probably reflects the influence of particle residence time (Jaenicke 1980). Aerosol particles show only limited concentration variations in the radius range of 0.1–1 μm. Particles in this radius range influence atmospheric radiation such as reduction in visibility and cloud formation.

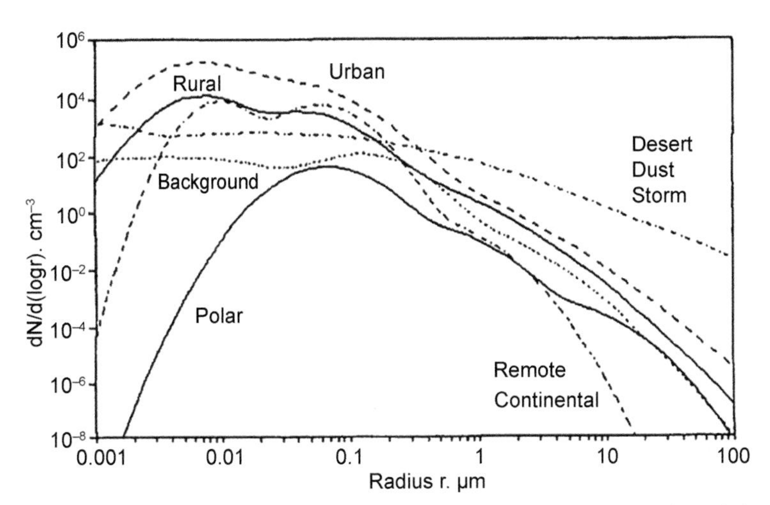

Fig. 2-2. Number size distributions of six aerosol types: polar, background, remote continental, desert dust storm, rural, and urban (Jaenicke 1993).

It is to be noted that particles having diameters < 50 nm (0.05 μm) and < 100 nm (0.1 μm) are also defined as nanoparticles and ultrafine particles, respectively. Nanoparticles in the atmosphere were reviewed by Buseck and Adachi (2008). There is evidence that several health effects are associated with the ultrafine particles with diameters of < 100 nm, and that they can penetrate the cell membranes, enter into the blood and even reach the brain (Oberdörster and Utell 2002; Nel 2005).

Aerosol spatial distribution

In this section, the horizontal distribution of aerosol particles is shown. One of the important researches on atmospheric radiation is the effect of aerosols on visibility. Conversely, the analysis of visibility is useful to evaluate the amount of aerosols in the atmosphere over land. Standard visibility, L (km), may be written as follows (Koschmieder 1924; Middleton 1935, 1952);

$$L = \frac{K}{b} \tag{2-1}$$

where b (km^{-1}) is the light extinction coefficient which is the sum of light scattering and absorption coefficients in ambient air. The constant K is 3.912. The extinction of light through the atmosphere is caused by the scattering and absorption of light due to aerosol particles and gaseous materials. For a wavelength of visible light equal to 0.55 μm, the scattering coefficient of air molecules is 0.01162 km^{-1} at standard pressure and temperature. Hence, the horizontal visibility determined by the Rayleigh scattering alone is estimated to be about 300 km. In particular, the light extinction by aerosol particles predominates over that by gaseous components. The light extinction coefficient obtained by visibility is nearly proportional to the mass concentration of aerosols (Charlson et al. 1967; Noll et al. 1968; Waggoner et al. 1981). Hence, visibility data obtained at surface meteorological observatories is useful to evaluate the horizontal distribution of aerosol mass loading.

Husar et al. (2000) obtained the global continental haze pattern evaluated based on daily average visibility data at about 7000 surface weather stations over five years, 1994–1998. The data used are obtained from the Global Summary of Day (SOD) database distributed by the National Climatic Data Center (NCDC).

Figure 2-3 shows the location density of visibility measurement stations used. As found in Fig. 2-3, the spatial coverage is highest in Europe, Russia, East Asia, and USA, where the stations are about 100–300 km apart. Throughout much of the remaining continents, the average station distance is 200–400 km. Low spatial coverage occurs over northern Canada, northern Siberia, western China, as well as over the central portions of South America, Africa, and Australia.

Although the global extinction coefficient was evaluated through a year, the results during the period from September to November are shown in Fig. 2-4. If there were no stations within 500 km, contour area was left blank (white). The results reveal that the continental haze is concentrated over distinct aerosol regions of the world. The haziest regions of Asia were found in the Indian subcontinent, eastern China, and Indochina where seasonal extinction coefficient exceeded 0.4 km^{-1}. In Africa, the highest year around extinction coefficient $>$ 0.4 km^{-1} was found over Mauritania, Mali, and Niger. During December, January, February, the savanna region of sub-Saharan Africa showed similar values.

The haziest region of South America was over Bolivia, adjacent to the Andes mountain range, with a peak during August–November (0.4–0.6 km^{-1}). In North America and Europe, there were isolated haze pockets, such as the San Joaquin Valley in California and the Po River Valley in northern Italy.

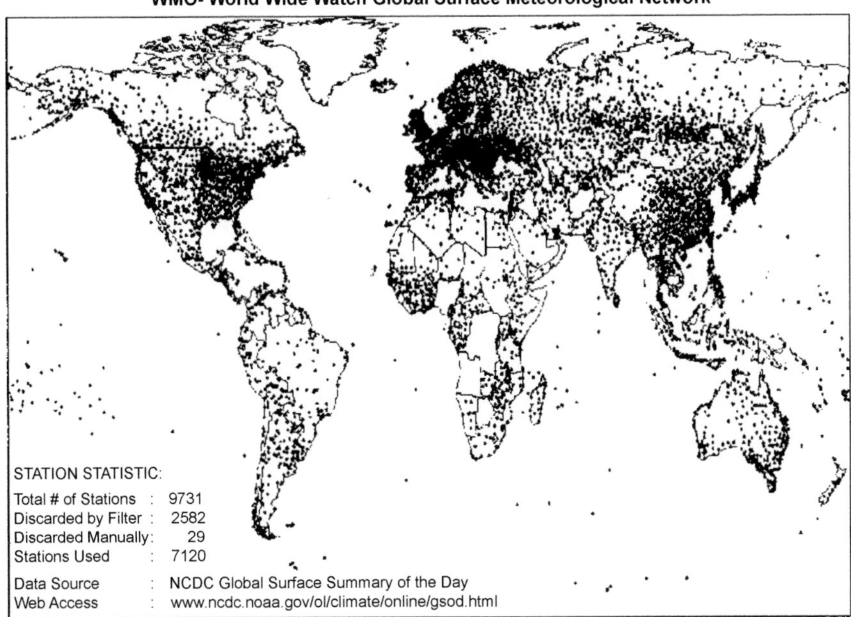

Fig. 2-3. Location density of visibility measurement stations used in the study of Husar et al. (2000).

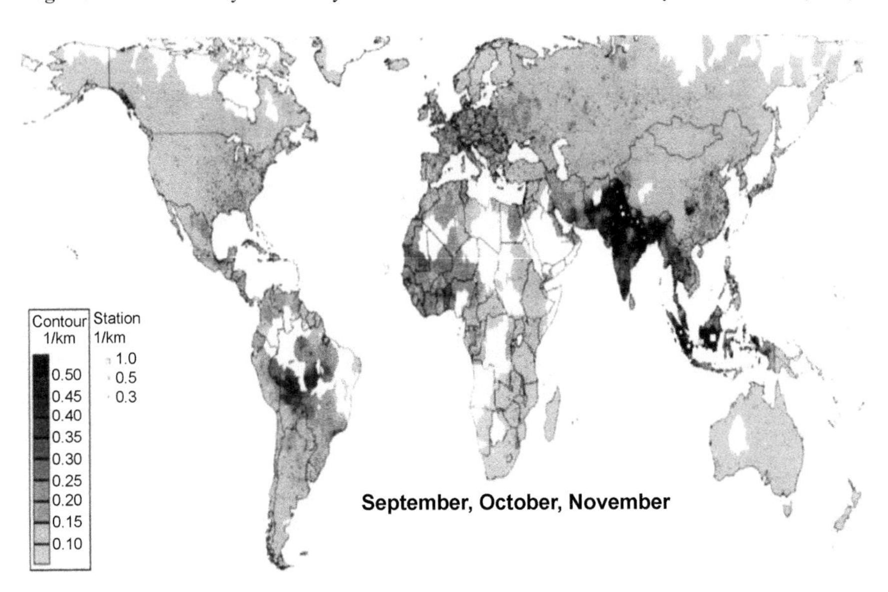

Fig. 2-4. Global extinction coefficient: the results during the period from September to November.

The analysis of light extinction coefficient from visibility data is also useful to study the trends in air pollution, associated with the combustion of fossil fuels (Husar et al. 1981; Okada and Isono 1982) and biomass burning (Field et al. 2009).

Although the horizontal distribution of aerosol optical depth over the ocean was studied by satellite observation (e.g., Husar et al. 1997), the results obtained by direct observations using ships are important to indicate aerosol concentration in the atmosphere near the surface. The horizontal distribution of Aitken particles over the ocean has been studied (e.g., Shiratori 1934; Podzimek 1980). Podzimek (1980) found that the concentrations are greatest close to land and, conversely, they are low values in the central parts of the ocean. As shown previously in Table 2-2, Aitken particles are present in high concentrations over the continent, mainly associated with the combustion processes. The Aitken number concentration decreases as air mass moves off the coast and it is several hundred cm^{-3} in the central parts of oceans.

Heintzenberg et al. (2000) reviewed the physical and chemical marine aerosol data in order to derive global-size distribution parameters and inorganic particle composition on a coarse 15° × 15° grid. Two submicrometer log-normal distributions were fitted to the number-size distribution data. The geometric mean diameters of both the Aitken and accumulation modes were larger in the Northern Hemisphere by about 25%. The combination of increased concentration and increased mean diameter for the accumulation mode results in a volume concentration ratio of about 2:1 between the Northern and Southern Hemispheres. The analysis of aerosol chemical data showed that the influence of anthropogenic aerosols was clearly recognized over the ocean in the Northern Hemisphere.

Chemical composition of aerosol particles

Chemical composition of fine particles with $D < 1$ μm collected near the surface was reviewed by Heintzenberg (1989). The aerosol data were classified into "urban (19 places)", "non-urban continent (14 places)", and "remote regions (11 places) by the sampling locations. Table 2-3 shows the total fine particle mass (TFP) and weight percentages of elemental carbon (EC), organic carbon (OC), NH_4^+, NO_3^-, and SO_4^{2-} at the urban, non-urban continent, and remote regions. Average urban TFP value is 33 μ gm^{-3}. The "non-urban" results were collected from experiments on the continents but away from large urban or industrial sources. Average TFP in non-urban continental areas is only a factor of two lower than the corresponding urban value. This result reflects the long residence times of submicrometer particles and the proximity to

Table 2-3. Main chemical composition of fine particles with $D < 1$ μm (Heintzenberg 1989).

Location	Mass concentration [μg m^{-3}]	Mass fraction [%]				
		EC	OC	NH_4^+	NO_3^-	SO_4^{2-}
Urban (19)*	32	9	31	8	6	28
Non-urban continental (14)*	15	5	24	11	4	37
Remote (11)*	4.8	0.3	11	7	3	22

*Number of places studied.

industrial regions. Consequently, no drastic changes in fine particle composition occur when moving away from the urbanized areas. In the "urban" and "non-urban continental" areas, 2/3 of the fine particle mass is composed of sulfates and carbon compounds while about 15% is nitrogen compounds. 1/4 to 1/6 of the carbonaceous material is elemental carbon (EC). It is this minor component that is responsible for most of the absorption of solar radiation.

The data sets from remote regions were taken on remote islands or research vessels. The Northern and Southern Hemispheres are covered roughly evenly. These mostly marine results are presented as a rough estimate of fine particle composition in the least-contaminated parts of the troposphere. The average TFP-value is lower by another factor of 3 compared to the non-urban continental level. On the average, 30% of TFP remained undetermined. Even here sulfate compounds are a major component. About 1/5 of the fine particle mass is sulfur compounds. Organic carbon is the next most important component with about 11% of TFP. EC as the most clearly identifiable anthropogenic component exhibits only 0.3% of total particle mass corresponding to an average concentration of 10 ng m^{-3}.

Aerosol particles produced by the mechanical disintegration of the earth's surface

Mineral particles

Strong winds over loose soil areas can produce large amounts of atmospheric mineral particles. It is well known that the dust storm occurred in the desert and arid regions of southwestern USA, Sahara in Africa, northwestern China, and Australia. Figure 2-5 shows the soil particle motions during wind erosion (Gillette 1980).

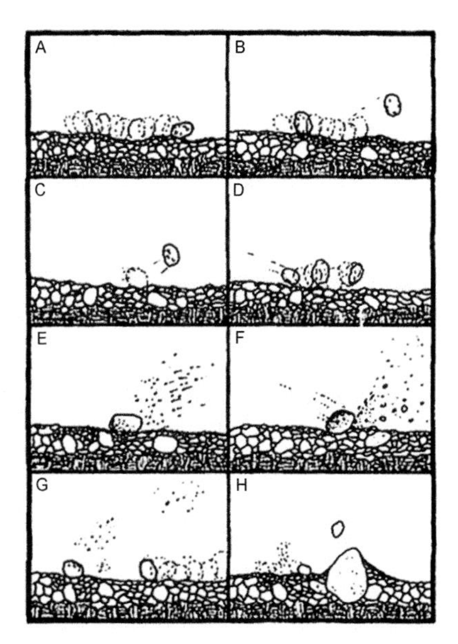

Fig. 2-5. Soil particle motions during wind erosion (Gillette 1980).

Figure 2-5a shows the creeping motion of a coarse particle close to threshold velocity. In Fig. 2-5b, the particle is lifted into the air by turbulent fluctuations, and in Fig. 2-5c, the particle collides with the surface and bounces. In Fig. 2-5D, a collision is followed by creeping and another lift. Figure 2-5E shows the breaking off of smaller particles that were encrusted on the colliding particle's surface. In Fig. 2-5F, a coarse particle collision is followed by splashing of fine particles into the air. Figures 2-5E and 2-5f illustrate sandblasting. In Fig. 2-5G, many of the above movements are combined, and the fine particles emitted are carried high into the air. Figure 2-5H shows an example of a nonerodible element giving protection to the soil downwind. Sources, distributions, and flux of mineral particles were reviewed by Duce (1995).

The number-size distributions of atmospheric mineral particles were measured in the radius range of 0.01–100 μm over the middle and southern Sahara desert (d'Almeida and Schütz 1983) (Fig. 2-6). The samples were examined by using a combination of a scanning electron microscope, optical microscope and sieving. The most western site is Matam, a city in Senegal at the northeastern border with Mauritania.

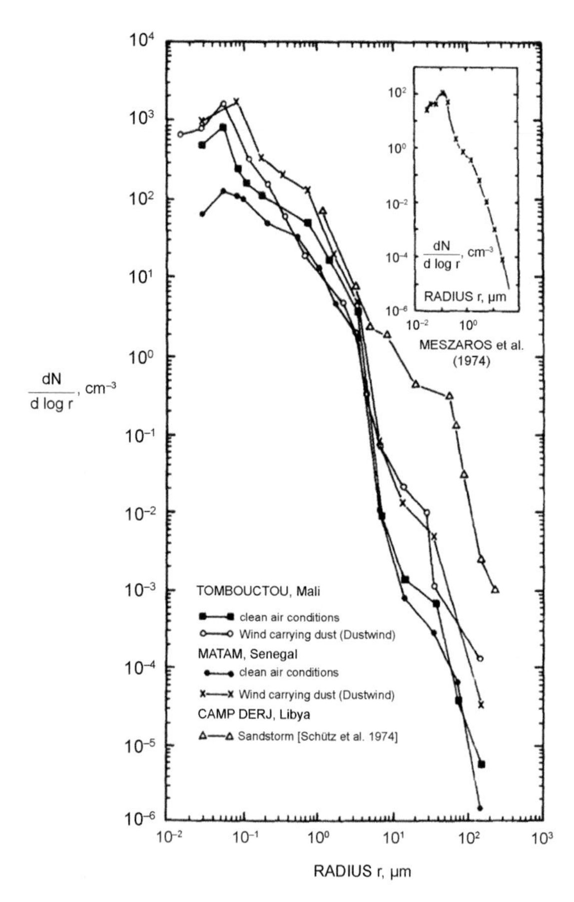

Fig. 2-6. Number-size distributions of atmospheric mineral particles over the middle and southern Sahara desert (d'Almeida and Schütz 1983).

Matam itself is located in the flood plains of the Senegal River and the surrounding area is of typical Sahelian character. Goundam and Dar Albeida are sampling sites in Mali, which are located southwest of Tombouctou. As found in Fig. 2-6, all size distributions of mineral particles show a distinct absolute maximum between 0.05 and 0.08 μm radius. This shows clearly that crustal generated particles are also found in the Aitken particle size range, which is usually the domain of "gas-to-particle-conversion" species, based on the general knowledge of the physicochemical behavior of these particles. From the point of view of aerosol mass or volume distributions, those mineral particles below 0.1 μm radius will not play an important role with respect to the total mass or volume of mineral particles.

Figure 2-7 shows the electron micrograph of mineral particles collected at Zhangye, China.

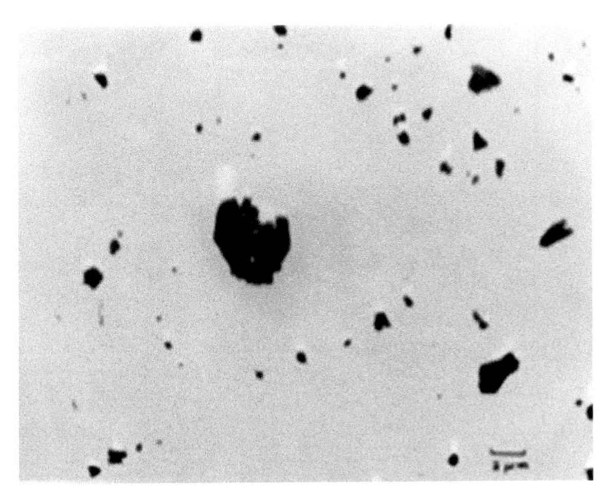

Fig. 2-7. Electron micrograph of mineral particles collected at Zhangye, China (Okada and Kai 1995).

As found in Fig. 2-7, the shape of atmospheric mineral particles shows irregular shapes. Okada et al. (2001) studied the shape of atmospheric mineral particles of 0.1–6 μm radius by electron microscopy applied to the samples collected in three arid-regions in China (Qira in the Taklamakan Desert, Zhangye near the southern border of the Badain-Jaran Desert, and Hohhot in northern China). In all three regions, the mineral particles showed irregular shapes with a median aspect ratio b/a (ratio of the longest dimension b to the orthogonal width a) of 1.4. Although the aspect ratio exhibited no clear size dependence, the circularity factor ($4\pi S/l^2$; S is surface area and l is periphery length) tended to decrease with increasing radius, suggesting the presence of aggregated mineral particles at larger sizes. The ratio of particle height-to-width h/a was also evaluated by measuring the shadow length. The median ratio h/a was 0.49 in Hohhot, 0.29 in Zhangye, and 0.23 in Qira. Analytical functions were fitted to the grand total of the frequency distributions of aspect ratios, height-to-width ratios, and circularity factors allowing parametric calculations of radiative effects and calculations of optical and sedimentation behavior of mineral particles.

Sea-salt particles

For many years, sea-salt particles are considered to be produced upon the evaporation of spray generated from the tops of breaking sea waves. However, salt particles formed through the sea spray are too large to be present in the air for long time. In the 1950s, it is known that the droplets are mainly formed by the breaking of many air bubbles as they reach the surface of the sea (e.g., Woodcock 1953; Blanchard 1954; Kientzler et al. 1954; Mason 1954; Blanchard and Woodcock 1957).

Figure 2-8 shows the schematic diagram of the production of sea-salt particles from the bursting of bubbles (Junge 1963). Day (1964) indicated that bubbles of < 300 μm bursting in seawater do not appear to produce any film drops, whereas a 2000 μm bubble will produce a maximum of about 100.

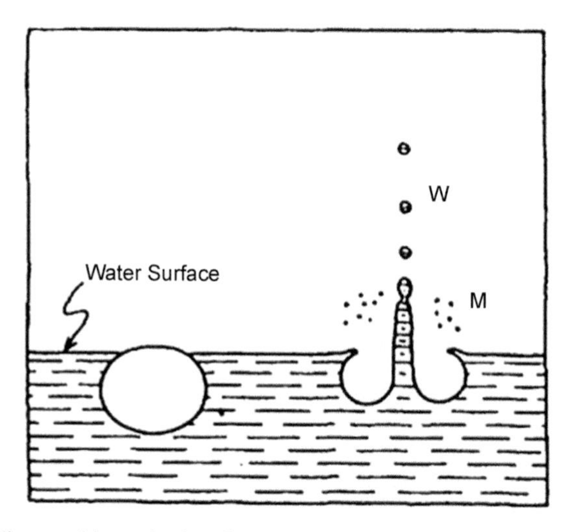

Fig. 2-8. Schematic diagram of the production of sea-salt particles from the bursting of bubbles (Junge 1963).

Figure 2-9 indicates that a large part of the aerosol particles between 0.2 and 1 μm radius on the collection surface (silicon oil; 25 Å thickness) were cubic in shape (Okada et al. 1983). These particles were collected at a coastal area of Nagoya in Japan during the strong summer-monsoon winds from the Pacific Ocean. Electron diffraction was applied to aerosol particles A and B in Fig. 2-9. The interplanar distances of the electron diffraction pattern from the particles agreed with the distance of sodium chloride (NaCl). It is found that other materials are present around the NaCl crystal.

In the dissolved components of seawater of salinity 35‰ (Berg and Winchester 1978), the weight ratios for major components are Cl/Na (1.7956), SO_4^{2-}/Na (0.2517), Mg^{2+}/Na (0.1201), Ca^{2+}/Na (0.0382), and K^+/Na (0.0370). Upon the evaporation of sea droplets, different materials re-crystallize separately due to each solubility product. Eugster et al. (1980) indicated that the evaporation of seawater produces NaCl as a major material, along with many alkali sulfates (e.g., $CaSO_4 \cdot 2H_2O$, $CaSO_4$, $Na_2Ca(SO_4)_2$, $K_2MgCa_2(SO_4) \cdot 2H_2O$, and $MgSO_4$. It is considered that the behavior of sea-salt particles is mainly determined by NaCl.

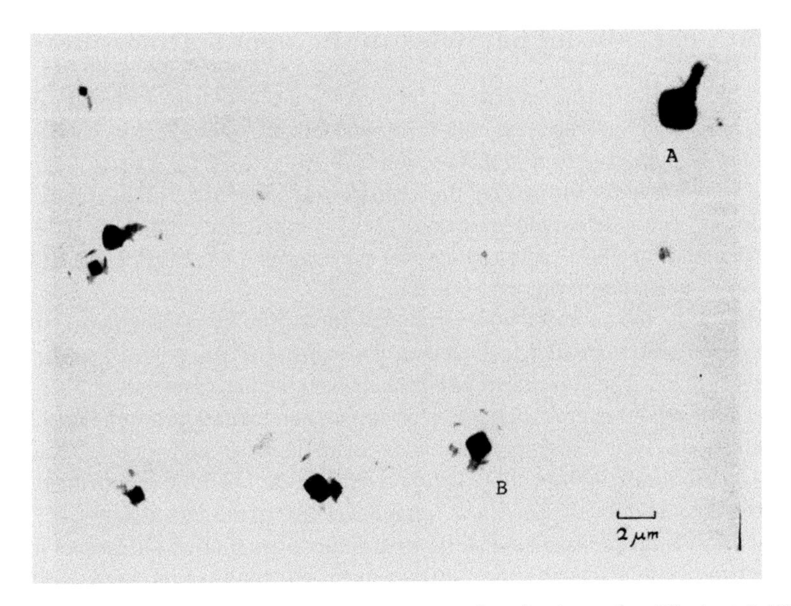

Fig. 2-9. Aerosol particles between 0.2 and 1 μm radius on the collection surface (Okada et al. 1983).

Emission of sea-salt particles from the ocean surface was studied in relation to wind speed. Lovett (1978) obtained the mass concentration of sea-salt particles from North Atlantic weather ships during 11 voyages during the period September 1974 to July 1975. Total sea-salt concentrations were measured by air sampling through membrane filters. It was found that total sea-salt concentration could be represented as a function of wind speed by the equation $\ln \theta = 0.16u + 1.45$ (θ = sea-salt concentration in μg m^{-3} and u = wind speed in m s^{-1}). This relation gives the mass concentration of sea-salt particles; that is, $\theta = 9.5$, 20, and 55 μg m^{-3} for wind speeds of 5, 10, 15 m s^{-1}, respectively.

Number-size distributions of sea-salt particles, including those with radii less than 0.1 μm, were measured over the ocean (Mészáros and Vissy 1974; Gras and Ayers 1983; O'Dowd and Smith 1993; Clarke et al. 2006). Their observed results indicated the presence of Aitken sea-salt particles. Bigg et al. (1995) also indicated the presence of the wind-produced particles in the Aitken size range.

Cipriano et al. (1983) carried out a laboratory experiment of generation of sea-salt particles by breaking wave or whitecap, and found that submicron- and even Aitken-sized particles were produced: the presence of salt particles of mass $< 10^{-17}$ g (radius < 0.01 μm) could be inferred. The evidence strongly suggests that the submicron fraction is composed of film drops, derived primarily from bubbles larger than 1 mm in diameter. Similar to mineral particles, sea-salt particles would be abundant in the Aitken size range over the ocean in the situation of strong winds.

Sulfate-containing particles in the urban atmosphere

Although aerosol particles can be formed by gas-to-particle conversion in the atmosphere over various areas, the formation of sulfate-containing particles in the urban atmosphere is shown in this section.

Sulfate aerosols formed by the oxidation of SO_2 are usually related to fine particles, or accumulation mode particles of 0.1–1 μm diameter (Whitby 1978). Hering and Friedlander (1982) collected aerosol particles at Los Angeles in California in 1976–1977 using a low-pressure impactor.

The collected samples were analyzed for sulfur by flash volatilization with subsequent flame photometric detection. Two characteristic types of sulfur aerosols were recognized. The Type I distributions had mass median diameters of 0.42–0.65 μm and were observed over a wide range of sulfate concentration, from 5 to 52 μg m^{-3}. In contrast, the Type II distributions had mass median diameters of 0.17–0.22 μm. Of the 69 distributions, 90% could be categorized as Type I or Type II with the remaining 10% showing a bimodal behavior. A "grand average" of the data is shown in Fig. 2-10, where the averaged normalized sulfur distributions are plotted for each of the types.

In the Type I distribution, sulfate mass concentrations were 3–9 μg m^{-3}. Calculations indicated that the Type I sulfates were formed from aerosol phase reactions, and the Type II sulfates were formed from gas phase reactions. The size of the Type I sulfates, which falls in the optimum light scattering range, accounts for the observed contribution of sulfates to visibility reduction.

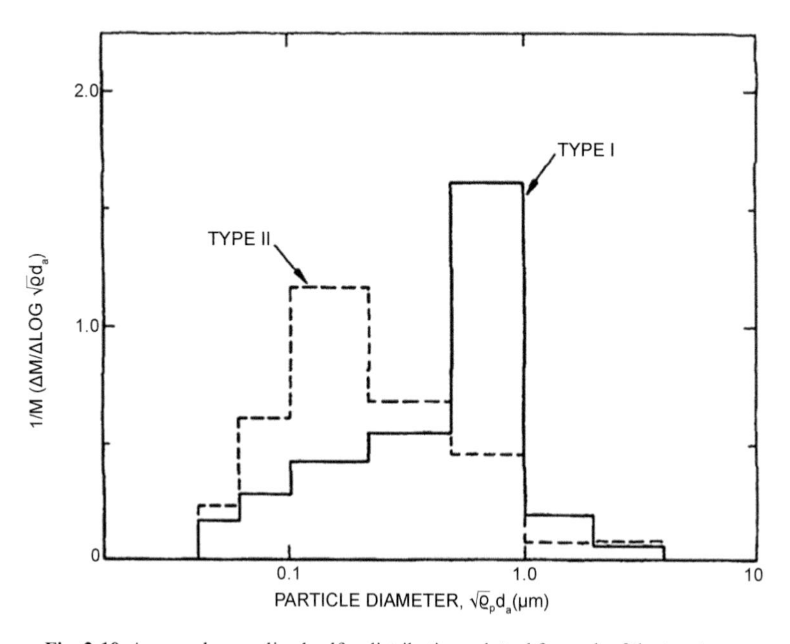

Fig. 2-10. Averaged normalized sulfur distributions plotted for each of the two types.

Similar findings were observed in the urban atmosphere of Nagoya in Japan by using electron microscopic examinations (Okada 1985). Sulfate ions in particles can be detected by the electron microscopy using a vapor-deposited thin film of barium chloride (Bigg et al. 1974; Ono et al. 1981). The procedures of the analysis are shown in the electron micrographs (Fig. 2-11). Individual aerosol particles were collected directly

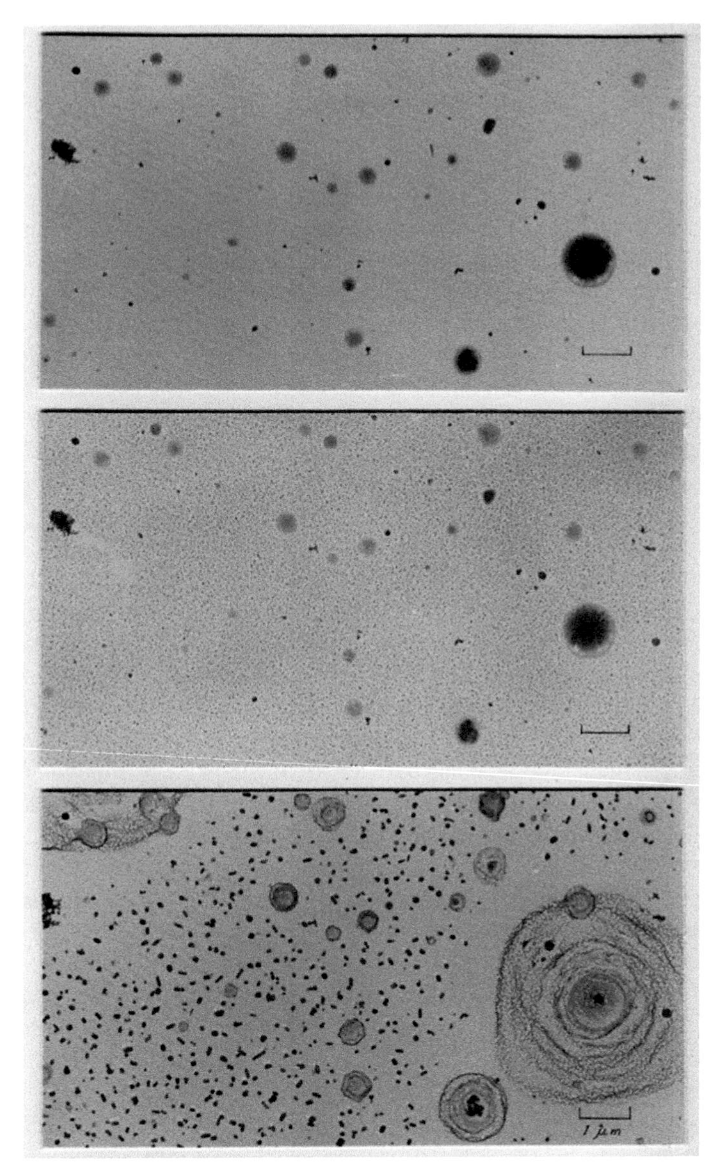

Fig. 2-11. Electron micrographs of aerosol particles collected at Yokkaichi, Japan (Okada et al. 1983b).

a) General view.
b) After evaporation of barium chloride.
c) After exposure to octanol vapor for 24 hours.

on a carbon-covered nitrocellulose film supported on an electron microscope grid by using an electrostatic aerosol sampler. The shape and apparent diameter of individual particles on a carbon film were investigated with a transmission electron microscope (Fig. 2-11a). After that, the same specimen was placed in a vacuum evaporator and coated with a vapor-deposited thin film of barium chloride (20 Å in thickness) with a shadowing angle arctan 0.5 in a high vacuum of about 4×10^{-6} torr (mmHg). It was taken out of the evaporator after the evaporation of barium chloride and kept in air of relative humidity less than 50%. This specimen was investigated with the electron microscope in order to estimate the shadow length of particles (Fig. 2-11b).

In the procedure mentioned above the specimen was irradiated by the electron beam at weak intensity to avoid the destruction of particles. Then it was placed in an atmosphere saturated with octanol vapor for 24 h at room temperature. Subsequently, it was investigated with the electron microscope (Fig. 2-11c). If Liesegang's rings of reaction product could be seen clearly around the individual particle, it was determined to be a sulfate-containing particle. The lower detection limit of sulfate particles was evaluated to be 0.016 μm radius.

Classification of the distributions into two characteristic types was carried out by using the number concentrations of particles of 0.005–0.009 μm radius N_1 and of 0.05–0.09 μm radius N_2. Distributions with the ratios N_1/N_2 more than 10 and less than 3 were termed A and B types, respectively.

Type A distribution was measured in the daytime on 30 July 1978 and 31 July 1981 (Fig. 2-12a).

The two days collection was carried out in the summer monsoon. During the collections the light scattering coefficient ranged from 0.4 to 0.7×10^{-4} m^{-1}, and visibility exceeded 20 km. The mass concentration of aerosols was estimated to be from 13 to 22 μg m^{-3}. In this type of distribution, both total and sulfate particles with radii more than 0.1 μm are present in a low number concentration, less than 2.5×10^3 cm^{-3} in $dN/d\log r$. However, the concentrations of both classes of particles tend to increase strongly with decreasing radius in the Aitken size range. High number concentrations of all particles with radii less than 0.01 μm are found in the distribution termed A.

Type B distribution was measured in the daytime on 7 August 1978 and 3 August 1981 as indicated in Fig. 2-12a. The collections of particles were carried out mainly in the situation of wind velocity less than 3 m s^{-1} and light scattering coefficient more than 3×10^{-4} m^{-1}. Visibility was less than 10 km during the collections. The mass concentrations of aerosols are estimated to be more than 100 μg m^{-3}. The distribution shows that total and sulfate-containing particles of 0.1 μm radius are present in high concentrations ranging from 7×10^3 to 2×10^4 cm^{-3} in $dN/d\log r$. The concentrations of both classes of particles with radii between 0.01 and 0.07 μm are also present in a concentration higher than 10^4 cm^{-3} in $dN/d\log r$. The concentration of all particles with radii less than 0.01 μm shown in distribution B seems to decrease and it is lower than that in distribution A. The distribution of sulfate-containing particles is found to be similar in shape to that of all particles.

The ratios of number concentration of sulfate-containing particles $N_{sulfate}$ to that of all particles N_{all} in the range of 0.016–0.28 μm radius is shown in Fig. 2-12b. Sulfate-containing particles are present in high number fractions in particles between 0.016 and 0.28 μm radius, from 0.66 to 0.97 in the type B distribution. On the other

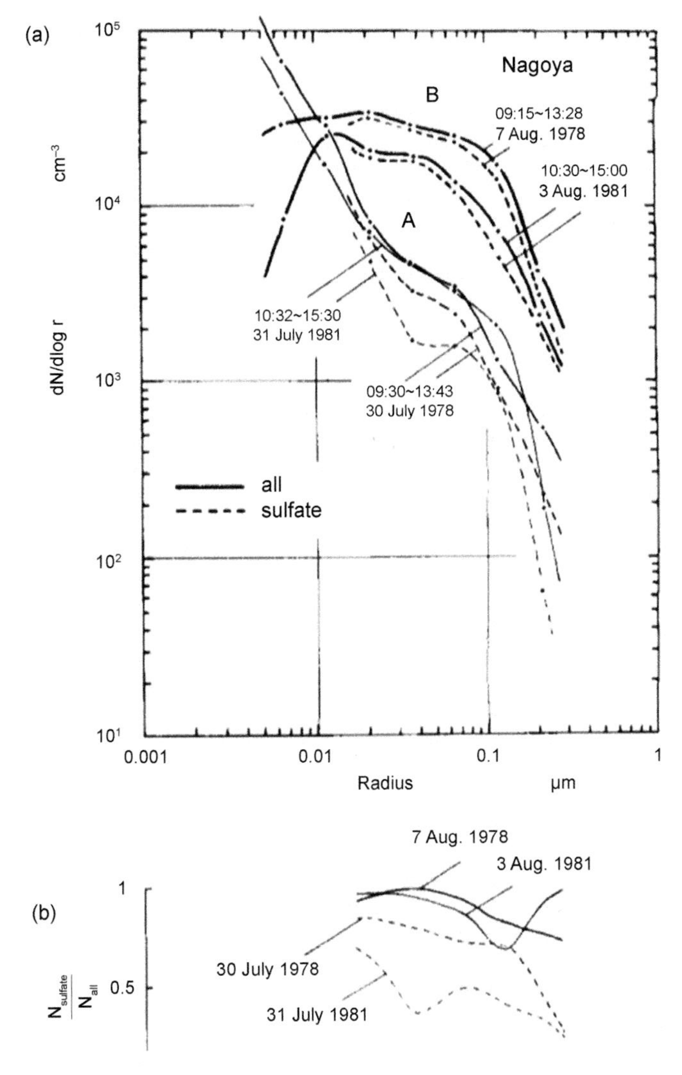

Fig. 2-12. (from Okada 1985).

a) Two characteristic types of number-size distribution in the daytime at Nagoya. Type A distributions (July 30, 1978 and July 31, 1981) and type B distribution (August 7, 1978 and August 3, 1981).

b) Ratios of number concentration of sulfate-containing particles $N_{sulfate}$ to that of all particles N_{all} in the range of 0.016–0.28 μm radius. Type A—dotted line, type B—solid line.

hand, sulfate-containing particles have ratios ranging from 0.33 to 0.83 in the type A distribution and the ratios have a tendency to increase with decreasing radius. In both types of distribution, most of the particles other than sulfate-containing particles were those made up of aggregations of electron-dense spherules, which are similar in morphological features to carbon particles.

Surface area of pre-existing particles has the depressing effect of the new particle formation (McMurry and Friedlander 1979; Friedlander 1982). As for the type A distribution, new particles would be formed in the daytime at a large rate by homogeneous nucleation of sulfuric acid vapor in the situation of low surface areas of particles and adequate solar radiation. Nagoya is one of the large cities of Japan, located near the sea. The atmosphere of Nagoya is often influenced by the invasion of maritime air. The invasion of maritime air with the strong monsoon persists for periods in excess of one day at a time. In such a condition, aerosol particles in the optically effective size range were present in persistently low concentrations. The maritime air with the strong monsoon, which would be mixed with the polluted air in Nagoya, passed too rapidly over the urban area to permit much growth of new particles formed in the urban atmosphere. Hence, the continuous supply of clean air from outside the urban area by strong winds would suppress the increase in the number concentration of aerosol particles in the "accumulation mode". These processes would be reflected in the number-size distribution classified as A type and a high appearance frequency of hygroscopic particles without water-insoluble inclusion in the Aitken size range.

On the other hand, a high number concentration of aerosol particles in "nuclei mode" was not measured in the daytime when type B distribution was observed. It is reasonable to say that the depressing effect of pre-existing particles on the formation of new particles would occur strongly in the situation of larger surface area than 10^3 μm^2 cm^{-3}. Type B distribution that exhibited the high concentrations sulfate-containing particles in the "accumulation mode" would be formed by heterogeneous processes on pre-existing particles in the air moving slowly over the urban area.

Type B distribution is important to decrease in visibility in the urban atmosphere. On the other hand, type A distribution provides significant information on the supply of small particles, formed in the urban atmosphere by homogeneous nucleation, to areas downwind from the urban area by strong monsoon winds. It is expected that these small particles would grow into larger particles during the transport and would play an important role in the increase in the amount of aerosol particles in places remote from air pollution sources.

Aerosol particles emitted by burning processes

Ambient aerosols, especially in the urban atmosphere, contain a significant fraction of carbonaceous material, which is mainly composed of organic carbon (OC) and elemental carbon (EC). EC is sometimes termed black carbon (BC) because of its specific absorption of solar radiation. Carbonaceous particles are emitted from the fuel combustion and biomass burning. Elemental carbon is released as primary particles, whereas organic carbon is also formed in the atmosphere as secondary organic carbon. This section mainly describes elemental carbon particles.

Shah et al. (1986) summarized the mass concentrations of carbonaceous aerosols at urban and rural sites in the United States. The samples were filter segments obtained from the National Air Surveillance Networks for the year 1975. Urban annual averages of OC ranged from 2.7 to 13.4 $\mu g\ m^{-3}$ and rural averages from 1.2 to account 3.4 $\mu g\ m^{-3}$. OC accounted for an average of 8% of the total aerosol mass at both urban and rural sites. Urban annual average of EC ranged from 0.9 to 7.7 $\mu g\ m^{-3}$ and rural averages ranged from 0.3 to 2.2 μg m^{-3}. EC accounted for an average of 8% of the urban aerosol mass concentration and 4% of the rural mass concentration.

BC aerosols were collected with cascade impactors in the two urban atmospheres of Vienna in Austria and Uji in Japan from September 1998 to November 1999 (Hitzenberger and Tohno 2001). The averages of BC in all samples in Uji and Vienna were 4.89 μg m^{-3} and 5.01 μg m^{-3}, respectively. BC accounted for an average of 14.4% of total mass in Uji and 9.81% of total mass in Vienna. The mass-size distributions of BC particles in Uji and Vienna measured in November are shown in Fig. 2-13, with the distribution of total aerosols in Vienna. The distribution in Uji shows two peaks at 0.15 μm and 0.39 μm diameter in the submicron range.

Parungo et al. (1994) studied long-range transport of BC aerosols from China to the downwind ocean. Lin-an station is situated on a mountain (140 m MSL) 200 km southwest of Shanghai City.

The mass concentration of BC aerosols measured at Lin-an station situated on a mountain (140 m MSL), 200 km southwest of Shanghai City, is suited for the use as a starting point for long-range transport from land to ocean.

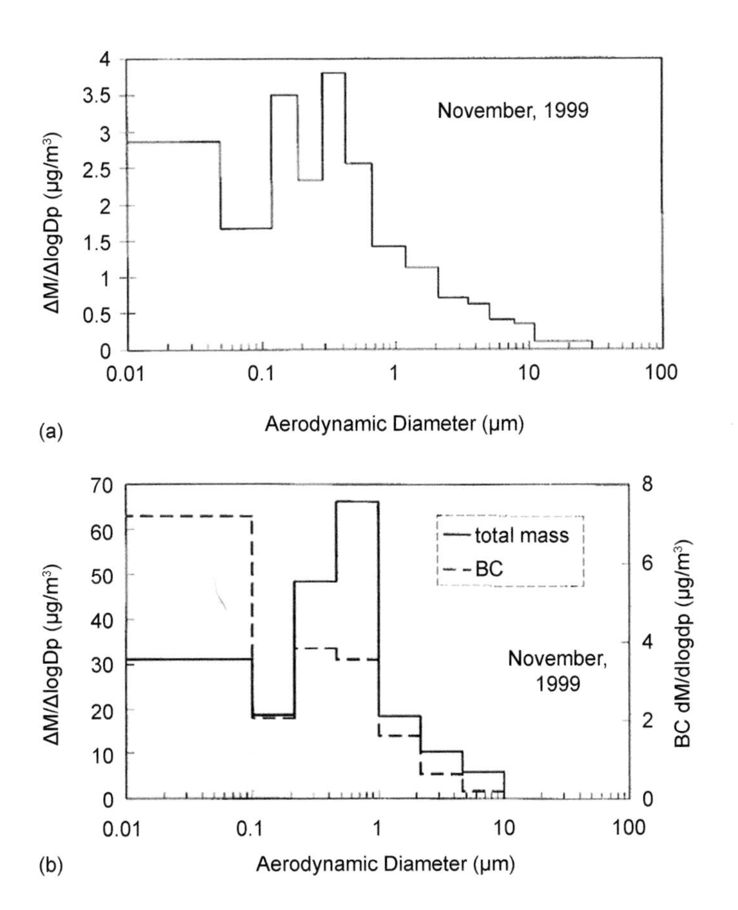

Fig. 2-13. Mass-size distributions of BC particles in Uji and Vienna (Hitzenberger and Tohno 2001).

The mean BC concentration was 2.07 μg m^{-3} (2070 ng m^{-3}) at Lin-an station. Figure 2-14 shows the relationship between observed BC mass concentration and distance from the China shore. Within 2000 km, an approximate linear relationship is found for the data points of Cruise 2 and 3, while wind speeds were ~ 5 m s^{-1}. An approximate linear relationship is also found for the data points of Cruise 1, which was conducted in winter (average wind speed was 7 m s^{-1}). The BC mass concentrations over the remote Pacific ranged from 7 to 20 ng m^{-3}. Measured BC concentrations in remote locations were summarized in Penner (1995).

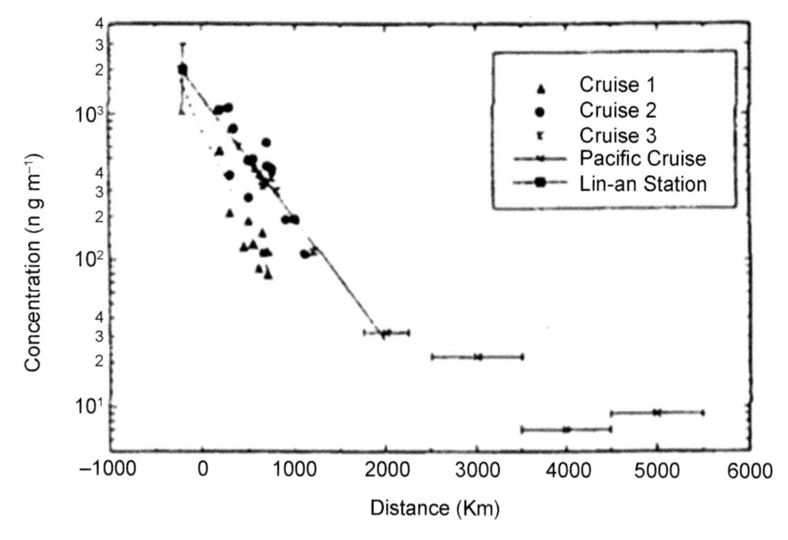

Fig. 2-14. Relationship between observed BC mass concentration and distance from China shore (Parungo et al. 1994).

Urban particulate pollution in the early 1950s in London is coal-originated pollution that effected human health and mortality. Change in fuel usage from coal to other fossil fuels (oil or natural gas) has made a large reduction of urban aerosols in cities. However, the emission of aerosol particles from automobiles increased because of urban population growth and increasing importance of traffic in urban areas. Due to their adverse health effects and their abundance in the vicinity of roads in urban areas, diesel particles have been of great concern.

Table 2-4 shows the emission factors for different sources of elemental (black) carbon (from Ogren and Charlson 1984). As found in Table 2-4, diesel-engine-powered vehicles or not well-optimized combustion processes such as fireplaces, have high emission factors.

Particles that were produced during combustion normally have sizes in the accumulation mode range. An example for soot particles emitted by a small diesel engine is shown in Fig. 2-15 (Horvath 1993).

The measurements were made during test runs, using a dilution tunnel, and a low-pressure impactor was used for sampling. The size range covered by the stages of the impactor is given by the horizontal bars. The solid line gives the mass of particles after

Table 2-4. Emission factors for different sources of elemental (black) carbon (Ogren and Charlson 1984).

Source	Elemental carbon /fuel [g kg^{-1}]
Vehicles, diesel engine	2
Fireplace, softwood	1.3
Jet engine	1
Fireplace, hardwood	0.39
Vehicles, gasoline engine	0.02
Solid fossil fuel	0.001
Natural gas	0.0003

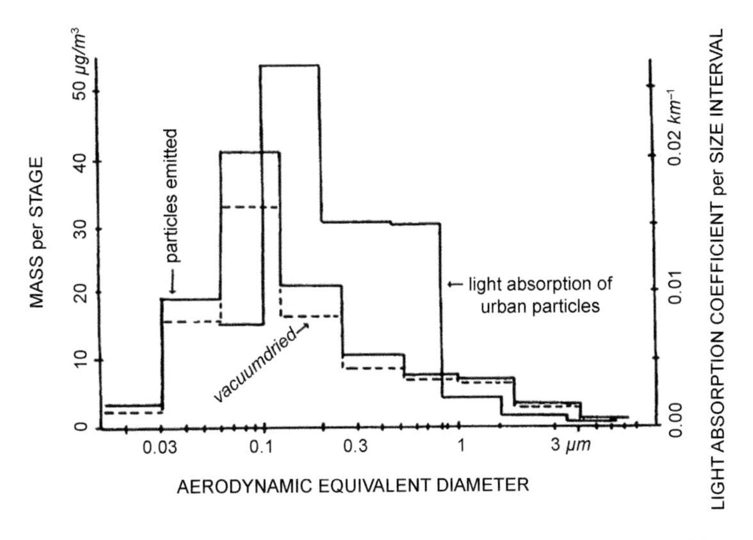

Fig. 2-15. Example of mass-size distribution and light absorption coefficient for soot particles emitted by a small diesel engine (Horvath 1993).

sampling, whereas the broken line after vacuum drying the collected aerosol particles. The majority of the particles are in the size range between 0.05 and 1 mm diameter. The reduction in the aerosol mass concentrations by vacuum drying indicates that the soot particles also contain volatile substances, which are organic carbon. This fraction can vary between 10% and 90% of the total mass depending on the quality and the operating conditions of the motor. For comparison the light absorption coefficient of a size-segregated atmospheric sample taken with a rotating impactor and subsequent analysis with an integrating plate in Vienna is also shown. The distributions show similarities, indicating that the light-absorbing atmospheric aerosols and the emitted diesel particles have similar size.

The diesel aerosols are mainly composed of highly agglomerated solid carbonaceous material and ash, volatile organic and sulfur compounds (Kittelson 1998). The structure is illustrated schematically in Fig. 2-16.

Solid carbon is formed during combustion in locally rich regions. Much of it is subsequently oxidized. The residue is exhausted in the form of solid carbon

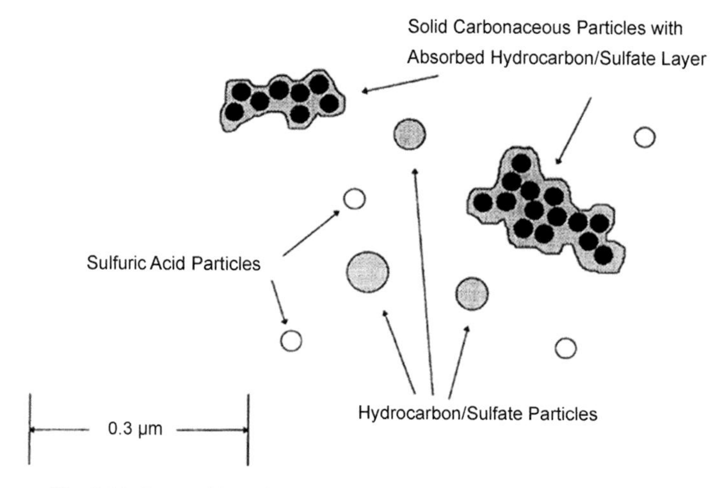

Fig. 2-16. Composition of particles emitted from a diesel engine (Kittelson 1998).

agglomerates. A tiny fraction of the fuel and atomized and evaporated lube oil escape oxidation and appear as volatile or soluble organic compounds (generally described as the soluble organic fraction, SOF) in the exhaust.

Figure 2-17 shows the idealized number- and mass-size distributions of diesel exhaust particles (Kittelson 1998). The distributions shown are trimodal, lognormal in form. Most of the particle mass exists in the so-called accumulation mode in the

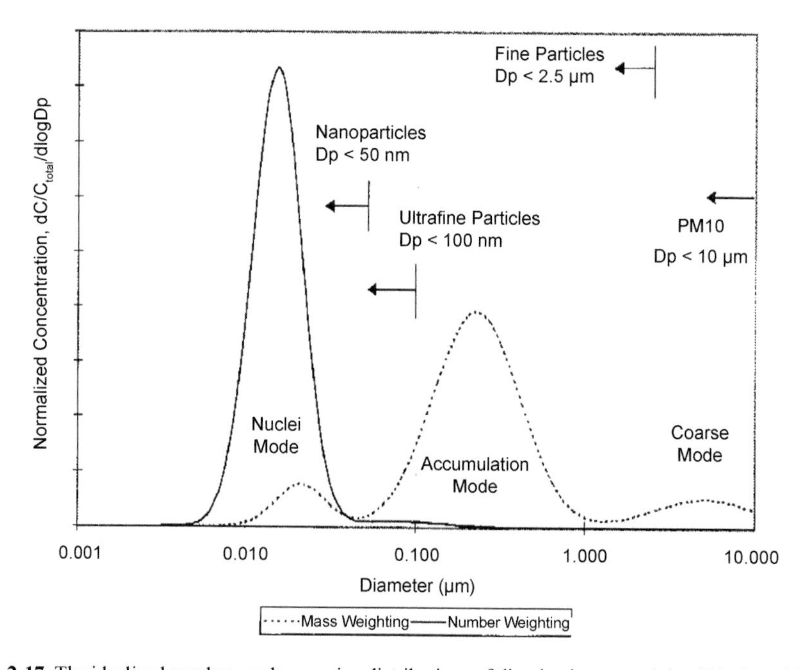

Fig. 2-17. The idealized number- and mass-size distributions of diesel exhaust particles (Kittelson 1998).

diameter range of 0.1–0.3 μm. This is where the carbonaceous agglomerates and associated adsorbed materials reside. The nuclei mode typically consists of particles in the diameter range of 0.005–0.05 μm. This mode usually consists of volatile organic and sulfur compounds that form during exhaust dilution and cooling, and may also contain solid carbon and metal compounds. The nuclei mode typically contains 1–20% of the particle mass and more than 90% of the particle number.

Because the hierarchy of carbon emission sources has been rapidly changing during the last two decades, studies of atmospheric elemental and black carbon are still very important, and therefore are still popular. For example, recently Zhi et al. (2014) published their results of the measurements on comparison of elemental and black carbon levels during normal and heavy haze periods in China. Also, Cheng et al. (2014) documented very interesting results of the five-year observation of correlations between black carbon mass and size-resolved particle number concentrations in the urban area in Taipei.

Mixing state of aerosols

Aerosols originating from anthropogenic sources play an important role in climate change through direct and indirect radiative forcing (Charlson et al. 1992). In order to clarify the effect of anthropogenic aerosol particles on atmospheric radiation and cloud processes, the composition and mixing properties of aerosol particles in the urban atmosphere have to be studied (e.g., Heintzenberg and Covert 1990). In particular, the volume fraction of water-soluble material (ε) in individual particles is important for assessing the nucleation scavenging properties of atmospheric aerosol particles (Junge and McLaren 1971; Fitzgerald 1973; Okada et al. 1983a; Hallberg et al. 1994) and its influence on the radiative properties of the atmosphere (Heintzenberg 1978; Mita 1982). Moreover, the mixing state and its associated hygroscopic growth may be important in assessing the deposition of particles in the human airway.

Most of the existing studies of the chemical composition of aerosol particles in the urban atmosphere used the size-segregated bulk samples. These studies have supplied information on particle composition as a function of size. However, the mixing properties of aerosol particles cannot be evaluated by the analysis of the bulk samples alone because atmospheric aerosols consist of many particles with different composition even in a narrow size range. Junge (1950) introduced the concept of "mixed nuclei" (mixture of water-soluble and insoluble material) to explain the hygroscopic growth of particles with relative humidity. Winkler (1973) proposed a definition of internal and external mixtures in aerosols. In an external mixture, aerosol particles with different pure compounds are present as separate particles. On the other hand, the state in which all particles consist of the same mixture of compounds is termed internal mixture. An intermediate situation between these extreme mixture types usually exists in the atmosphere.

The mixing properties of aerosols have been studied in the size-segregated samples using their hygroscopic growth (Ahlberg et al. 1978; Covert and Heintzenberg 1984; Heintzenberg and Covert 1987) and by the samples processed through the nucleation scavenging (Harrison 1985; Hallberg et al. 1994; Hitzenberger et al. 2000). Moreover, several measurements of aerosol hygroscopic growth using a tandem differential mobility analyzer (TDMA) system showed a variability of the hygroscopic properties of submicrometer aerosol particles (Sekigawa 1983; McMurry and Stolzenburg 1989; Svenningsson et al. 1992; Zhang et al. 1993). In addition to the above studies, a technique that has the capability to resolve external mixtures in aerosols with respect to their optical properties was developed by Covert et al. (1990).

Single particle analysis is a useful method for evaluating mixed properties of aerosol particles. Husar et al. (1976) obtained electron micrographs of aerosol particles collected at 390 m altitude over Pasadena and found that the submicrometer liquid-like particles contained a single or multiple solid nuclei. They suggested heterogeneous nucleation as the formation mechanism for the aerosol. More explicit measurements of the mixing properties of individual particles with respect to the degree of internal and external mixtures of water-soluble and -insoluble material were carried out by dialysis with water. Okada (1983a,b) measured the mixed state of individual particles with radii between 0.03 and 0.35 μm in the urban atmospheres of Yokkaichi and Nagoya in Japan using a dialysis method for the extraction of water-soluble material by electron microscopy and found that more than 80% of the aerosol particles were hygroscopic.

Figure 2-18 shows the example of electron micrographs. The results also showed that 34% of the hygroscopic particles in the Aitken size range (0.03–0.1 μm radius) and 67% in the large size range (0.1–0.35 μm radius) were mixed particles. Differences in the mixed properties were found to be associated with the formation process of aerosol particles by gas-to-particle conversion (Okada 1985). Pastuszka and Okada (1995) examined mixing properties of individual particles collected at Katowice in Poland and found the large abundance of mixed particles in air influenced by anthropogenic sources.

Individual aerosol particles were collected on five days with different meteorological conditions in March, April, and June 1991 in the urban atmosphere of Vienna in Austria (Okada and Hitzenberger 2001). The samples collected with an impactor were examined by electron microscopy. The mixing properties of submicrometer aerosol particles with radii between 0.1 and 1 μm were studied by using the dialysis (extraction) of water-soluble material. The averaged results showed that more than 85% of particles with radii between 0.1 and 0.7 μm were hygroscopic. However, more than 50% of particles with radii larger than 0.2 μm were mixed particles (hygroscopic particles with water-insoluble inclusions), and they were dominant (80%) in the size range 0.5–0.7 μm radius. The results also showed that the number proportion of mixed particles increased with increasing radius and the abundance increased with increasing particle loading in the atmosphere. The volume fraction of water-soluble material (ε) in mixed particles tended to decrease with increasing radius, implying the formation of mixed particles by heterogeneous processes such as condensation and/or surface reaction.

Naoe and Okada (2001) collected individual aerosol particles with different meteorological conditions in June 2000 in the urban atmosphere of Tsukuba, Japan. The samples collected with an electrostatic aerosol sampler were examined by electron microscopy. The mixing properties of submicrometer aerosol particles of 0.02–0.2 μm radius were studied using the dialysis (extraction) of water-soluble material. By their morphological appearance, soot-containing particles were classified into two types,

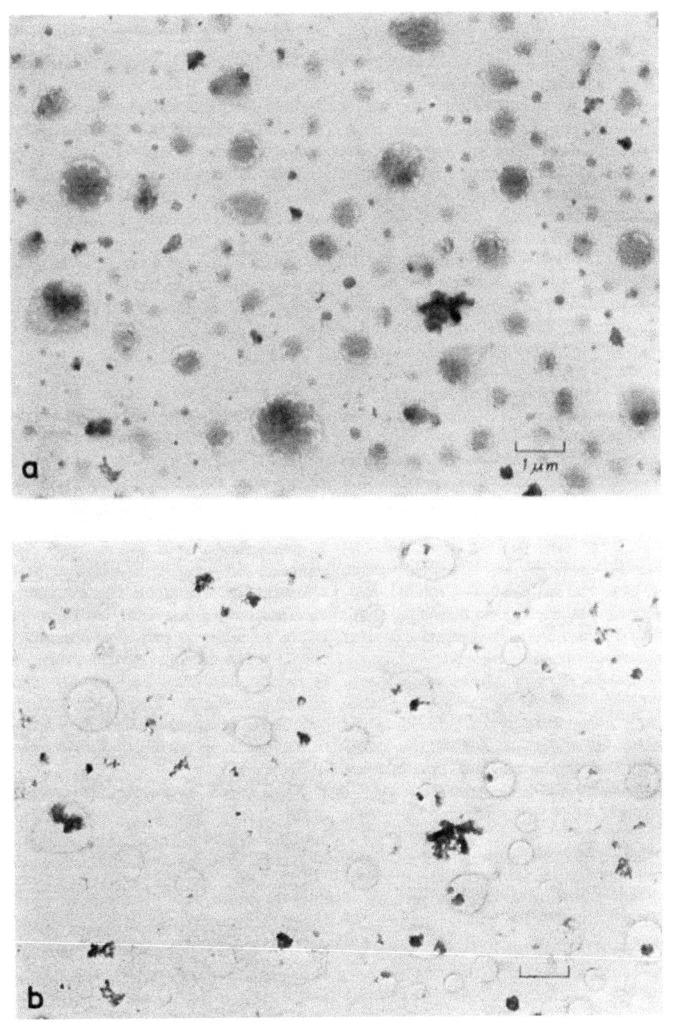

Fig. 2-18. Example of electron micrographs of individual particles with radii between 0.03 and 0.35 μm in the urban atmospheres of Yokkaichi, Japan before and after the dialysis (Okada 1983a,b).

i.e., externally mixed soot-particles and internally mixed soot-particles. The number fractions of internally mixed soot-particles increased with increasing radius. In a "polluted" case, the sample showed a dominant number fraction (75%) of internally mixed soot-particles in the larger radius range of 0.1–0.2 μm.

The mixing state of soot particles was also studied by electron microscopy (Hasegawa and Ohta 2002; Adachi et al. 2010).

Finally, it should be noted that the analysis of features and elemental composition of individual atmospheric particles is still a very efficient tool in aerosol science and is useful for obtaining the information about airborne particles in various areas of the world. For example, Cong et al. (2010) recently published their study on the elemental and individual particle analysis of atmospheric aerosols in the high Himalayas.

References

Adachi, K., S.H. Chung and P.R. Buseck. 2010. Shapes of soot aerosol particles and implications for their effects on climate. J. Geophys. Res. 115: D15206, doi:10.1029/2009JD012868.

Ahlberg, M.S., A.C.D. Leslie and J.W. Winchester. 1978. Characteristics of sulfur aerosol in Florida as determined by PIXE analysis. Atmos. Environ. 12: 773–777.

Aitken, J. 1881. On dust, fogs, and clouds, Trans. Roy. Soc. Edinb. 30: 337–368.

Aitken, J. 1923. Collected Scientific Papers of John Aitken. G.G. Knott [ed.]. Cambridge University Press, London.

Andreae, M.O. 1995. Climate effects of changing atmospheric levels. pp. 347–398. *In*: A. Henderson-Sellers [ed.]. World Survey of Climatology, Vol. 16, Future Climates of the World. Elsevier Science B.V., Amsterdam, the Netherlands.

Berg, W.W. and J.W. Winchester. 1978. Aerosol chemistry of the marine atmosphere. pp. 173–231. *In*: J.P. Riley and R. Chester [eds.]. Chemical Oceanography, Vol. 7 (2nd edition). Academic Press, London.

Bigg, E.K., A. Ono and J. Williams. 1974. Chemical tests for individual submicron aerosol particles. Atmos. Environ. 8: 1–13.

Bigg, E.K., J.L. Gras and D.J.C. Mossop. 1995. Wind-produced submicron particles in the marine atmosphere. Atmos. Res. 36: 55–68.

Blanchard, D.C. 1954. Bursting of bubbles at an air-water interface. Nature 173: 1048.

Blanchard, D.C. and A.H. Woodcock. 1957. Bubble formation and modification of the sea and its meteorological significance. Tellus 9: 145–158.

Buseck, P.R. and K. Adachi. 2008. Nanoparticles in the atmosphere. Elements 4: 389–394.

Charlson, R.J., H. Horvath and R.F. Pueschel. 1967. The direct measurement of atmospheric light scattering coefficient for studies of visibility and pollution. Atmos. Environ. 1: 469–478.

Charlson, R.J., S.E. Schwartz, J.M. Hales, R.D. Cess, J.A. Coakley, J.E. Hansen and D.J. Hofmann. 1992. Climate forcing by anthropogenic aerosols. Science 255: 423–430.

Cheng, Y.-H., Y.-Y. Kao and J.-J. Liu. 2014. Correlations between black carbon mass and size-resolved particle number concentrations in the Taipei urban areas: a five-year long-term observation. Atmos. Poll. Res. 5: 62–72.

Cipriano, R.J., D.C. Blanchard, A.W. Hogan and G.G. Lala. 1983. On the production of Aitken nuclei from breaking waves and their role in the atmosphere. J. Atmos. Sci. 40: 469–479.

Clark, W.E. and K.T. Whitby. 1967. Concentration and size distribution measurements of atmospheric aerosols and a test of the theory of self-preserving size distributions. J. Atmos. Sci. 24: 677–687.

Clarke, A.D., S.R. Owens and J. Zhou. 2006. An ultrafine sea-salt flux from breaking waves: Implications for cloud condensation nuclei in the remote marine atmosphere. J. Geophys. Res. 111, D06202, doi:10.1029/2005JD006565.

Cong, Z., S. Kang, S. Dong, X. Liu and D. Qin. 2010. Elemental and individual particle analysis of atmospheric aerosols from high Himalayas. Environ. Monit. Assess. 160: 323–335.

Coulier, M. 1885a. Note sur une nouvelle propriété de l'air. Journal de Pharmacie et de Chimie 22: 165–173.

Coulier, M. 1885b. Note sur une nouvelle propriété de l'air. Journal de Pharmacie et de Chimie 22: 254–255.

Covert, D.S. and J. Heintzenberg. 1984. Measurement of the degree of internal/external mixing of hygroscopic compounds and soot in atmospheric aerosols. Sci. Total Environ. 36: 347–352.

Covert, D.S., J. Heintzenberg and H.-C. Hansson. 1990. Electro-optical detection of external mixtures in aerosols. Aerosol Sci. Technol. 12: 446–456.

d'Almeida, G.A. and L. Schütz. 1983. Number, mass and volume distributions of mineral aerosol and soil of the Sahara. J. Climate Appl. Meteor. 22: 233–243.

Duce, R.A. 1995. Sources, distributions, and fluxes of mineral aerosols and their relationship to climate. pp. 43–72. *In*: R.J. Charlson and J. Heintzenberg [eds.]. Aerosol Forcing of Climate. John Wiley & Sons Ltd., West Sussex, England.

Eugster, H.P., C.E. Harvie and J.H. Weare. 1980. Mineral equilibria in a six-component seawater system, Na-K-Mg-Ca-SO_4-Cl-H_2O at 25°C. Geochim. Cosmochim. Acta 44: 1335–1347.

Field, R.D., G. van der Werf and S.S.P. Shen. 2009. Human amplification of drought-induced biomass burning in Indonesia since 1960. Nature Geosci. 2: 185–188.

Fitzgerald, J.W. 1973. Dependence of the supersaturation spectrum of CCN on aerosol size distribution and composition. J. Atmos. Sci. 30: 628–634.

Friedlander, S.K. 1982. The behavior of constant rate aerosol reactors. J. Aerosol Sci. Technol. 1: 3–13.

Gillette, D. 1980. Major contributions of natural primary continental aerosols: Source mechanisms. Ann. N. Y. Acad. Sci. 338: 348–358.

Gras, J.L. and G.P. Ayers. 1983. Marine aerosol at southern mid-latitude. J. Geophys. Res. 88: 10,661–10,666.

Hallberg, A., J.A. Ogren, K.J. Noone, K. Okada, J. Heintzenberg and I.B. Svenningsson. 1994. The influence of aerosol particle composition on cloud droplet formation. J. Atmos. Chem. 19: 153–171.

Harrison, L. 1985. The segregation of aerosols by cloud-nucleating activity. Part II: observation of an urban aerosol. J. Climate Appl. Meteor. 24: 312–321.

Hasegawa, S. and S. Ohta. 2001. Some measurements of the mixing state of soot-containing particles at urban and non-urban sites. Atmos. Environ. 36: 3899–3908.

Heintzenberg, J. 1978. Light scattering parameters of internal and external mixtures of soot and non-absorbing material in atmospheric aerosols, 278–281. *In*: The Proceedings of the Conference on Carbonaceous Particles in the Atmosphere. CONF-7803101, Berkeley, CA, USA.

Heintzenberg, J. 1989. Fine particles in global troposphere. A review. Tellus. 41B: 149–160.

Heintzenberg, J. and D.S. Covert. 1987. Chemically resolved submicrometric size distribution and external mixing of the Arctic haze aerosols. Tellus 39B: 374–382.

Heintzenberg, J. and D.S. Covert. 1990. On the distribution of chemical and physical particle properties in the atmospheric aerosol. J. Atmos. Chem. 10: 383–397.

Heintzenberg, J., D.S. Covert and R. Van Dingenen. 2000. Size distribution and chemical composition of marine aerosols: A compilation and review. Tellus 52B: 1104–1122.

Hering, S.V. and S.K. Friedlander. 1982. Origins of aerosol sulfur size distributions in the Los Angeles basin. Atmos. Environ. 16: 2647–2656.

Hitzenberger, R. and S. Tohno. 2001. Comparison of black carbon (BC) aerosols in two urban areas—concentration and size distribution. Atmos. Environ. 35: 2153–2167.

Hitzenberger, R., A. Berner, R. Kromp, A. Kasper-Giebl, A. Limbeck, W. Tscherwenka and H. Puxbaum. 2000. Black carbon and other species at a high-elevation European site (Mount Sonnblick, 3106 m, Austria): Concentrations and scavenging efficiencies. J. Geophys. Res. 105: 24,637–24,645.

Horvath, H. 1993. Atmospheric light absorption—a review. Atmos. Environ. 27A: 293–317.

Husar, R.B., W.H. White and D.L. Blumenthal. 1976. Direct evidence of heterogeneous aerosol formation in Los Angeles smog. Environ. Sci. Technol. 10: 490–491.

Husar, R.B., J.M. Prospero and L.L. Stowe. 1997. Characterization of tropospheric aerosols over the oceans with the NOAA advanced very high resolution radiometer optical thickness operational product. J. Geophys. Res. 102: 16,889–16,909.

Husar, R.B., J.M. Holloway, D.E. Patterson and W.E. Wilson. 1981. Spatial and temporal pattern of eastern U.S. haziness: A summary. Atmos. Environ. 15: 119–1928.

Husar, R.B., J.D. Husar and L. Martin. 2000. Distribution of continental surface aerosol extinction based on visual range data. Atmos. Environ. 34: 5067–5078.

Jaenicke, R. 1980. Atmospheric aerosols and global climate. Atmos. Environ. 11: 577–588.

Jaenicke, R. 1993. Tropospheric aerosols. pp. 1–31. *In*: P.V. Hobbs [ed.]. Aerosol-Cloud-Climate Interactions. Academic Press, San Diego.

Junge, C. 1950. Das Wachstum der Kondensationskerne mit der relativen Feuchtigkeit. Annalen der Meteorologie 3: 129–135.

Junge, C. 1963. Air Chemistry and Radioactivity. Academic Press, New York and London.

Junge, C. and E. McLaren. 1971. Relationship of cloud nuclei spectra to aerosol size distribution and composition. J. Atmos. Sci. 28: 382–390.

Kiehl, J.T. and H. Rodhe. 1995. Modeling geographical and seasonal forcing due to aerosols. pp. 281–296. *In*: R.J. Charlson and J. Heintzenberg [eds.]. Aerosol Forcing of Climate. John Wiley & Sons Ltd., West Sussex, England.

Kientzler, C.F., A.B. Arons, D.C. Blanchard and A.H. Woodcock. 1954. Photographic investigation of the projection of droplets by bubble bursting at a water surface. Tellus 6: 1–7.

Kittelson, D.B. 1998. Engines and nanoparticles: A review. J. Aerosol Sci. 29: 575–588.

Koschmieder, H. 1924. Theorie der Horizontalen Sichweit. Beitr. Phys. Atmos. 12: 33–53, 171–181.

Landsberg, H. 1938. Atmospheric condensation nuclei. Ergebnis. der Kosm. Physik 3: 155–252.

Lovett, R.F. 1978. Quantitative measurement of airborne sea-salt in the North Atlantic. Tellus 30: 358–364.

Mason, B.J. 1954. Bursting of air bubbles at the surface of sea water. Nature 174: 470–471.

McMurry, P.H. 2000. The history of condensation nucleus counters. Aerosol Sci. Technol. 33: 297–322.

McMurry, P.H. and S.K. Friedlander. 1979. New particle formation in the presence of an aerosol. Atmos. Environ. 13: 1635–1651.

McMurry, P.H. and M.R. Stolzenburg. 1989. On the sensitivity of particle size to relative humidity for Los Angeles aerosols. Atmos. Environ. 23: 497–507.

Mészáros, A. and K. Vissy. 1974. Concentration, size distribution and chemical nature of atmospheric aerosol particles in remote oceanic areas. J. Aerosol Sci. 5: 101–109.

Middleton, W.E.K. 1935. Visibility in Meteorology. The University of Toronto Press, Toronto, Canada.

Middleton, W.E.K. 1952. Vision through the Atmosphere. The University of Toronto Press, Toronto, Canada.

Mita, A. 1982. Light absorption properties of inhomogeneous spherical aerosol particles. J. Meteor. Soc. Japan 60: 763–776.

Naoe, H. and K. Okada. 2001. Mixing properties of submicrometer aerosol particles in the urban atmosphere—with regard to soot particles. Atmos. Enviro. 35: 5765–5722.

Nel, A. 2005. Air pollution-related illness: Effects of particles. Science 308: 804–806.

Noll, K.E., P.K. Mueller and M. Imada. 1968. Visibility and aerosol concentration in urban air. Atmos. Environ. 2: 465–475.

Oberdörster, G. and M.J. Utell. 2002. Ultrafine particles in the urban air: To the respiratory tract—and beyond? Environ. Health. Perspect. 110: A440–A441.

O'Dowd, C.D. and M.H. Smith. 1993. Physicochemical properties of aerosols over the northeast Atlantic: Evidence for wind-speed-related submicron sea-salt aerosol production. J. Geophys. Res. 98: 1137–1149.

Ogren, J.A. and R.J. Chalson. 1984. Wet deposition of elemental carbon and sulfate in Sweden. Tellus 36B: 262–271.

Okada, K. 1983a. Volume fraction of water-soluble material in individual aerosol particles. J. Aerosol Sci. 14: 301–302.

Okada, K. 1983b. Nature of individual hygroscopic particles in the urban atmosphere. J. Meteor. Soc. Japan 61: 727–736.

Okada, K. 1985. Number-size distribution and formation process of submicrometer sulfate-containing particles in the urban atmosphere of Nagoya. Atmos. Environ. 19: 743–757.

Okada, K. and K. Isono. 1982. Trends in visibility in the urban atmosphere—a case study in Nagoya, Japan. J. Meteor. Soc. Japan 60: 777–786.

Okada, K. and K. Kai. 1995. Features and elemental composition of mineral particles collected in Zhangye, China. J. Meteor. Soc. Japan 73: 947–957.

Okada, K. and R. Hitzenberger. 2001. Mixing properties of individual submicrometer aerosol particles in Vienna. Atmos. Environ. 35: 5617–5628.

Okada, K., A. Kobayashi and N. Kuba. 1983. Features of light scattering coefficient and aerosol particles in the urban atmosphere of Nagoya. Atmos. Environ. 17: 2087–2092.

Okada, K., J. Heintzenberg, K. Kai and Y. Qin. 2001. Shape of atmospheric mineral particles collected in three Chinese arid-regions. Geophys. Res. Lett. 28: 3123–3126.

Ono, A., K. Okada and K. Akagawa. 1981. On the validity of the vapor-deposited thin film of barium chloride for the detection of sulfate in individual atmospheric particles. J. Meteor. Soc. Japan 59: 417–422.

Parungo, F., C. Nagamoto, M.-Y. Zhou, A.D.A. Hansen and J. Harris. 1994. Aeolian transport of aerosol black carbon from China to the ocean. Atmos. Environ. 28: 3251–3260.

Pastuszka, J.S. and K. Okada. 1995. Features of atmospheric aerosol particles in Katowice, Poland. Sci. Total Environ. 175: 179–188.

Penner, J.E. 1995. Carbonaceous aerosols influencing atmospheric radiation: Black and organic carbon. pp. 91–108. *In*: R.J. Charlson and J. Heintzenberg [eds.]. Aerosol Forcing of Climate. John Wiley & Sons Ltd., West Sussex, England.

Podzimek, J. 1980. Advances in marine aerosol research. J. Rech. Atmos. 14: 35–61.

Sekigawa, K. 1983. Estimation of the volume fraction of water soluble material in submicron aerosols in the atmosphere. J. Meteor. Soc. Japan 61: 359–366.

Shah, J.J., R.L. Johnson, E.K. Heyerdahl and J.J. Huntzicker. 1986. Carbonaceous aerosol at urban and rural sites in the United States. JAPCA 36: 254–257.

Shiratori, K. 1934. Ionic balance in air and nuclei over ocean. Mem. Fac. Sci. Agr. Taihoku Imp. Univ. 10: 175–202.

Spurny, K.R. 1999. Methods of aerosol measurement before the 1960s. pp. 3–21. *In*: K.R. Spurny [ed.]. Analytical Chemistry of Aerosols. Lewis Publishers, New York.

Spurny, K.R. 2000. Atmospheric condensation nuclei: P.J. Coulier 1875 and J. Aitken 1880 (historical review). Aerosol Sci. Technol. 32: 243–248.

Svenningsson, I.B., H.-C. Hansson, A. Wiedensohler, J.A. Ogren, K.J. Noone and A. Hallberg. 1992. Hygroscopic growth of aerosol particles in the Po Valley. Tellus 44B: 556–569.

Waggoner, A.P., R.E. Weiss, N.C. Ahlquist, D.S. Covert, S. Will and R.J. Charlson. 1981. Atmos. Environ. 15: 1891–1909.

Whitby, K.T. 1978. The physical characteristics of sulfur aerosols. Atmos. Environ. 12: 135–159.

Whitby, K.T. and W.E. Clark. 1966. Electric aerosol particle counting and size distribution measuring system for the 0.015 to 1 μ size range. Tellus 18: 573–586.

Whitby, K.T., R.B. Husar and B.Y.H. Liu. 1972. The aerosol size distribution of Los Angeles smog. pp. 237–264. *In*: G.M. Hidy [ed.]. Aerosols and Atmospheric Chemistry. Academic Press, New York and London.

Willeke, K. and F.A. Baron. 1993. Aerosol Measurement. Van Nostrand Reinhold, New York.

Winkler, P. 1973. The growth of atmospheric aerosol particles as a function of the relative humidity—II. An improved concept of mixed nuclei. J. Aerosol Sci. 4: 373–387.

Woodcock, A.H. 1953. Salt nuclei in marine air as a function of altitude and wind force. J. Meteor. 10: 362–371.

Zhang, X.Q., P.H. McMurry, S.V. Hering and G.S. Gasuccio. 1993. Mixing characteristics and water content of submicron aerosols measured in Los Angeles and at the Grand Canyon. Atmos. Environ. 27A: 1593–1607.

Zhi, G., Y. Chen, Z. Xue, F. Meng, J. Cai, G. Sheng and J. Fu. 2014. Comparison of elemental and black carbon measurements during normal and heavy haze periods: implications for research. Environ. Monit. Assess. 186: 6097–6106.

3

PAH and Heavy Metals in Ambient Particulate Matter: A Review of Up-to-Date Worldwide Data

Wioletta Rogula-Kozłowska

Introduction

It is impossible to tell which ones in the whole tangle of the properties of atmospheric aerosol particles (ambient particulate matter, PM) are crucial to the PM environmental and health hazardousness. It seems that there is not much reason in attributing the PM adverse effects to some selected, "main", intrinsic PM properties, because they easily lose or acquire the importance depending on time and place, the PM effects being site and time dependent. Furthermore, this attribution may be erroneous because the observable health or environmental effects are usually not due to PM only; they rather derive from the concordant actions of PM and accompanying circumstances, the latter in general unknown.

The importance of some PM properties arises from their alleged effects on the human health and environment as well as the ability to measure and control their concentrations. From this point of view, the most important PM properties are the PM concentration in the air and PM chemical composition (Harrison and Yin 2000; Englert 2004). These PM properties are synergistic with each other and with particle size distribution in causing health hazard to humans. The finest particles can be most hazardous, depending on the chemical composition of PM (Dreher 2000; Pope and Dockery 2006; Ostro et al. 2007; Valavanidis et al. 2008; Karlsson et al. 2009; Daher et al. 2012, 2014).

Polish Academy of Sciences, Institute of Environmental Engineering, 34 M. Skłodowska-Curie St., 41-819 Zabrze, Poland.
E-mail: wioletta@ipis.zabrze.pl

Roughly, PM consists of elemental and organic carbon (carbonaceous matter or carbonaceous aerosol), ammonia, nitrates, sulfates (secondary inorganic aerosol), mineral particles (soil or crustal matter), sodium and chlorates (often linked to sea salt), trace elements, and water (Table 3-1). Their mutual proportions vary dynamically in time and between localities.

Despite their relatively small mass share in PM (Table 3-1), trace elements and polycyclic aromatic hydrocarbons (PAH), being by weight a tiny fraction of the total organic matter content of PM, are the most intensely investigated PM components.

The elemental analysis of PM partly characterizes the PM chemical composition; it usually suffices for understanding the temporal and spatial relationship between the source and the receptors of PM and for the assessment of the efficiency of emission abatement methods (e.g., Chow 1995; Viana et al. 2008a; Zhang et al. 2010). It also provides some information on the potential PM health and environmental effects (e.g., Swaine 2000; Duvall et al. 2008; Wiseman and Zereini 2009; Lettino et al. 2012; Baxter et al. 2013).

When other data are inadequate, the Polycyclic Aromatic Hydrocarbon (PAH) content of PM may be useful for finding the PM origin, like the PM elemental composition, but in general, they are investigated because of their adverse health effects (e.g., Durant et al. 1996, 1999; Nielsen et al. 1996; Hernández et al. 2010; Darpa et al. 2014).

Polycyclic organic matter (organic compounds with more than one benzene ring that have boiling point greater than or equal to 100°C) and some elements (heavy metals) are considered hazardous by the US EPA and the International Agency for Research on Cancer (IARC).

The hazard from PM-bound PAH and heavy metals to humans depends on the PAH or metal ambient concentrations, their mass distribution respect to particle size, and PM physicochemical properties; the health condition and habitat of the population may enhance or suppress the PM toxicity.

However, despite the many-factor dependence between PM and the PM health effects, the concentrations of only few PM components are controlled in the atmosphere. Usually, they are controlled by setting the limits, which vary between countries. The yearly permissible ambient concentrations of PM10-bound benzo(a) pyrene, As, Ni, and Cd in Europe are 1, 6, 20, and 5 ng m^{-3}, respectively (WHO 2000; EC 2004). Such an air protection policy does not need data more sophisticated than ambient concentrations of the PM-bound chemicals, but they can only give a general view on the health hazard from PM.

The PM concentrations, size distribution, and chemical composition are always observed within some limited area. The area is usually of a definite kind, the measurements consist in analyzing PM samples taken at a single monitoring (sampling) point, and the point is representative of the PM situation within the whole area. The areas are categorized by the law or are easy to be defined (rural site, urban background, suburban background, regional background, urban area, residential area, city center, street canyon), and a sampling point is representative of the PM situation in the area if the point and the area are in the relation defined by the regulations. The difficulties with the selection of such a point consist in the sensitivity of the concentrations and chemical composition of PM to even small shifts of the point location.

Table 3-1. Percentages of PM components in the PM mass in selected European localities.

City (Country), averaging period; PM fraction (Reference)	Carbonaceous matter	Secondary inorganic aerosol	Mineral matter	Sea salt	Trace elements
Rural/regional background sites					
Streithofen (Austria), year; $PM_{2.5}$ (Puxbaum et al. 2004)	36	46	-	1	< 1
Chaumont (Switzerland), year; $PM_{2.5}$ (Hüeglin et al. 2005)	26	48.2	8.8	-	3.6
Bemantes (Spain), year; $PM_{2.5}$ (Salvador et al. 2007)	30	33	11	7	< 1
Villar Arzobispo (Spain), year; $PM_{2.5}$ (Viana et al. 2008b)	16	28	29	2	< 1
Montseny (Spain), year; $PM_{2.5}$ (Pey et al. 2009)	28	36	9	2	-
Bologna (Italy), [1]summer/winter; PM_1 (Carbone et al. 2010)	48/43	48/56	2/0	2/0	-
Melpitz (Germany), [2]summer/winter; PM_1 (Spindler et al. 2010)	20/18	< 35/55	-	-	-
Melpitz (Germany), [3]summer/winter; PM_1 (Spindler et al. 2010)	27/25	< 32/41	-	-	-
Racibórz (Poland), summer/winter; PM_1 (Rogula-Kozłowska and Klejnowski 2013)	38/65	24/18	2/1	10/6	*4.0/2.1*
Diabla Góra (Poland), year; $PM_{2.5}$ (Rogula-Kozłowska et al. 2014)	32.7	36.3	7.9	13.5	0.1
Urban background sites					
Duisburg (Germany), autumn; $PM_{2.5}$ (Sillanpää et al. 2006)	40	39.5	5.6	7.9	1.2
Prague (Czech Republic), winter; $PM_{2.5}$ (Sillanpää et al. 2006)	59.7	34.6	2.5	1.1	0.47
Amsterdam (Holland), winter; $PM_{2.5}$ (Sillanpää et al. 2006)	28.4	39	2.6	10	0.32
Helsinki (Finland), spring; $PM_{2.5}$ (Sillanpää et al. 2006)	54.4	39.3	6.4	5.8	0.35
Barcelona (Spain), spring; $PM_{2.5}$ (Sillanpää et al. 2006)	28.6	38.7	4.4	6.0	0.57
Athens (Italy), summer; $PM_{2.5}$ (Sillanpää et al. 2006)	41.6	41.4	5.1	1.3	0.34
Barcelona (Spain), year; $PM_{2.5}$ (Pérez et al. 2008)	34	30	16	2	-

Table 3-1. contd....

Table 3-1. contd.

City (Country), averaging period; PM fraction (Reference)	Carbonaceous matter	Secondary inorganic aerosol	Mineral matter	Sea salt	Trace elements
Birmingham (UK), spring/winter; PM_1 (Yin and Harrison 2008)	31/49	55/41	5/8	1/2	-
Milan (Italy), winter/summer; PM_1 (Vecchi et al. 2008)	*39/35*	*41/39*	*1/1*	-	*< 1/< 1*
Florence (Italy), winter/summer; PM_1 (Vecchi et al. 2008)	*57/27*	*23/45*	*< 1/2*	-	*< 1/< 1*
Genoa (Italy), winter/summer; PM_1 (Vecchi et al. 2008)	*50/33*	*20/38*	*2/2*	-	*< 1/< 1*
Zabrze (Poland), summer/winter; PM_1 (Rogula-Kozłowska and Klejnowski 2013)	51/64	24/15	3/1	6/7	*6.6/4.0*
Gdańsk (Poland), year; $PM_{2.5}$ (Rogula-Kozłowska et al. 2014)	31.2	25.4	7.7	9.2	0.2

italics—data extracted from a chart.
[1]samples taken during daytime.
[2]air inflow from western sector.
[3]air inflow from eastern sector.

Heavy metals

All heavy metals in PM are considered toxic (Costa and Dreher 1997; Swaine 2000). They accumulate in body tissues (bones, kidneys, brain). Exposure to their salts or oxides can cause acute or chronic poisonings, tumors, diseases of the cardiovascular and nervous systems, and of kidneys (Lippmann 2008); some heavy metals can weaken the immune system in humans (Goyer 1986). The adverse effects of lead are well known (Chow 1995). Some heavy metals are not bioavailable in their elemental forms. Recent studies show that PM-bound heavy metals are bioavailable; majority occurs in well water-soluble compounds (e.g., Na and Cocker 2009; Wiseman and Zereini 2009; Rogula-Kozłowska et al. 2013a).

Many heavy metals may be found in PM (Schroeder et al. 1987). Each of them can be released by a variety of sources within any area (Nriagu 1979; Pacyna 1984; Nriagu and Pacyna 1988); so they are ubiquitous and their ambient concentrations are strongly site-dependent. Their concentrations should be measured cautiously to avoid the dominating effect of accidental emissions. Although these metals can be extremely toxic (lead), and although their ambient concentrations should be thoroughly controlled, the selection of the area-representative PM sampling points is not usually heavy metal-oriented, except for special purpose measurements.

The metals bound to fine particles (PM2.5-bound) come from incomplete combustion of substances containing carbon in car engines, power plants, waste incinerators, domestic ovens, etc. Some groups of metals are characteristic of specific sources of PM2.5 (Table 3-2).

Table 3-2. Elements and some chemical compounds in particles from various sources (Chow 1995).

Source Type	Mass percentage			
≤ 2.5 µm	< 0.1%	0.1–1%	1–10%	> 10%
[1]Motor Vehicles	Cr,Ni,Y,Sr,Ba	Si,Cl,Al,P,Ca,Mn,Fe, Zn,Br,Pb	$SO_4^{=}$,NH_4^{+} S,Cl^-,NO_3	[2]OC, [3]EC
Vegetative Burning	Ca,Fe,Mn,Zn Br,Rb,Pb	NO_3^-, $SO_4^{=}$, NH_4^{+} Na^+,S	Cl^-,K^+,Cl,K	OC,EC
Residual Oil Combustion	K^+,OC,Cl,Ti Cr,Co,Ga,Se	NH_4^{+}, Na^+,Zn Fe,Si	V,OC,EC,Ni	S,$SO_4^{=}$
Incinerator	V,Mn,Cu,Ag,Sn	K^+,Al,Ti Zn,Hg	NO_3^-,Na^+,EC Si,S,Ca,Fe, Br,La,Pb	$SO_4^{=}$, NH_4^{+} OC,Cl
Coal-Fired Boiler	Cl,Cr,Mn,Ga, As,Se,Br,Rb,Zr	NH_4^{+},P,K,Ti,V Ni,Zn,Sr,Ba,Pb	$SO_4^{=}$,OC,EC Al.,S,Ca,Fe	Si
Oil Fired Power Plant	V,Ni,Se,As,Br,Ba	Al,Si,P,K,Zn	NH_4^{+},OC,EC Na,Ca,Pb	S, $SO_4^{=}$
Smelter Fine	V,Mn,Sb,Cr,Ti	Cd,Zn,Mg,Na Ca,K,Se	Fe,Cu,As,Pb	S
Antimony Roaster	V,Cl,Ni,Mn	$SO_4^{=}$,Sb,Pb	S	–
2.5–10 µm	< 0.1%	0.1–1%	1–10%	> 10%
Paved Road Dust	Cr,Sr,PB,Zr	$SO_4^{=}$,Na^+,K^+,P,S,Cl,Mn,Zn, Ba,Ti	EC,Al,K,Ca, Fe	OC,Si
Unpaved Road Dust	NO_3^-,NH_4^{+},P,Zn, Sr,Ba	$SO_4^{=}$,Na^+,K^+,P,S,Cl,Mn,Ba, Ti	OC,Al,K,Ca, Fe	Si
Construction	Cr,Mn,Zn,Sr, Ba	$SO_4^{=}$,K^+,S,Ti Ca,Fe	OC,Al,K	Si
Agriculture Soil	NO_3^-,NH_4^{+},Cr Zn,Sr,Cl,Mn,Ba,Ti	$SO_4^{=}$,Na^+,K^+,S,Ca,Fe	OC,Al,K	Si
Natural Soil	Cr,Mn,Zn,Sr Zn,Ba	Cl^-,Na^+,EC,P S,Cl,Ti	OC,Al,Mg K,Ca,Fe	Si
Lake Bed	Mn,Sr,Ba	K^+,Ti	$SO_4^{=}$,Na^+ OC,Al,S,Cl K,Ca,Fe	Si
≤ 10 µm	< 0.1%	0.1–1%	1–10%	> 10%
Marine	Ti,V,Ni,Sr,Zr,Pd Ag,Sn,Sb,Pb	Al,Si,K,Ca,Fe Cu,Zn,Ba,La	NO_3^-,$SO_4^{=}$ OC,EC	Cl^-,Na Na,Cl

[1] names of the emission sources are taken from the original papers.

In general, coarse dust PM2.5–10 contains metals from natural sources. A variety of sources emit PM2.5–10 abundant in iron (Table 3-2). Iron is the most ubiquitous component of the Earth crust, and its greater ambient concentrations are associated with the occurrence of PM2.5–10 rather than PM2.5, like those of silicone or aluminum, also of the natural origin. However, fly dust from a power station can contain Al, Ca, and Fe in its mineral components as well (Table 3-2, Coal-Fired Boiler).

Road traffic is an important source of ambient metals. Before lead had ceased to be a gasoline additive, cars were the main source of airborne lead (Pacyna 1984; Nriagu and Pacyna 1988; Richter et al. 2007). Crude oil contains sulfur, vanadium, nickel, iron, sodium, potassium, magnesium, zinc; they are not removed in the refining process and they pass on to PM2.5 originating from oil combustion in car engines (Table 3-2).

The PM2.5–10 from traffic contains metal compounds from corrosion of car bodies, wear of brake linings, clutches, and tires, or from road surface (Harrison et al. 2012; Pant and Harrison 2013); airborne Mo, Pd, and Rh come from catalytic converters in cars (Dias da Silva et al. 2008).

Table 3-3 presents the concentrations of PM and the most often investigated PM-bound metals (V, Fe, Cu, Zn, Ba, As, Cd, Co, Cr, Mn, Ni, and Pb) at background sites (rural site/suburban/regional background) and at urban sites (urban background/urban area/residential area/city center) in various regions in the world; they characterize in some way the areas they come from, but they are rather incomparable.

The highest ambient metal concentrations at background sites (background concentrations) occur in Asia. While the concentrations of PM-bound V, Cr, Mn, Co, Ni, Cu, Zn, As, Cd, and Pb at the majority of the European background sites do not exceed 10 ng/m^3, their concentrations in China reach 500 ng/m^3; they are very high in India, and Korea as well (Table 3-3). In Europe, the highest background concentrations occur in Poland in winter and in Greece in summer. Ireland, Switzerland, France, and some regions in Italy are the least metal-polluted European background regions.

The seasonal variations of the heavy metal concentrations occur at the sites where both the PM and the PM-bound metal background concentrations are high; in Poland and China, the PM and PM-bound metal background concentrations are highest. In Diabla Góra (northern Poland) and Racibórz (southern Poland), in the hot season, they are low and comparable with the concentrations elsewhere in Europe. In the cold season they are several times higher. There is neither dense traffic nor industry at these two Polish sites, and PM-bound metals come from the combustion of fossil fuels for heat and power production in the cities and agglomerations in neighboring areas. Therefore, relatively remote traffic or industrial sources may contribute significantly to the background concentrations of heavy metals, as at sites in China, where background concentrations are affected by industry and traffic from the surrounding agglomerations during whole year and, additionally, by emissions from heating in winter, or as in Diabla Góra and Racibórz (Poland), where the heavy metals may come exclusively from combustion of coal for energy production. Iron is an exception; being bound to coarse PM, it comes mainly from natural sources.

In some Asian background areas, heavy metals brought in by winds with coarse PM from north and northwest deserts may occur along with the heavy metals coming from energy production, industry, and traffic in surrounding big urban agglomerations.

Table 3-3. Concentrations of the twelve trace elements associated with various PM fractions in various regions across the world.[1]

Reference	City (Country)	Averaging period	Fraction	Ambient concentration, ng m^{-3}												
				PM µg m^{-3}	V	Cr	Mn	Fe	Co	Ni	Cu	Zn	As	Cd	Ba	Pb
urban background/urban area/residential area/city center																
(Saliba et al. 2010)	Vienna (Austria)	I–XII 2004	PM_{10}	27.7	1.4	5.5	13	780	1.8	5.7	21	40	0.9	0.5	12	11
(von Schneidemesser et al. 2010)	Zonguldak (Turkey)	XII 2004–X 2005	$PM_{2.5}$	29.6	-	3.8	8.0	130.0	-	3.0	61.0	58.0	-	-	-	11.9
	Zonguldak (Turkey)	XII 2004–X 2005	$PM_{2.5-10}$	24.9	-	3.7	12.0	352.0	-	2.9	60.0	26.0	-	-	-	7.3
(Theodosi et al. 2010)	Istanbul (Turkey)	XI 2007–VI 2009	PM_{10}	*39.1*	*0.014*	*0.004*	*0.02*	*0.70*	-	*0.007*	*0.020*	*0.24*	-	*0.001*	-	*0.07*
(Byrd et al. 2010)	County Cork (Ireland)	2005 S	PM_{10}	14.0	-	4.0	4.0	203.0	bld	12.0	4.0	80.0	-	bld	bld	8.0
(Sánchez-Jiménez et al. 2012)	Glasgow (UK)	VIII 2006	PM_{10}	21.6	0.183	0.141	1.57	26.0	-	0.473	3.06	-	0.040	0.054	-	0.714
	London (UK)	VIII 2006	PM_{10}	25.5	1.49	-	0.352	5.04	-	0.803	1.56	7.73	0.524	0.061	-	2.44
(Vercauteren et al. 2011)	Aarschot (Belgium)	IX 2006–IX 2007	PM_{10}	27.6	4.7	3.4	7.2	495	-	2.6	10.7	74	1.2	-	-	16.0
	Mechelen (Belgium)	IX 2006–IX 2007	PM_{10}	28.2	6.3	3.7	8.3	674	-	3.5	15.8	66	1.2	-	-	19.0
	Borgerhout (Belgium)	IX 2006–IX 2007	PM_{10}	33.7	9.6	4.5	10.5	1040.0	-	4.6	24.6	72.0	17.5	-	-	27.0
(Gu et al. 2011)	Augsburg (Germany)	2006/2007 W	PM_{10}	31.775	-	9.4	15.7	1261.0	0.23	3.9	43.9	47.4	-	0.19	-	8.8
(Contini et al. 2010)	Lecce (Italy)	2007–2008 S	PM_{10}	27.2	2.1	2.5	9.8	365.8	-	3.6	13.0	19.5	< 0.9	-	-	6.2
	Lecce (Italy)	2007–2008 W	PM_{10}	25.2	1.0	1.8	5.8	236.1	-	2.3	12.8	31.3	< 0.9	-	-	9.3

(Moroni et al. 2012)	Terni (Italy)	XII 2008–XI 2009 W	$PM_{1.3}$	47.0	5.2	21.0	9.5	170.0	-	16.0	23.0	50.0	-	-	-	27.0
	Terni (Italy)	XII 2008–XI 2009 W	$PM_{1.3-10}$	23.0	2.1	16.0	16.0	940.0	-	11.0	28.0	47.0	-	-	-	9.9
	Terni (Italy)	XII 2008–XI 2009 S	$PM_{1.3}$	13.0	0.5	5.4	2.1	36.0	-	2.2	15.0	25.0	-	-	-	5.0
	Terni (Italy)	XII 2008–XI 2009 S	$PM_{1.3-10}$	16.0	0.2	5.5	3.0	115.0	-	2.4	5.6	25.0	-	-	-	2.4
(Cuccia et al. 2013)	Genoa (Italy)	V 2009–V 2010	PM_{10}	22.0	13.0	7.0	10.0	780.0	-	9.0	34.0	34.0	-	-	29.0	10.0
	Genoa (Italy)	V 2009–V 2010	$PM_{2.5}$	15.0	12.0	3.0	4.0	135.0	-	7.0	5.0	17.0	-	-	17.0	8.0
(Lettino et al. 2012)	Tito Scalo (Italy)	IV 2010	$PM_{2.5}$	9.0	-	47.0	9.0	195.0	3.0	16.0	8.0	420.0	-	2.0	-	34.0
(Szoboszlai et al. 2012)	Debrecen (Hungary)	X 2008	$PM_{2.5}$	13.7	-	-	11.1	655.4	-	-	34.6	16.9	-	-	-	6.8
	Debrecen (Hungary)	V 2009	$PM_{2.5-10}$	9.3	-	-	5.1	222.2	-	-	8.5	3.3	-	-	-	1.5
	Debrecen (Hungary)	X 2008	$PM_{2.5}$	11.4	-	-	7.9	353.8	-	-	15.5	38.4	-	-	-	30.2
	Debrecen (Hungary)	V 2009	$PM_{2.5}$	7.5	-	-	2.1	111.6	-	-	8.0	5.3	-	-	-	2.4
(Aldabe et al. 2011)	Navarra (Spain)	I–XII 2008	PM_{10}	25.91	-	2.81	6.88	-	1.20	2.21	26.79	29.25	0.21	0.04	18.91	3.33
	Navarra (Spain)	I–XII 2008	$PM_{2.5}$	15.38	-	2.39	2.57	-	0.99	1.31	11.98	17.98	0.16	0.05	12.08	2.29
(Fernández-Camacho et al. 2012)	Huelva (Spain)	IV 2008–XII 2009	PM_{10}	32.7	5.3	2.3	9.6	600.0	0.3	3.7	45.3	47.4	6.2	0.7	31.9	14.4
	Huelva (Spain)	IV 2008–XII 2009	$PM_{2.5}$	19.3	3.4	1.6	4.0	200.0	0.2	2.3	31.2	37.3	5.1	0.6	19.7	10.8
	Huelva (Spain)	IV 2008–XII 2009	$PM_{2.5-10}$	13.4	1.9	0.7	5.6	400.0	0.1	1.4	14.1	10.1	1.1	0.1	12.2	3.6
(Aldabe et al. 2011)	Navarra (Spain)	I–XII 2008	PM_{10}	25.91	-	2.81	6.88	-	1.20	2.21	26.79	29.25	0.21	0.04	18.91	3.33
	Navarra (Spain)	I–XII 2008	$PM_{2.5}$	15.38	-	2.39	2.57	-	0.99	1.31	11.98	17.98	0.16	0.05	12.08	2.29

Table 3-3. contd....

Table 3-3. contd.

Reference	City (Country)	Averaging period	Fraction	Ambient concentration, ng m^{-3}												
				PM $\mu g\ m^{-3}$	V	Cr	Mn	Fe	Co	Ni	Cu	Zn	As	Cd	Ba	Pb
urban background/urban area/residential area/city center																
(Fernández-Camacho et al. 2012)	Huelva (Spain)	IV 2008–XII 2009	PM_{10}	32.7	5.3	2.3	9.6	600.0	0.3	3.7	45.3	47.4	6.2	0.7	31.9	14.4
	Huelva (Spain)	IV 2008–XII 2009	$PM_{2.5}$	19.3	3.4	1.6	4.0	200.0	0.2	2.3	31.2	37.3	5.1	0.6	19.7	10.8
	Huelva (Spain)	IV 2008–XII 2009	$PM_{2.5-10}$	13.4	1.9	0.7	5.6	400.0	0.1	1.4	14.1	10.1	1.1	0.1	12.2	3.6
(Gianini et al. 2012)	Zurich (Switzerland)	VIII 2008–VII 2009	PM_{10}	20.7	0.6	2.0	5.6	420.0	-	1.0	20.6	28.2	0.52	0.12	3.7	5.2
(Zwoździak et al. 2013)	Wrocław (Poland)	XII 2009–X 2010 W	$PM_{2.5}$	55.0	-	4.3	23.0	215.0	-	4.0	40.0	227.0	4.9	-	-	81.0
	Wrocław (Poland)	XII 2009–X 2010 S	$PM_{2.5}$	11.0	-	1.6	9.0	78.0	-	0.7	20.0	43.0	1.2	-	-	27.0
(Pastuszka et al. 2010)	Zabrze (Poland)	I 2006	$PM_{2.5}$	187.3	-	24	25	-	-	18	36	-	-	6	-	-
(Rogula-Kozłowska and Klejnowski 2013)	Zabrze (Poland)	VIII 2009	PM_1	16.66	0.56	0.92	11.9	196	Bld	0.87	5.06	56.37	3.71	1.75	3.71	15.51
	Zabrze (Poland)	XII 2009	PM_1	50.03	0.45	5.15	7.47	96.07	9.26	1.22	6.45	131.59	14.19	4.58	6.79	42.03
(Rogula-Kozłowska et al. 2014)	Katowice (Poland)	2010 W	$PM_{2.5}$	82.97	-	-	-	-	-	3.12	-	-	3.68	2.92	-	83.46
	Katowice (Poland)	2010 S	$PM_{2.5}$	17.29	-	-	-	-	-	3.34	-	-	1.39	1.54	-	22.92
	Gdańsk (Poland)	2010 W	$PM_{2.5}$	42.75	-	-	-	-	-	1.26	-	-	1.33	2.09		43.57
	Gdańsk (Poland)	2010 S	$PM_{2.5}$	13.09	-	-	-	-	-	2.22	-	-	0.33	0.35		5.49
(Friend et al. 2011)	Hong Kong (China)	XI 2004–X 2005	$PM_{2.5}$	-	*0.02*	*0.002*	*0.02*	*0.19*	*0.0003*	*0.01*	*0.01*	*0.22*	*0.01*	-	-	*0.05*

(Xu et al. 2012)	Fuzhou (China)	IV 2007–I 2008 S	$PM_{2.5}$	23.58	2.6	5.0	16.9	219.0	-	3.2	73.8	151.1	7.2	-	-	23.1
	Fuzhou (China)	IV 2007–I 2008 W	$PM_{2.5}$	59.81	3.7	10.1	47.6	563.2	-	3.4	164.9	232.6	16.8	-	-	46.6
(Yang et al. 2012)	Jinan (China)	III 2006–II 2007 S	$PM_{2.5}$	129.04	*0.01*	*0.02*	*0.05*	*0.99*	*0.01*	*0.01*	*0.02*	*0.56*	*0.03*	*0.01*	*0.04*	*0.28*
	Jinan (China)	III 2006–II 2007 W	$PM_{2.5}$	204.89	-	*0.03*	*0.16*	*2.02*	*0.02*	*0.01*	*0.05*	*0.99*	*0.03*	-	*0.07*	*0.43*
(Schleicher et al. 2011)	Beijing (China)	VI 2005–V 2008	TSP	373	2.4	1.9	129	137	1.0	2.7	30	625	19	5.3	-	24
	Beijing (China)	VI 2005–V 2008	$PM_{2.5}$	63	1.3	1.9	33	55	0.4	1.6	16	336	13	2.5	-	32
(Wang et al. 2013)	Shanghai (China)	VI 2009–IX 2010	PM_{10}	97.44	-	22	92	2660	-	11	22	303	-	2	-	71
	Shanghai (China)	VI 2009–IX 2010	$PM_{2.5}$	62.25	-	9	66	1328	-	9	15	236	-	1	-	59
(Saliba et al. 2010)	Beirut (Lebanon)	XII 2006–VIII 2007	PM_{10}	77.10	-	-	*0.01*	*1.06*	-	-	*0.01*	*0.08*	-	-	-	*0.02*
(von Schneidemesser et al. 2010)	Lahore (Pakistan)	2007	PM_{10}	340	*0.021*	*0.03*	*0.3*	*8.2*	*0.0031*	*0.018*	*0.073*	*11*	*0.022*	*0.077*	*0.12*	*4.4*
(Mansha et al. 2012)	Karachi (Pakistan)	I 2006–I 2008 W	$PM_{2.5}$	98.44	-	*0.015*	*0.053*	*3.706*	-	-	*0.039*	*2.31*	*0.05*	-	*0.136*	*0.13*
	Karachi (Pakistan)	I 2006–I 2008 S	$PM_{2.5}$	55.89	-	*0.031*	*0.053*	*3.36*	-	-	*0.056*	*2.89*	*0.071*	-	*0.707*	*0.119*
(Singh et al. 2011)	Delhi (India)	IV–VI 2008 S	PM_{10}	95.1	2.2	144.7	19.1	2021	-	31.2	22.3	431.5	-	1.8	-	210.5
	Delhi (India)	XI 2007–II 2008 W	PM_{10}	182	60.4	128.5	44.7	2884	-	15.3	36.2	612.3	-	9.2	-	420.7
	Delhi (India)	IV–VI 2008 S	$PM_{2.5}$	39.4	1.8	74.1	33.7	612.7	-	29.5	65.2	477.2	-	2.4	-	500.2
	Delhi (India)	XI 2007–II 2008 W	$PM_{2.5}$	61.8	5.6	80.5	25.6	2446	-	15.3	125.3	825.1	-	12.6	-	630.8
(Safai et al. 2010)	Pune city (India)	2007–2008	TSP	171	-	-	0.02	3.07	-	-	5.73	0.42	-	-	-	-
(Alolayan et al. 2013)	Kuwait City (Kuwait)	II 2004–X 2005 W	$PM_{2.5}$	29	*0.01*	-	*0.03*	*1.5*	-	*0.00*	*0.02*	*0.08*	-	-	-	*0.03*
	Kuwait City (Kuwait)	II 2004–X 2005 S	$PM_{2.5}$	57	*0.01*	-	*0.04*	*2.1*	-	*0.01*	*0.01*	*0.05*	-	-	-	*0.02*

Table 3-3. contd....

Table 3-3. contd.

Reference	City (Country)	Averaging period	Fraction	Ambient concentration, ng m^{-3}												
				PM µg m^{-3}	V	Cr	Mn	Fe	Co	Ni	Cu	Zn	As	Cd	Ba	Pb
urban background/urban area/residential area/city center																
(Lestari and Mauliadi 2009)	Bandung City (Indonesia)	2001–2007 DS	$PM_{2.5}$	48	-	*0.06*	*0.01*	*0.33*	-	*0.01*	*0.02*	*0.46*	-	-	-	*0.03*
	Bandung City (Indonesia)	2001–2007 DS	$PM_{2.5-10}$	19	-	*0.04*	*0.02*	*0.31*	-	*0.01*	*0.02*	*0.40*	-	-	-	*0.01*
	Bandung City (Indonesia)	2001–2007 WS	$PM_{2.5}$	39	-	*0.02*	*0.02*	*0.51*	-	*0.03*	*0.01*	*0.36*	-	-	-	*0.03*
	Bandung City (Indonesia)	2001–2007 WS	$PM_{2.5-10}$	16	-	*0.02*	*0.02*	*0.50*	-	*0.05*	*0.01*	*0.30*	-	-	-	*0.01*
(Halek et al. 2010)	Tehran (Iran)	2007 W	PM_{10}	336.7	-	*0.33*	*0.42*	*7.87*	-	-	*1.92*	*26.60*	*0.27*	-	-	*0.94*
	Tehran (Iran)	2007 W	$PM_{2.5}$	210.5	-	*0.30*	*0.32*	*3.73*	-	-	*1.37*	*24.26*	*0.20*	-	-	*0.58*
(Cong et al. 2011)	Lhasa (Tibet)	IX 2007–VIII 2008	PM_{10}	51.8	4.8	19	27	1034	1.8	7.2	9.1	81	1.8	0.52	12	37
(Moreno et al. 2012)	Kuamamoto (Japan)	IV–V 2010	PM_{10}	41.25	4.31	2.50	24.71	840	0.36	2.22	5.70	51.43	2.38	0.44	16.48	22.57
(Cohen et al. 2010)	Hanoi (Vietnam)	IV 2001–XII 2008	$PM_{2.5}$	54.2	*0.003*	*0.005*	*0.061*	*0.394*	*0.002*	*0.004*	*0.010*	*0.487*	-	-	-	*0.236*
(Murillo et al. 2012)	Salamanca (Mexico)	XI 2006–XI 2007	$PM_{2.5}$	45	*0.15*	*0.018*	-	*1.01*	-	*0.012*	*0.044*	*0.430*	-	-	-	*0.123*
(Saffari et al. 2013)	Los Angeles (USA)	IV 2008–III 2009	$PM_{0.25}$	9.15	3.72	1.22	1.78	111	0.05	1.38	9.02	8.05	0.22	0.06	6.22	1.79
(Wong et al. 2011)	Fresno (California, USA)	VI 2006	$PM_{1.8}$	17.3	-	-	-	12.17	-	-	1.59	2.246	0.104	-	-	-
	Fresno (California, USA)	VI 2006	$PM_{0.1}$	0.399	-	-	-	0.138	-	-	0.233	0.189	0.018	-	-	-

(Osornio-Vargas et al. 2011)	Mexicali (USA)	X 2005–III 2006	$PM_{2.5}$	-	*0.13*	*0.09*	*0.04*	*1.94*	-	*0.01*	*0.04*	*0.05*	-	-	-	-
	Mexicali (USA)	X 2005–III 2006	PM_{10}	-	*0.10*	*0.08*	*0.03*	*1.13*	-	*0.01*	*0.02*	*0.04*	-	-	-	-
	Mexicali (USA)	X 2005–III 2006	$PM_{2.5}$	-	*0.10*	*0.07*	*0.05*	*1.57*	-	*0.01*	*0.08*	*0.16*	-	-	-	-
	Mexicali (USA)	X 2005–III 2006	PM_{10}	-	*0.13*	*0.12*	*0.07*	*1.60*	-	*0.03*	*0.09*	*0.18*	-	-	-	-
(Chiou et al. 2009)	Hamshire (USA)	VII 2003–VIII 2005	$PM_{2.5}$	10.986	3.3	0.95	1.4	64	-	0.92	1.33	8.4	0.90	-	10.5	2.4
(Pancras et al. 2013)	Dearborn (USA)	VII–VIII 2007	$PM_{2.5}$	15.66	1.03	-	6.70	37.55	-	0.36	3.08	34.53	0.94	0.30	3.45	4.04
(Saffari et al. 2013)	Long Beach (USA)	IV 2008–III 2009	$PM_{0.25}$	11.5	8.54	1.43	3.07	148	0.14	2.45	5.84	17.0	0.24	0.06	6.44	2.01
(Davis et al. 2011)	Atlanta (USA)	2005–2007	$PM_{2.5}$	-	1.2	0.9	1.4	88.2	-	0.5	9.0	10.1	1.5	-	-	3.0
	Chicago (USA)	2005–2007	$PM_{2.5}$	-	1.4	2.0	3.2	105.2	-	1.0	5.6	24.6	1.0	-	-	5.0
	Cleveland (USA)	2005–2007	$PM_{2.5}$	-	2.7	4.0	10.5	336.5	-	2.7	9.0	52.1	2.8	-	-	11.7
	Dallas (USA)	2005–2007	$PM_{2.5}$	-	1.3	0.6	2.7	120.0	-	0.4	3.7	9.9	0.4	-	-	2.7
	Detroit (USA)	2005–2007	$PM_{2.5}$	-	3.1	3.2	9.1	241.8	-	1.7	9.4	54.0	1.6	-	-	6.7
	Houston (USA)	2005–2007	$PM_{2.5}$	-	5.1	0.8	3.0	106.9	-	1.6	5.5	14.0	1.2	-	-	2.3
	Los Angeles (USA)	2005–2007	$PM_{2.5}$	-	4.8	1.4	3.4	168.6	-	2.5	10.2	13.6	0.6	-	-	3.4
	Miami (USA)	2005–2007	$PM_{2.5}$	-	5.8	7.5	2.5	150.2	-	3.4	4.4	9.2	0.6	-	-	2.0
	Minneapolis (USA)	2005–2007	$PM_{2.5}$	-	3.3	1.8	1.7	56.2	-	1.7	3.6	9.7	0.9	-	-	2.7
	New York (USA)	2005–2007	$PM_{2.5}$	-	7.1	1.8	3.8	124.6	-	11.7	5.1	31.8	0.8	-	-	4.6
	Philadelphia (USA)	2005–2007	$PM_{2.5}$	-	4.5	3.0	2.2	93.3	-	4.3	7.3	15.1	1.0	-	-	4.2
	Phoenix (USA)	2005–2007	$PM_{2.5}$	-	3.2	2.8	4.0	196.1	-	1.2	6.5	10.3	0.6	-	-	2.8
	Pittsburgh (USA)	2005–2007	$PM_{2.5}$	-	1.7	3.7	5.4	142.3	-	2.0	6.8	35.4	3.1	-	-	12.5
	Raleigh (USA)	2005–2007	$PM_{2.5}$	-	1.9	2.7	0.9	74.3	-	1.4	4.3	7.6	0.8	-	-	2.5

Table 3-3. contd....

Table 3-3. contd.

Reference	City (Country)	Averaging period	Fraction	Ambient concentration, ng m^{-3}												
				PM µg m^{-3}	V	Cr	Mn	Fe	Co	Ni	Cu	Zn	As	Cd	Ba	Pb
urban background/urban area/residential area/city center																
	Riverside (USA)	2005–2007	$PM_{2.5}$	-	6.2	4.1	3.5	181.9	-	2.3	8.8	18.5	0.8	-	-	4.3
	San Diego (USA)	2005–2007	$PM_{2.5}$	-	3.6	2.7	2.1	97.2	-	1.9	9.3	7.4	0.8	-	-	2.9
	San Jose (USA)	2005–2007	$PM_{2.5}$	-	2.2	2.0	1.1	79.6	-	1.3	4.7	6.8	0.6	-	-	2.4
	Seattle (USA)	2005–2007	$PM_{2.5}$	-	5.9	2.3	7.6	99.4	-	2.6	5.1	15.2	1.0	-	-	3.8
(Herrera–Murillo et al. 2013)	San Jose (Costa Rica)	VI 2010–VII 2011	PM_{10}	37	3.7	11.5	58	361	-	4.8	47	-	-	-	-	11.8
			$PM_{2.5}$	26	2.9	7.0	61	257	-	3.86	43	-	-	-	-	7.6
	Moravia (Costa Rica)		PM_{10}	25	1.2	5.7	84	189	-	3.4	86	-	-	-	-	8.6
			$PM_{2.5}$	18	0.9	5.1	44	141	-	2.94	57	-	-	-	-	7.29
(Jorquera and Barraza 2012)	Santiago (Chile)	2004	$PM_{2.5}$	32.3	0.5	1.7	11.9	244	-	0.4	19.3	46	10.3	-	7.9	18.9
(Jorquera and Barraza 2013)	Antofagasta (Chile)	XII 2007–I 2008	PM_{10}	*161*	*0.014*	*0.004*	*0.049*	*2.097*	-	*0.003*	*0.582*	*0.093*	*0.076*	-	-	*0.021*
	Antofagasta (Chile)	XII 2007–I 2008	$PM_{2.5}$	*42*	*0.007*	-	*0.010*	*0.354*	-	*0.002*	*0.068*	*0.060*	*0.050*	-	-	*0.014*
(Miranda and Tomaz 2008)	Saõ Paulo (Brazil)	VIII 2003–VIII 2004	PM_{10}	20.85	3.4	4.2	14.8	585	-	4.3	18.1	80	-	-	-	3.6
(Lemos et al. 2012)	Rio Grande (Brazil)	X 2009	TSP	45.94	1.43	bld	-	2.38	-	bld	2.38	bld	bld	bld	-	bld
	Rio Grande (Brazil)	I 2010	$PM_{2.5}$	41.17	0.70	bld	-	4.18	-	1.39	4.18	bld	bld	bld	-	bld
	Rio Grande (Brazil)	X 2009	TSP	12.43	0.79	bld	-	bld	-	0.79	1.19	14.69	bld	bld	-	0.40

	Rio Grande (Brazil)	I 2010	$PM_{2.5}$	8.15	0.86	bld	-	0.86	-	0.86	0.86	35.14	bld	bld	-	bld
(Dieme et al. 2012)	Dakar (Senegal)	VII–IX 2009	$PM_{2.5}$	75.1	*0.002*	*0.002*	*0.027*	*3.027*	bld	*0.001*	*0.005*	*0.024*	bld	-	*0.026*	*0.007*
	Dakar (Senegal)	VII–IX 2009	$PM_{2.5}$	105.4	*0.004*	*0.003*	*0.036*	*4.567*	*0.001*	*0.002*	*0.006*	*0.054*	bld	-	*0.040*	*0.019*
rural site/suburban/regional background																
(Pey et al. 2009)	Villar Arzobispo (Spain)	2004–2005	PM_{10}	21.0	2.6	2.1	3.8	-	0.1	3	2	11	0.2	0.1	4	5
	Villar Arzobispo (Spain)	2004–2005	$PM_{2.5}$	18.0	2.8	2.0	3.2	-	0.1	4	1	12	0.3	0.1	5	6
	Montseny (Spain)	2004–2007	PM_{10}	16.2	3.5	1.0	4.8	-	0.1	1.5	3.6	12	0.3	0.1	6	4
(Aldabe et al. 2011)	Navarra (Spain)	I–XII 2008	PM_{10}	15.23	-	1.32	2.20	-	0.12	0.90	2.03	17.34	0.10	0.05	20.03	2.88
(Dongarrà et al. 2007)	Palermo (Italy)	II–XI 2005	PM_{10}	25	10	3.1	6.6	298	0.2	4.6	9.9	17	1.8	-	23	9.8
(Caggiano et al. 2011)	Tito Scalo (Italy)	VI–VII 2006	$PM_{2.5}$	16	-	68	8	190	-	11	11	34	-	2	-	14
(Perrone et al. 2013)	Salento (Italy)	VII 2008–V 2010	$PM_{2.5}$	25	4	4	2	100	-	5	4	-	-	0.2	5	7
	Salento (Italy)	VII 2008–V 2010	PM_{1}	15	3	4	1	60	-	3	2	-	-	0.1	4	4
	Narni (Italy)	XII 2008–XI 2009 W	$PM_{1.3}$	16	1.0	6.0	4.7	97	-	2.6	18	33	-	-	-	6.4
	Narni (Italy)	XII 2008–XI 2009 W	$PM_{1.3-10}$	17	0.7	6.0	7.7	535	-	1.8	16	15	-	-	-	2.0
	Narni (Italy)	XII 2008–XI 2009 S	$PM_{1.3}$	7.0	0.4	3.6	1.5	30	-	1.1	9.3	14	-	-	-	3.2
	Narni (Italy)	XII 2008–XI 2009 S	$PM_{1.3-10}$	12	0.1	2.6	1.5	82	-	0.7	2.4	7.1	-	-	-	1.2
(Byrd et al. 2010)	County Cork (Ireland)	2005 A	PM_{10}	11	-	5	5	150	1	9	5	141	-	1	bld	11

Table 3-3. contd....

Table 3-3. contd.

Reference	City (Country)	Averaging period	Fraction	Ambient concentration, ng m^{-3}												
				PM μg m^{-3}	V	Cr	Mn	Fe	Co	Ni	Cu	Zn	As	Cd	Ba	Pb
rural site/suburban/regional background																
	County Cork (Ireland)	2006/2007 W	PM_{10}	8	-	7	3	85	1	5	7	39	-	1	bld	12
(Kopanakis et al. 2012)	Akrotiri (Greece)	VIII 2007	PM_{10}	28.3	-	64.4	2254.0	-	-	1.9	4.8	-	-	-	-	69.3
	Akrotiri (Greece)	VIII 2007	$PM_{2.1}$	12.1	-	6.5	842.3	-	-	1.3	1.5	-	-	-	-	24.7
	Basel (Switzerland)	VIII 2008–VII 2009	PM_{10}	18.8	0.7	0.9	3.7	181	-	0.9	6.7	24.6	0.39	0.17	2.2	5.0
	Payerne (Switzerland)	VIII 2008–VII 2009	PM_{10}	19.1	0.5	0.7	2.9	119	-	0.7	4.2	19.4	0.51	0.09	1.7	3.7
	Magadino (Switzerland)	VIII 2008–VII 2009	PM_{10}	20.9	0.6	1.4	5.1	315	-	0.9	8.6	21.3	0.66	0.14	3.1	4.0
(Vercauteren et al. 2011)	Aarschot (Belgium)	IX 2006–IX 2007	PM_{10}	27.6	4.7	3.4	7.2	495	-	2.6	10.7	74	1.2	-	-	16.0
	Mechelen (Belgium)	IX 2006–IX 2007	PM_{10}	28.2	6.3	3.7	8.3	674	-	3.5	15.8	66	1.2	-	-	19.0
	Hasselt (Belgium)	IX 2006–IX 2007	PM_{10}	27.2	3.5	5.0	8.3	618	-	2.5	16.1	101	0.9	-	-	17.0
	Houtem (Belgium)	IX 2006–IX 2007	PM_{10}	27.2	10.6	1.6	6.9	286	-	3.8	4.6	37	0.7	-	-	12.0
(Tran et al. 2012)	small village (France)	V–VII; 2009 OP	PM_{1}	13.3	3.1	0.98	2.3	56	0.23	1.8	1.19	11.6	0.34	0.16	2.56	4.4
	small village (France)	V–VII; 2009 OP	$PM_{2.5}$	16.3	3.3	1.25	3.7	100	0.31	1.8	1.92	19.0	0.38	0.20	5.78	5.4
	small village (France)	V–VII; 2009 OP	PM_{10}	29.6	3.5	2.15	5.7	161	0.45	2.2	2.70	23.1	0.41	0.22	7.64	6.0

	small village (France)	V–VII; 2009 UP	PM_1	11.6	4.3	0.33	0.9	14	0.06	1.9	0.37	3.3	0.26	0.06	0.19	2.7
	small village (France)	V–VII; 2009 UP	$PM_{2.5}$	19.0	4.6	1.25	2.2	43	0.10	2.5	1.12	6.3	0.31	0.07	3.42	3.0
	small village (France)	V–VII; 2009 UP	PM_{10}	23.1	4.8	1.35	3.3	65	0.12	2.6	1.51	7.4	0.33	0.08	4.79	3.1
(Makkonen et al. 2010)	Virolahti (Finland)	VIII 2007	PM_{10}	31.15	2.7	-	7.3	270	0.31	1.3	2.6	20	0.85	0.31	-	13
(Rogula-Kozłowska and Klejnowski 2013)	Racibórz (Poland)	VIII 2009	PM_1	13.32	0.67	5.07	4.94	61.97	Bld	1.36	2.36	12.94	1.23	1.15	3.69	2.04
	Racibórz (Poland)	XII 2009	PM_1	57.27	0.71	26.47	8.67	211.25	0.03	4.54	8.25	91.7	11.02	5.23	7.07	35.24
(Rogula-Kozłowska et al. 2014)	Diabla Góra (Poland)	2010 W	$PM_{2.5}$	25.93	-	-	-	-	-	0.54	-	-	0.79	0.88	-	19.95
	Diabla Góra (Poland)	2010 S	$PM_{2.5}$	8.74	-	-	-	-	-	1.05	-	-	0.23	0.20	-	2.27
(Lim et al. 2010)	Ansan-shi (Korea)	V 2004–I 2006	PM_{10}	73.9	5.7	7.7	60.2	816	-	12.0	-	264	5.8	3.8	37.3	196
(Yang et al. 2012)	Jinan (China)	III 2006–II 2007 W	$PM_{2.5}$	146.80	*0.1*	*0.01*	*0.07*	*0.61*	*0.01*	-	*0.02*	*0.51*	*0.01*	*0.01*	*0.03*	*0.26*
	Jinan (China)	III 2006–II 2007 S	$PM_{2.5}$	69.56	-	*0.02*	*0.03*	*0.50*	*0.01*	-	*0.02*	*0.22*	*0.01*	*0.01*	*0.04*	*0.29*
(Shridhar et al. 2010)	Delhi (India)	IX 2003–VIII 2004	SPM	479.08	*0.31*	*0.10*	*0.36*	*10.49*	-	*0.07*	*1.01*	*1.85*	-	bld	-	*0.25*
(Safai et al. 2010)	Pune city (India)	III–V 2007 S	TSP	335	-	-	*0.03*	*3.14*	-	-	*1.63*	*0.07*	-	-	-	-
	Pune city (India)	VI–IX 2007 M	TSP	135	-	-	bld	*1.55*	-	-	*3.61*	*0.08*	-	-	-	-
	Pune city (India)	X–XI 2007 pM	TSP	100	-	-	*0.02*	*2.94*	-	-	*5.97*	*0.66*	-	-	-	-
	Pune city (India)	XII 2007–II 2008 W	TSP	101	-	-	*0.04*	*4.32*	-	-	*10.28*	*0.83*	-	-	-	-
(Khare and Baruah 2010)	Jorhat City (India)	I 2007–I 2008 R	$PM_{2.5}$	108	106	15.3	168	359	47	30.9	23.5	33.3	-	10.7	-	22
	Jorhat City (India)	I 2007–I 2008 W	$PM_{2.5}$	143.25	50.71	16.16	16.3	495	48.4	51.3	23.7	17.57	-	39.7	-	26

Table 3-3. contd....

Table 3-3. contd.

Reference	City (Country)	Averaging period	Fraction	Ambient concentration, ng m^{-3}												
				PM µg m^{-3}	V	Cr	Mn	Fe	Co	Ni	Cu	Zn	As	Cd	Ba	Pb
rural site/suburban/regional background																
(Massey et al. 2013)	Agra (India)	X 2007–III 2009	$PM_{2.5}$	143.07	-	*0.6*	*0.1*	*1.9*	-	*0.2*	*0.2*	-	-	*0.5*	-	*0.4*
(Saliba et al. 2010)	Beirut (Lebanon)	I 2004–I 2005	PM_{10}	103.78	-	-	*0.03*	*1.86*	-	*0.006*	*0.05*	*0.08*	-	-	-	*0.08*
	Beirut (Lebanon)	I 2004–I 2005	$PM_{2.5}$	41.40	-	-	*0.01*	*0.26*	-	*0.005*	*0.01*	*0.10*	-	-	-	*0.11*
(Dieme et al. 2012)	Ngaparou (Senegal)	VII–IX 2009	$PM_{2.5}$	16.9	bld	bld	*0.004*	*0.393*	bld	bld	*0.001*	*0.002*	bld	-	*0.004*	bld

italics—data in µg/m^3
W—winter
S—summer
A—autumn
DS.—dry season
WS—wet season
OP—occupied period
UP—unoccupied period
M—monsoon
pM—post-monsoon
R—rainy
bld—below limit of detection

[1] Table 3-3 presents some results from urban sites, background sites or/and rural sites in the last decade (or earlier if the averaging period and the decade overlap). The symbols for the measuring periods and fractions, decimal places in the numbers (concentrations of dust and compounds), and units are as in the source texts.

In India, the concentrations of coarse PM reach hundreds of μg m^{-3}, and the ambient concentrations of the related (crustal) iron reach 10 μg m^{-3}.

In almost every European and Asian background locality, if one of V, Cr, Mn, Co, Ni, Cu, Zn, As, Cd, or Pb has high ambient concentration, the concentrations of the rest are also high (Table 3-3). The metal concentrations are also highly correlated with the PM concentrations, and where the PM concentrations are high, the metal concentrations are high as well. Therefore, if a background locality in Europe or Asia is affected by industry, traffic, or combustion of coal for energy production in surrounding urban agglomerations, the PM concentration and the concentration of each of these metals can indicate the presence and the level of airborne heavy metals, becoming an air quality index on a wider than local scale (Harrison et al. 2003; Shah and Shaheen 2007).

It is slightly different at urban sites. Although the high PM concentrations are accompanied by the high concentrations of almost all considered heavy metals in majority of the cities (it refers especially to lead), it happens that in big agglomerations the low concentrations of PM co-occur with high concentrations of some heavy metals. Despite low concentrations of PM, the ambient concentrations of V in Genoa and of Cr, Zn, and Pb in Tito Scalo (Italy) are high. As for Asia, the concentrations of PM2.5 in Delhi are low, but the heavy metal concentrations are high.

Majority of the metal concentrations are low, whilst the PM concentrations are moderate or high, in Glasgow and London (UK), Augsburg (Germany), Navarra (Spain) in Europe, Hanoi (Vietnam), Beijing, Shanghai (China), Beirut (Lebanon) in Asia, Salamanca (Mexico) in America, Dakar (Senegal) in Africa.

The ambient heavy metal concentrations are highest in Asia, the lowest in the cities of the USA (Table 3-3); in Europe, the highest concentrations are in the Polish cities, especially in winter. In Europe, the highest ambient concentrations of As, the tracer for hard coal combustion, occur in Zabrze and Katowice (Poland) and Borgehout (Belgium). Similar high (or even higher) concentrations of As were reported in Beijing (China), and in Lahore and Karachi (Pakistan).

Almost all heavy metal concentrations vary seasonally; the variability is greater at urban that at background sites. At many urban sites, the seasonal variability occurs of the mass proportions of fine to coarse PM and of fine- to coarse PM-bound metals, as in Terni (Italy). However, more information on the distribution of PM-bound metal mass with respect to particle size is needed, especially at the sites where ambient metal concentrations are high. The available, not numerous, works on the PM-bound metal mass size distribution confirm the tendency of PM-bound heavy metals to accumulate in the finest PM particles (e.g., Fernandez et al. 2001; Daher et al. 2012, 2014; Rogula-Kozłowska et al. 2013b).

On the other hand, there are urban sites where more metals are in coarse than in fine particles (Table 3-3); at some sites the ambient concentrations of both fine- and coarse PM-bound heavy metals are high (Delhi, India or Teheran, Iran; Table 3-3). The direct influence of the coarse PM-bound heavy metals on humans, e.g., through the respiratory system, is of minor importance, but they affect the human health indirectly, through the environment: coarse PM-bound metals contaminate soil and water; they accumulate in crops.

It is rather obvious that the data on the PM-bound chemicals, their distributions among the PM fractions, and their fraction related concentrations and physicochemical properties are far too poor to picture the hazard from PM worldwide. The chemical forms of the element occurrence in PM may be numerous; in general they are unknown, but they decide the mechanisms of the element toxicity.

Polycyclic aromatic hydrocarbons

Polycyclic aromatic hydrocarbons (PAHs) are another components of PM, extremely different from heavy metals and equally important. Their congeners are complex, they are built of several (at least two) benzene rings. The variety of structures that can arise from combining these rings, multiplied by the possibility of attaching side chains, makes the PAHs group very numerous (more than 100 PAHs). 16 PAHs investigated most often in ambient air are listed below in Table 3-4. It should be noted that the physicochemical properties of individual PAHs are very different from one another. In the atmospheric air, they can occur in solid, vapor, and gas phases; the phases usually differ in their toxicity (Patri et al. 2009; Bae et al. 2010; Hanzalova et al. 2010; Hassan and Khoder et al. 2012).

PAHs have low saturated vapor pressure and they are not easily water-soluble. The PAHs' distribution among the phases depends chiefly on the number of benzene rings, but also on air temperature and humidity, PM concentration, molecular weight, and vapor pressure of particular PAHs. The lightest, two- or three-ring PAHs can only be vaporous in the air, while four-ring PAHs can change their phase depending on weather conditions. Five- or more ring PAHs occur only in the PM-bound phase (Khalili 1995; Ravindra et al. 2008; Tobiszewski and Namieśnik 2012; Hassan and Khoder 2012).

According to the International Agency for Research on Cancer (IARC), 48 PAHs can be human or animal carcinogens. They are the first air pollutants that have been identified as carcinogens. The strength of their carcinogenicity grows with their molecular weight.

After entering the human body, PAHs are transported by blood to various organs, where they are metabolized (Nikolao et al. 1984; White 2002). The metabolites, bound covalently to DNA or RNA, cause neoplasm and affect replications, transcriptions, and biosynthesis of proteins (Skupińska et al. 2009). All PAHs can accumulate in the body tissues.

However, it is unknown so far if the carcinogenicity of PAHs, always a mixture, never a single compound in the air, may be ascribed to individual hydrocarbons or if it is due to the concerted effects of some number of PAHs. The carcinogenicity of a PM-bound PAH may be enhanced or suppressed by other PM components.

Benzo(a)pyrene (BaP) is a well-studied five-ring hydrocarbon; it is of special importance to environmental toxicology. It is one of the most mutagenic and carcinogenic hydrocarbons known (Nikolao et al. 1984; Durant et al. 1996; Saunders et al. 2002; Ravindra et al. 2008). In exposed humans, it penetrates the whole body accumulating in lungs, liver, spleen, kidneys, heart, muscles, fat (Elovaara et al. 2007). Its importance consists also in being a basis for defining the toxic equivalence factor (TEF) and the toxic equivalent (TEQ) for other PAH. Namely, TEF for a PAH

Table 3-4. PAHs diagnostic ratios for various emission sources.

Ratio	Range	Source Type	Reference
ΣLMW/ΣHMW	< 1	Pyrogenic	(Zhang et al. 2008)
	> 1	Petrogenic	
CPAH/TPAH	~ 1	Combustion	(Prahl et al. 1984; Takada et al. 1990; Mantis et al. 2005; Ravindra et al. 2006a, 2008; Gogou et al. 1996)
FL/(FL + PYR)	< 0.5	Petrol emissions	(Ravindra et al. 2008)
	> 0.5	Diesel emissions	
ANT/(ANT + PHE)	< 0.1	Petrogenic	(Pies et al. 2008)
	> 0.1	Pyrogenic	
	0.24	Coal burning	(Guo et al. 2003; Kong et al. 2010)
PHE/(PHE + ANT)	0.50	Gasoline	(Alves et al. 2001)
	0.65	Diesel	
	> 0.70	Crude oil	(Sicre et al. 1987)
	0.76	Coal	(Alves et al. 2001)
FLA/(FLA + PYR)	< 0.4	Unburned petroleum	(De La Torre-Roche et al. 2009)
	0.4–0.5	Liquid fossil fuels	
	> 0.5	Coal, grass, wood	
	0.53	Coal/coke	(Saarnio et al. 2008; Kong et al. 2010)
	0.57	Coal burning	(Galarneau 2008; Kong et al. 2010)
	0.40	Steel manufacture	(Manoli et al. 2004)
	0.50	Cement production	(Manoli et al. 2004)
FLA/PYR	0.6	Vehicles	(Neilson 1998)
BaA/(BaA + CHR)	0.2–0.35	Coal combustion	(Akyüz and Cabuk 2010)
	> 0.35	Vehicular emissions	
	< 0.2	Petrogenic	(Yunker et al. 2002)
	> 0.35	Combustion	
	0.5	Coal/coke	(Tang et al. 2005; Kong et al. 2010)
	0.46	Coal burning	(Galarneau 2008; Kong et al. 2010)
	0.48	Steel manufacture	(Manoli et al. 2004)
	0.30	Cement production	
BaP/(BaP + CHR)	0.5	Diesel	(Khalili et al. 1995; Guo et al. 2003)
	0.73	Gasoline	

Table 3-4. contd....

Table 3-4. contd.

Ratio	Range	Source Type	Reference
BaA/BaP	0.5	Gasoline	(Li and Kamens 1993)
	1	Diesel	
	1	Wood combustion	
IcdP/(IcdP + BghiP)	< 0.2	Petrogenic	(Yunker et al. 2002)
	0.2–0.5	Petroleum combustion	
	> 0.5	Grass, wood and coal combustion	
	0.18	Gasoline engine	(Grimmer et al. 1983; Ravindra et al. 2006a,b; Kavouras et al. 2001)
	0.33	Coal/coke	(Tang et al. 2005; Kong et al. 2010)
	0.56	Coal burning	(Ravindra et al. 2008; Kong et al. 2010)
	0.37	Steel manufacture	(Manoli et al. 2004)
	0.90	Cement production	(Manoli et al. 2004)
	0.62	Wood combustion	
	0.35–0.70	Diesel engine	
IcdP/BghiP	< 0.4	Gasoline	(Caricchia et al. 1999)
	~ 1	Diesel	
BbF/BkF	2.5–2.9	Aluminium smelter emission	(Callen et al. 2011)
	> 0.5	Diesel	(Pandey et al. 1999; Park et al. 2002)
BaP/BghiP	< 0.6	Non-traffic emissions	(Katsoyannis et al. 2007)
	> 0.6	Traffic emissions	
	0.5–0.6	Gasoline engine	(Pandey et al. 1999; Park et al. 2002)
	> 1.25	Coal/coke	(Ravindra et al. 2008; Kong et al. 2010)
	0.9–6.6	Wood combustion	(Akyüz and Cabuk 2008; Kong et al. 2010)
	0.71	Steel manufacture	(Manoli et al. 2004)
	0.85	Cement production	(Manoli et al. 2004)
	> 1.25	Brown coal	(Pandey et al. 1999)
PYR/BaP	~ 10	Diesel	(Oda et al. 2001)
	~ 1	Gasoline	

ΣLMW—total concentration of two- and three-ring PAH.
ΣHMW—total concentration of four- and five-ring PAH.
CPAH—total concentration of more important non-alkylene PAH (FL, PYR, BaA, CHR, BbF, BkF, BaP, IcdP, BghiP).
TPAH—total PAH concentration.
Naftalen: NP, Acenaftylen: ACY, Acenaften: ACE, Fluoren: FL, Fenantren: PHE, Antracen: ANT, Fluoranten: FLA, Piren: PYR, Benzo(a)antracen: BaA, Chryzen: CHR, Benzo(b)fluoranten: BbF, Benzo(k)fluoranten: BkF, Benzo(a)pyrene: BaP, Dibenzo(a,h)anthracene: DahA, Benzo(g,h,i)perylen: BghiP, Indeno(1,2,3-c,d) piren: IcdP.

is defined relative to the TEF of BaP, assumed to be 1; TEQ of a group of PAHs is the linear combination of the TEF and the ambient concentrations of the PAH from this group. TEF expresses the absolute toxicity of a particular PAH, while TEQ expresses the toxicity of a group of PAHs in the air (Nisbet and LaGoy 1992).

For a long time, airborne BaP, or rather its concentration, has been used to indicate the health hazard from PAH. However, PAHs being more carcinogenic than BaP have been identified recently, such as dibenzo(a,h)anthracene and dibenzo(a,l)pyrene, which are at least five times more carcinogenic than BaP (Nisbet and LaGoy 1992), and which probably can represent the carcinogenicity of PAHs mixtures better. Unfortunately, they are not as well-studied as BaP. It should be noted that PM10-bound BaP is routinely monitored in many countries (the most common limit for its yearly concentrations is 1 ng m^{-3}), and before they replace BaP, they will need to be investigated further.

PAHs arise from pyrolysis and pyrosynthesis; secondary ambient PAHs arise from photolysis in the atmosphere. In general, each process involving heating or (incomplete) combustion of organic compounds can yield PAHs. Chemical industry (production of coal and oil derivatives), combustion of materials containing carbon are the most important sources of PAHs within urbanized areas (Zou et al. 2003; Kozielska and Konieczyński 2007, 2008; Ravindra et al. 2008; Tobiszewski and Namieśnik 2012). Combustion of coal, waste, or biomass in domestic ovens, waste incineration, and combustion of gasoline and oil in car engines are everywhere occurring PAHs sources (e.g., Nikolao et al. 1984; Kristensson et al. 2004; Ravindra et al. 2006a,b; Brown and Brown 2012). Cigarette smoke is an important source of PAHs because of its specific role in human life (Li et al. 2003).

Despite relatively low mass contribution to PM, PAHs may be used to trace the origin of PM; a group of PAHs in PM can be characteristic of the PM source (Harrison et al. 1996; Fang et al. 2004). They may be used as markers or diagnostic ratios may be computed based on their concentrations. The markers of a source are the PAHs that are characteristic of the source. The diagnostic ratios are the mutual proportions of the ambient concentrations of a single PAH or groups of PAHs that have similar physicochemical properties (Tobiszewski and Namieśnik 2012). In Table 3-4, some diagnostic ratios for the most often investigated 16 PAHs are presented. Although the diagnostic ratios are better to use than markers, they must be used cautiously because they are sensitive to changes of the atmospheric conditions and may be the same for various PAHs sources (Dvorská et al. 2011; Křůmal et al. 2013).

Ambient PAHs take part in complex physicochemical processes, and they transform in the atmosphere (e.g., Shiraiwa et al. 2009; Lammel et al. 2009). Their persistence in the air is strongly dependent on solar radiation, air humidity and temperature, and precipitation. Half-lives of ambient PAHs may vary from hours (high density of solar radiation, moderate temperatures, and humidity) to days or even weeks (low insolation, low temperatures, and low humidity) (Kamens et al. 1988; Eiguren-Fernandez et al. 2004). Some PAHs are transported with air masses to such remote areas as the Arctic, particularly those bound to fine PM (Valerio et al. 1984; Van Jaarsveld 1997; Mc Veety et al. 1988). PAHs may be removed from the air by PM deposition, wet or dry. PM-bound PAHs are more persistent in the air than gaseous. The PM2.5 and PM10-bound PAHs are investigated most often; in many countries the PM2.5- and PM10-bound PAHs are measured routinely in networks of air quality monitoring stations. The gaseous PAHs, having two, three, sometimes four rings, are less significant. Their ambient concentrations are not routinely measured because they are very site-dependent, especially in urban conditions. Also, they are considered less harmful than more-ring PAHs. They may be measured to monitor special areas or sites exposed to special pollution sources.

In Table 3-5, the sums (∑PAH) of the ambient concentrations of PAHs are presented. Each ∑PAH is the sum of a number of the concentrations of PAHs found in a single PM fraction. The PAHs are the 16 EPA PAHs (Table 3-4, below) and other important PAHs, such as cyclopenta(cd)pyrene and coronene, which are markers of traffic emissions. The PM comes from various sites in the world, and in general, the number of the PM-bound PAHs, whose concentrations are summed up, depends on the laboratory capability; in Table 3-5 it ranges from 7 to 28 PAHs. The numbers 15 and 16 occur most often.

At many sites, road traffic is a prominent PAHs source, and in general ∑PAH are higher at the traffic localities than at the urban background sites in the same city, e.g., in Florence and Venice-Mestre (Italy), Paris (France), Madrid (Spain). But in Katowice (Poland) the concentrations of PM2.5-related PAHs are almost two times greater at the urban background site than at the traffic sampling point located on a highway shoulder (Table 3-5). The concentrations in Katowice are not short-term results (daily or weekly), they both are averages over the same long period (almost a half-year), and so the tendency is rather persistent.

In Europe, such high concentrations of ∑PAH as in Poland, reaching 200 ng m^{-3} at the background site in Katowice, occur only in Belgium and at the industrial site in Turkey in winter. In Asia, however, in China, India, Taiwan, and Afghanistan, such high ∑PAH concentrations as in Poland, and much higher, are not only noted at traffic or affected by industry sites, but also at typical urban sites (Table 3-5). Therefore, although traffic contributes to the PM-related PAHs everywhere where the PAHs concentrations are high, it may not be the main PAHs source. In the industrialized Asian cities, in southern Poland, or in other densely populated areas, where biomass and/or fossil fuels are combusted to produce energy, the traffic emissions have a small share in the PM-bound PAHs (Kozielska et al. 2013). Also, where the ∑PAH concentrations are high (several tens or hundreds of ng/m^3) their seasonal variations are very high; where the concentrations are low (several ng m^{-3} or less than 1 ng m^{-3}), their seasonal variations are also low. In the latter case, the seasonal variability may not only be due to the additional PAHs emissions from the local fossil fuel combustion in winter, but also, or maybe mainly, to the transport of PM-bound PAHs from other, more polluted, regions. The winter meteorological condition favor the PAH adsorption on PM particles and occurrence of PAHs in solid phase. In Upper Silesia, Poland, the PM-bound PAHs content in the air is, on average, more than ten times higher in winter than in summer (Bodzek et al. 1993). In other regions it is similar (Gaga and Ari 2011; Hassan and Khoder 2012). The difference in the gaseous PAHs content of the air between winter and summer is much smaller (Hassan and Khoder 2012; Zhou et al. 2013).

Table 3-5. Concentrations of benzo(a)pyrene (BaP) and sums of PAHs (∑PAH) associated with various fractions of PM from various regions in the world.[2]

Reference	City (Country)	Averaging period	Sampling site	Fraction	Concentration		
					PM, µg m^{-3}	BaP, ng m^{-3}	∑PAH (n), ng m^{-3}
(Rogula-Kozłowska et al. 2012)	Zabrze (Poland)	2007 S	U	PM_1	9.65	0.17	3.48 (15)
				$PM_{2.5}$	12.82	0.31	4.49 (15)
				PM_{10}	18.90	0.65	5.86 (15)
		2007 W		PM_1	33.65	16.08	128.10 (15)
				$PM_{2.5}$	41.31	19.19	153.10 (15)
				PM_{10}	46.29	19.32	156.11 (15)
(Rogula-Kozłowska et al. 2013c)	Katowice (Poland)	VIII 2009–XII 2010	U	PM_1	30.77	7.72	84.50 (15)
				$PM_{2.5}$	49.60	14.14	196.80 (15)
	Katowice (Poland)		T	PM_1	29.98	9.50	121.07 (15)
				$PM_{2.5}$	38.98	9.74	107.26 (15)
	Złoty Potok (Poland)		R	PM_1	13.19	3.24	20.82 (15)
				$PM_{2.5}$	25.48	4.72	40.29 (15)
(Křůmal et al. 2013)	Brno (Czech Republic)	2009 W	U	PM_1	19.9	13.9	22.20 (16)
		2009 S			13.2	0.35	5.01 (16)
		2010 W			33.4	2.82	39.8 (16)
		2010 S			12.9	0.15	1.68 (16)
	Šlapanice (Czech Republic)	2009 W			19.2	1.65	21.00 (16)
		2009 S			13.2	0.39	5.41 (16)
		2010 W			35.1	2.84	38.70 (16)
		2010 S			11.6	0.15	1.52 (16)
(Delhomme and Millet 2012)	Strasbourg (France)	IV 2006–I 2007	U	PM_{10}	-	2.1	12.6 (10)
	Besançon (France)				-	-	9.5 (10)
	Spicheren (France)				-	-	8.9 (10)
(Di Filippo et al. 2010)	Rome (Italy)	X 2007–II 2008	sU	PM_1	-	-	6.70 (14)
				$PM_{2.5}$	-	-	7.77 (14)
				PM_{10}	-	-	7.98 (14)

Table 3-5. contd....

Table 3-5. contd.

Reference	City (Country)	Averaging period	Sampling site	Fraction	Concentration		
					PM, µg m^{-3}	BaP, ng m^{-3}	∑PAH (n), ng m^{-3}
(Martellini et al. 2012)	Florence (Italy)	III 2009–III 2010 c	T	$PM_{2.5}$	29.6	1.0	13.0 (16)
		w			26.2	0.21	7.2 (16)
		c	U		18.2	0.47	6.7 (16)
		w			16.4	0.049	1.7 (16)
	Livorno (Italy)	c	sU		13.6	0.20	3.6 (16)
		c			15.4	0.02	0.92 (16)
(Makkonen et al. 2010)	Virolahti (Finland)	V–IX 2006	R	PM_1	4.3	< 0.08	1.77 (11)
				$PM_{2.5}$	8.7	< 0.08	1.77 (11)
				PM_{10}	10.3	< 0.08	1.78 (11)
		II–IV 2006 and X 2006–I 2007		PM_1	3.8	0.52	7.54 (11)
				$PM_{2.5}$	6.3	0.69	14.53 (11)
				PM_{10}	8.2	0.73	13.9 (11)
(Masiol et al. 2013)	Venice-Mestre (Italy)	2011	R	PM_{10}	35	-	4.2 (8)
			U		39	-	5.9 (8)
	Padova (Italy)		T		46	-	10.3 (8)
			R		43	-	13.1 (8)
	Verona (Italy)		U		44	-	9.7 (8)
			T		42	-	8.3 (8)
	Vicenza (Italy)		I		45	-	8.5 (8)
			sU		35	-	5.1 (8)
	Belluno (Italy)		U		48	-	4.7 (8)
			U		46	-	6.6 (8)
	Feltre (Italy)		U		23	-	7.1 (8)
			sU		28	-	11.8 (8)
(Vercauteren et al. 2011)	Zelzate (Belgium)	X 2006–III 2007	I	PM_{10}	35.0	0.95	209.95 (17) 10.5 (7)
		IV–IX 2007				0.37	156.37 (17) 4.5 (7)
	Borgerhout (Belgium)	X 2006–III 2007	U		35.7	1.18	257.18 (17) 12.0 (7)

Table 3-5. contd....

Table 3-5. contd.

Reference	City (Country)	Averaging period	Sampling site	Fraction	Concentration		
					PM, μg m^{-3}	BaP, ng m^{-3}	∑PAH (n), ng m^{-3}
		IV–IX 2007				0.51	241.51 (17) 6.0 (7)
	Aarschot (Belgium)	X 2006–III 2007	R		28.2	0.81	155.81 (17) 9.1 (7)
		IV–IX 2007				0.41	111.41 (17) 4.9 (7)
(Chrysikou et al. 2009)	Thessaloniki (Greece)	I–II 2006	T	$PM_{1.5-3.0}$	-	0.04	1.41 (13)
(Evagelopoulos et al. 2010)	Kozani (Greeece)	XII 2005–X 2006	U	$PM_{2.5}$	-	0.38	4.77 (15)
(Slezakova et al. 2011)	Oporto (Portugal)	2008 A/W	T	$PM_{2.5}$	25.2	1.88	19.2 (17)
				PM_{10}	36.2	2.02	20.8 (17)
(Dejean et al. 2009)	Toulouse (France)	IV–V 2006	T	PM_{10}	17.87	0.17	19.7 (16)
			U		19.53	0.13	12.3 (16)
			I		18.77	0.81	21.6 (16)
(Kliucininkas et al. 2011)	Kaunas (Lithuania)	I–II 2009 1	U	$PM_{2.5}$	34.5	3.2	32.7 (13)
		2			36.7	6.2	75.1 (13)
(Pietrogrande et al. 2011)	Augsburg (Germany)	VIII–IX 2007	U	$PM_{2.5}$	-	0.08	1.34 (11)
		II–III 2008			-	0.83	11 (11)
(Bari et al. 2009)	Dettenhausen (Germany)	XI 2005–III 2006	sU	PM_{10}	60–95	1.62	22.9 (21)
(Arruti et al. 2012)	Santander Spain	2008–2009	U	$PM_{2.5}$	13	0.02	1.20 (16)
				PM_{10}	29	0.04	1.47 (16)
	Castro Urdiales,			PM_{10}	22	0.07	1.55 (16)
	Reinosa,			PM_{10}	20	0.13	2.82 (16)
	Los Tojos,		R	PM_{10}	-	< 0.02	1.64 (16)
(Ringuet et al. 2012)	Paris (France)	2009–2010 S d	T	PM_{10}	40	0.14	6.71 (17)
		n			39	0.19	7.05 (17)
		d	sU		14	0.02	0.53 (17)
		n			15	0.03	0.61 (17)
(Mirante et al. 2013)	Madrid (Spain)	VI 2009	T	PM_1	-	0.034	0.70 (16)
				$PM_{2.5}$	-	0.039	1.00 (16)

Table 3-5. contd....

Table 3-5. contd.

Reference	City (Country)	Averaging period	Sampling site	Fraction	Concentration		
					PM, µg m^{-3}	BaP, ng m^{-3}	∑PAH (n), ng m^{-3}
				PM_{10}	22–49	0.057	1.501 (16)
		II 2010		PM_1	-	0.24	4.10 (16)
				$PM_{2.5}$	-	0.256	4.35 (16)
				PM_{10}	24–57	0.261	4.46 (16)
		VI 2009	U	PM_1	-	0.022	0.28 (16)
				$PM_{2.5}$	-	0.053	0.89 (16)
				PM_{10}	14–30	0.065	1.08 (16)
		II 2010		PM_1	-	0.054	0.775 (16)
				$PM_{2.5}$	-	0.05571	0.790 (16)
				PM_{10}	8–48	0.05641	0.816(16)
(Akyüz and Çabuk 2008)	Zonguldak (Turcja)	2007 S	I	$PM_{2.5}$	-	0.40	5.70 (14)
				$PM_{2.5-10.0}$	-	0.20	1.60 (14)
		2007 W		$PM_{2.5}$	-	15.70	152.70 (14)
				$PM_{2.5-10.0}$	-	0.70	10.50 (14)
(Wingfors et al. 2011)	Kabul (Afghanistan)	2009 A	U	$PM_{2.5}$	86	25	223 (15)
	Mazar-e Sharif (Afghanistan)			$PM_{2.5}$	68	6.7	64 (15)
(Singh et al. 2011)	Delhi (India)	XI 2007–II 2008	U	$PM_{2.5}$	50.6	9.9	96 (16)
				PM_{10}	138.5	6.9	81.5 (16)
(Lakhani 2012)	Agra (India)	V 2006–XII 2009	I	TSP	348	12.3	72.7 (16)
(Rajput and Lakhani 2012)	Agra (India)	IV–IX 2006	I/U	TSP	400	140	269 (14)
(Singla et al. 2012)	Angra (India)	VII–XII 2010	I/sU/T	$PM_{2.5}$	-	18.2	779.4 (13)
				PM_{10}	-	31.4	1445.5 (13)
(Kaur et al. 2013)	Amritsar (India)	XI 2011	U	PM_{10}	-	1.2	154 (16)
(Mohanraj et al. 2011a)	Tiruchirappalli (India)	III 2009–II 2010	U	$PM_{2.5}$	13.9–132.3	-	333.7 (9)
			sU			-	202.6 (9)
(Mohanraj et al. 2011b)	Chennai (India)	III 2009–II 2010	U	$PM_{2.5}$		-	582.9 (11)
			sU			-	325.7 (11)
			U/sU			-	330.7 (11)
			I			-	790.8 (11)

Table 3-5. contd....

Table 3-5. contd.

Reference	City (Country)	Averaging period	Sampling site	Fraction	Concentration		
					PM, µg m^{-3}	BaP, ng m^{-3}	∑PAH (n), ng m^{-3}
(Kuo et al. 2013)	Taichung County (Tajwan)	VIII 2008	T	$PM_{2.5}$	30.0	-	96.7 (28)
				PM_{10}	46.2	-	107.3 (28)
	Taichung City (Tajwan)	6–10 IX 2008	T	$PM_{2.5}$	34.3	-	130.0 (28)
				PM_{10}	48.9	-	142.2 (28)
(Kong et al. 2010)	Fushun (China)	2004–2005	U	$PM_{2.5}$	-	10.71	261.82 (13)
				$PM_{2.5-10}$	-	1.98	72.44 (13)
			sU	$PM_{2.5}$	-	48.44	1899.36 (13)
				$PM_{2.5-10}$	-	3.44	166.92 (13)
	Jinzhou (China)		U	$PM_{2.5}$	-	13.61	190.86 (13)
				$PM_{2.5-10}$	-	1.13	20.78 (13)
			sU	$PM_{2.5}$	-	6.61	106.94 (13)
				$PM_{2.5-10}$	-	0.77	16.40 (13)
(Wang et al. 2009)	Beijing (China)	2005–2007 S	sU	$PM_{2.5}$	-	1.19	39.28 (15)
				$PM_{2.5-10}$	-	1.14	24.87 (15)
		2005–2007 W		$PM_{2.5}$	-	19.82	360.71 (15)
				$PM_{2.5-10}$	-	5.09	102.00 (15)
(Hong et al. 2007)	Xiamen (China)	2005 S	I/T	PM_{10}	-	0.1	5.24 (15)
		2005 A	T	PM_{10}	-	1.4	37.10 (15)
		2005 S	sU	PM_{10}	-	< 0.1	1.69 (15)
		2005 A		PM_{10}	-	1.5	26.60 (15)
(Dong et al. 2012)	Beijing (China)	XII 2005–I 2006	U	$PM_{2.5}$	-	-	30.1 (16)
				PM_{10}	-	-	44.1 (16)
			T	$PM_{2.5}$	-	-	80.1 (16)
				PM_{10}	-	-	99.8 (16)
(Hien et al. 2007)	Ho Chi Minh (Vietnam)	I–II 2005 DS	T	TSP	470	2.00	39.80 (9)
		VII 2005 WS		TSP	418	5.70	58.70 (9)
(Ladji et al. 2009)	Bumardas (Algeria)	X 2006	U/T	$PM_{1.0-10.0}$	-	0.02	0.33 (14)
	Rouiba-Réghaia (Algeria)		I	$PM_{1.0-10.0}$	-	0.04	0.59 (14)

Table 3-5. contd....

Table 3-5. contd.

Reference	City (Country)	Averaging period	Sampling site	Fraction	Concentration		
					PM, µg m^{-3}	BaP, ng m^{-3}	∑PAH (n), ng m^{-3}
	Chréa National Park (Algeria)		R	$PM_{1.0–10.0}$	-	0.002	0.10 (14)
(Ding et al. 2009)	Golden (British Columbia, Canada)	2006 Sg	sU	$PM_{2.5}$	12.40	0.14	1.76 (15)
		2007 W		$PM_{2.5}$		2.67	31.39 (15)
(Li et al. 2009)	Atlanta (USA)	XII 2003	U	$PM_{2.5}$	-	0.37	3.16 (19)
			sU		-		4.13 (19)
			R		-		3.40 (19)
		VI 2004	U		-	0.04	0.60 (19)
			sU		-		0.74 (19)
			R		-		0.24 (19)
(Krudysz et al. 2009)	Long Beach. California (USA)	I–III 2005	sU	$PM_{0.25}$	-	0.17	0.94 (10)
				$PM_{0.25–2.5}$	-	0.03	0.43 (10)
				$PM_{2.5–10.0}$	-	0.00	0.01 (10)
(Phuleria et al. 2007)	Los Angeles (USA)	I 2005	R	$PM_{0.108–2.5}$	-	0.03	0.31 (9)
				$PM_{>2.5}$	-	0.16	1.45 (9)
			T	$PM_{0.108–2.5}$	-	0.04	0.38 (9)
				$PM_{>2.5}$	-	0.17	1.82 (9)

n—number of PAH in the sum
U—urban background, urban area, sU: suburban, campus site, downtown, commercial, residential, urban site, city center; T: traffic, highway, road site; R: rural, regional background, background; I: industrial, industrial region
W—winter
S—summer
A—autumn
Sg—spring
c—cold
w—warm
DS—dry season
WS—wet season
n—night
d—day

[2] Table 3-5 presents some results from the last decade (or earlier if the averaging period and the decade overlap). The symbols for the measuring periods and fractions, decimal places in numbers (concentrations of dust and compounds), and units are as in the source texts. The symbols for the sampling points are explained in the legend below the table.

The particulate phase of ambient aerosol is diversified. It consists of particles of soot, fly-ash, metal oxides, salts (ammonium nitrate, ammonium sulfate, etc.), their amounts depending on the source contributions to the aerosol. The particle capability to adsorb PAHs depends on the physicochemical properties of the particle, therefore on its origin. In the same conditions, particles of soot are capable of adsorbing more PAHs than metal oxides or mineral particles. PM1 at the traffic sites consists mainly of soot, whose adsorptive capacity is very high. In Katowice, at both urban and traffic sites (Table 3-5), the PM1 concentrations were almost the same, but the concentrations of ∑PAH are greater at the traffic than at the urban site. At the traffic site in Madrid (Spain), a half of the PM-bound ∑PAH mass is in PM1, while at the urban sites it is less than 30%. In winter, when the concentrations of PM1-bound ∑PAH are several times higher than in summer, and the elevation of the ∑PAH concentrations is due to domestic heating, the PAH distribution among fractions is more even (Table 3-5).

Within the areas affected by traffic, industry, energy production, where the ambient PAHs concentrations are highest, the PAHs absorb mainly onto sub-micron particles of soot that dominate the PM mass. Although in some regions the ∑PAH concentrations are not high (Europe, the USA, Table 3-5), they can exert strong health effects because of the synergistic action of their own toxicity and the sub-micron particle capability of deep-penetration into human body. On the other hand, it is not clear if in such regions as Agra (India), where the majority of PAHs is in coarse PM, the greater hazard comes from coarse than fine PM-bound PAHs. The fall velocity of an ambient particle having aerodynamic diameter 0.1 μm is 8.65×10^{-5} cm s^{-1}, of a particle with the diameter 1 μm—3.48×10^{-3} cm s^{-1}, of a 10 μm particle—3.06×10^{-1} cm s^{-1} (Monn 2001). Coarse PM can settle close to its source, contaminating water, soil, and crops with the absorbed PAHs. The sub-micron particles, contaminated with PAHs, can travel long distances in the atmosphere and the prime source of exposure to PAHs is not inhalation, it is rather water, or food.

Almost everywhere the PAHs concentrations are high, the BaP concentrations are also high; where the BaP concentrations are low, the PAHs concentrations are low too (Table 3-5). However, the dependence is not linear; it cannot be easily drawn from the available data, which are very incomparable. Roughly, where the air pollution by PAHs is high, the ∑PAH mass contribution to PM can reach 0.5%, and the BaP contribution to ∑PAH is greater than 10 or even 12%; where ∑PAH concentrations are low, the mass contribution of ∑PAH to PM is usually less than 0.1%, and of the BaP concentrations to ∑PAH only a few percent. But it is a simplification, and it does not hold in every region: the former, for example, is not true for China, where the BaP mass contribution to ∑PAH is only 2 to 8%; the latter for the USA, where the ∑PAH concentrations are low and BaP is more than 10% of the PAHs mass.

Summary and conclusions

All domestic, mobile, industrial, agricultural, and natural air pollutants sources can emit PM, metals and PAHs. The former four kinds, which are anthropogenic, can be controlled; the fifth one is rather uncontrollable. The hazard from these sources depends on their site in the world because of the synergy among PM components and environmental factors.

The concentrations of PM and PM-bound air contaminants are measured or monitored to identify the sources and to take actions against air pollution. The only way to limit the exposure to air pollution seems to be limiting anthropogenic emissions—but it is expensive.

In the USA and in western Europe, the concentrations of PM-bound heavy metals and PAHs (including BaP) have been low in the last decade. Despite some hot spots, the exposure to high concentrations of heavy metals and PAHs ceased to be a problem in these areas. It is a result of the fifty-year effort at reducing industrial and municipal emissions; the dominating source of PM-bound PAHs and heavy metals is road traffic and, probably, more polluted neighboring regions.

In central, central-eastern, and eastern Europe, on the contrary, traffic is not the prime PAHs or heavy metal source; the PM-bound PAHs and heavy metals come mainly from combustion of fossil fuels for heat and power production, and therefore the concentration levels of these pollutants are high. In Poland, besides hard coal, waste is burnt in the domestic ovens. To decrease the hazard from PM-bound metals and PAHs, the heat and power producers should stop relying primarily on fossil fuels and adopt cleaner technologies, nuclear for example.

There are very scarce data on PM-bound PAHs and heavy metals in Bosnia, Herzegovina, Serbia, Kosovo, Albania, and Romania. They have very weak economies; supposedly, the air pollution is not thoroughly controlled in these countries.

In Asia, PM-bound PAH and heavy metals come equally from industry, traffic, and vast amounts of combusted fossil fuels. Their concentrations are very diversified across regions. Air pollution in big cities may be very high, and, at the same time, very low in suburban regions. The specific problem of coarse PM-bound heavy metals travelling long distances in the periods of strong winds occurs in such countries as India. Even if industrial emissions were reduced, the domestic fires (food preparation) would keep the metal and PAHs levels in the air high enough to be hazardous to inhabitants of populated urban regions.

It is impossible to entirely eliminate combustion, the main source of anthropogenic air pollution. But it is possible to limit emissions of some selected hazardous air contaminants coming from combustion, or their groups, like metals or PAHs. There are many ways to do so; in general, it is a matter of costs. Because the air protection policy consists in balancing the costs of the emission control against the severity of air pollution effects, the geography of industrial or urban regions with clean air is the geography of strong economies.

References

Akyüz, M. and H. Çabuk. 2008. Particle-associated polycyclic aromatic hydrocarbons in the atmospheric environment of Zonguldak, Turkey. Sci. Total. Environ. 405: 62–70.

Akyüz, M. and H. Çabuk. 2010. Gaseparticle partitioning and seasonal variation of polycyclic aromatic hydrocarbons in the atmosphere of Zonguldak, Turkey. Sci. Total. Environ. 408: 5550–5558.

Aldabe, J., D. Elustondo, C. Santamaría, E. Lasheras, M. Pandolfi, A. Alastuey, X. Querol and J.M. Santamaría. 2011. Chemical characterization and source apportionment of PM2.5 and PM10 at rural, urban and traffic sites in Navarra (North of Spain). Atmos. Res. 102: 191–205.

Alolayan, M., K.W. Brown, J.S. Evans, W.S. Bouhamra and P. Koutrakis. 2013. Source apportionment of fine particles in Kuwait City. Sci. Total. Environ. 448: 4–25.

Alves, C., C. Pio and A. Duarte. 2001. Composition of extractable organic matter of air particles from rural and urban Portuguese areas. Atmos. Environ. 35: 5485–5496.

Arruti, A., I. Fernández-Olmo and Á. Irabien. 2012. Evaluation of the urban/rural particle-bound PAH and PCB levels in the northern Spain (Cantabria region). Environ. Monit. Assess. 184: 6513–6526.

Bae, S., X.-Ch. Pan, S.-Y. Kim, K. Park, Y.-H. Kim and H. Kim. 2010. Exposure to particulate matter and polycyclic aromatic hydrocarbons and oxidative stress in school children. Environ. Health Perspect. 118: 579–583.

Bari, Md. A., G. Baumbach, B. Kuch and G. Scheffknecht. 2009. Wood smoke as a source of particle-phase organic compounds in residential areas. Atmos. Environ. 43: 4722–4732.

Baxter, L.K., R.M. Duvall and J. Sacks. 2013. Examining the effects of air pollution composition on within region differences in $PM_{2.5}$ mortality risk estimates. J. Expo. Sci. Environ. Epidemiol. 23: 457–465.

Bodzek, D., K. Luks-Betlej and L. Warzecha. 1993. Determination of particle-associated polycyclic aromatic hydrocarbons in ambient air samples from the Upper Silesia region of Poland. Atmos. Environ. 27A: 759–764.

Brown, A.S. and R.J.C. Brown. 2012. Correlations in polycyclic aromatic hydrocarbon (PAH) concentrations in UK ambient air and implications for source apportionment. J. Environ. Monit. 14: 2072–2082.

Byrd, T., M. Stack and A. Furey. 2010. The assessment of the presence and main constituents of particulate matter ten microns (PM_{10}) in Irish, rural and urban air. Atmos. Environ. 44: 75–87.

Caggiano, R., S. Fiore, A. Lettino, M. Macchiato, S. Sabia and S. Trippetta. 2011. PM2.5 measurements in a Mediterranean site: two typical cases. Atmos. Res. 102: 157–166.

Callén, M.S., M.T. de la Cruz, J.M. López and A.M. Mastral. 2011. PAH in airborne particulate matter. Carcinogenic character of PM10 samples and assessment of the energy generation impact. Fuel Process. Technol. 92: 176–182.

Carbone, C., S. Decesari, M. Mircea, L. Giulianelli, E. Finessi, M. Rinaldi, S. Fuzzi, A. Marinoni, R. Duchi, C. Perrino, T. Sargolini, M. Vardè, F. Sprovieri, G.P. Gobbi, F. Angelini and M.C. Facchini. 2010. Size-resolved aerosol chemical composition over the Italian Peninsula during typical summer and winter conditions. Atmos. Environ. 44: 5269–5278.

Caricchia, A.M., S. Chiavarini and M. Pezza. 1999. Polycyclic aromatic hydrocarbons in the urban atmospheric particulate matter in the city of Naples (Italy). Atmos. Environ. 33: 3731–3738.

Chiou, P., W. Tang, C.-J. Lin, H.-W. Chu and T.C. Ho. 2009. Atmospheric aerosol over a southeastern region of Texas: chemical composition and possible sources. Environ. Model. Assess. 14: 333–350.

Chow, J.C. 1995. Measurement methods to determine compliance with ambient air quality standards for suspended particles. J. Air Waste Manag. Assoc. 45: 320–382.

Chrysikou, L.P., P.G. Gemenetzis and C.A. Samara. 2009. Wintertime size distribution of polycyclic aromatic hydrocarbons (PAHs), polychlorinated biphenyls (PCBs) and organochlorine pesticides (OCPs) in the urban environment: Street- vs rooftop-level measurements. Atmos. Environ. 43: 290–300.

Cohen, D.D., J. Crawford, E. Stelcer and V.T. Bac. 2010. Characterisation and source apportionment of fine particulate sources at Hanoi from 2001 to 2008. Atmos. Environ. 44: 320–328.

Cong, Z., S. Kang, C. Luo, Q. Li, J. Huang, S. Gao and X. Li. 2011. Trace elements and lead isotopic composition of PM_{10} in Lhasa, Tibet. Atmos. Environ. 45: 6210–6215.

Contini, D., A. Genga, D. Cesari, M. Siciliano, A. Donateo, M.C. Bove and M.R. Guascito. 2010. Characterization and source apportionment of PM10 in an urban background site in Lecce. Atmos. Res. 95: 40–54.

Costa, D.L. and K.L. Dreher. 1997. Bioavailable transition metals in particulate matter mediate cardiopulmonary injury in healthy and compromised animal models. Environ. Health Perspect. 105: 1053–1060.

Cuccia, E., D. Massabò, V. Ariola, M.C. Bove, P. Fermo, A. Piazzalunga and P. Prati. 2013. Size-resolved comprehensive characterization of airborne particulate matter. Atmos. Environ. 67: 14–26.

Daher, N., A. Ruprecht, G. Invernizzi, C. De Marco, J. Miller-Schulze, J. Bae Heo, M.M. Shafer, B.R. Shelton, J.J. Schauer and C. Sioutas. 2012. Characterization, sources and redox activity of fine and coarse particulate matter in Milan, Italy. Atmos. Environ. 49: 130–141.

Daher, N., N.A. Saliba, A.L. Shihadeh, M. Jaafar, R. Baalbaki, M.M. Shafer, J.J. Schauer and C. Sioutas. 2014. Oxidative potential and chemical speciation of size-resolved particulate matter (PM) at near-freeway and urban background sites in the greater Beirut area. Sci. Total Environ. 470-471: 417–26.

Darpa, S.J., P.S. Khillare and S. Sayantan. 2014. Risk assessment of inhalation exposure to polycyclic aromatic hydrocarbons in school children. Environ. Sci. Pollut. Res. 21: 366–378.

Davis, J.A., Q. Meng, J.D. Sacks, S.J. Dutton, W.E. Wilson and J.P. Pinto. 2011. Regional variations in particulate matter composition and the ability of monitoring data to represent population exposures. Sci. Total Environ. 409: 5129–5135.

De La Torre-Roche, R.J., W.-Y. Lee and S.I. Campos-Díaz. 2009. Soil-borne polycyclic aromatic hydrocarbons in El Paso. Texas: analysis of a potential problem in the United States/Mexico border region. J. Hazard. Mater. 163: 946–958.

Dejean, S., C. Raynaud, M. Meybeck, J.P. della Massa and V. Simon. 2009. Polycyclic aromatic hydrocarbons (PAHs) in atmospheric urban area: monitoring on various types of sites. Environ. Monit. Assess. 148: 27–37.

Delhomme, O. and M. Millet. 2012. Characterization of particulate polycyclic aromatic hydrocarbons in the east of France urban areas. Environ. Sci. Pollut. R. 19: 1791–1799.

Di Filippo, P., C. Riccardi, D. Pomata, C. Gariazzo and F. Buiarelli. 2010. Seasonal abundance of particle-phase organic pollutants in an urban/industrial atmosphere, Water Air Soil Pollut. 211: 231–250.

Dias da Silva, L.I., J.E. de Souza Sarkis, F.M. Zanon Zotin, M. Castro Carneiro, A.A. Neto, A. dos Santos, A.G. da Silva, M.J. Baldini Cardoso and M.I. Couto Monteiro. 2008. Traffic and catalytic convert—related atmospheric contamination in the metropolitan region of the city of Rio de Janeiro, Brazil. Chemosphere 71: 677–684.

Dieme, D., M. Cabral-Ndior, G. Garçon, A. Verdin, S. Billet, F. Cazier, D. Courcot, A. Diouff and P. Shirali. 2012. Relationship between physicochemical characterization and toxicity of fine particulate matter ($PM_{2.5}$) collected in Dakar city (Senegal). Environ. Res. 113: 1–13.

Ding, L.C., F. Ke, D.K.W. Wang, T. Dann and C.C. Austin. 2009. A new direct thermal desorption-GC/MS method: Organic speciation of ambient particulate matter collected in Golden, BC. Atmos. Environ. 43: 4894–4902.

Dong, X., D. Liu and S. Gao. 2012. Characterization of PM2.5- and PM10-bound polycyclic aromatic hydrocarbons in urban and rural areas in Beijing during the winter. Adv. Mat. Res. 518-523: 1479–1491.

Dongarrà, G., E. Manno, D. Varrica and M. Vultaggio. 2007. Mass levels, crustal component and trace elements in PM_{10} in Palermo, Italy. Atmos. Environ. 41: 7977–7986.

Dreher, K.L. 2000. Particulate matter physicochemistry and toxicology: in search of causality—a critical perspective. Inhalation Toxicology 12: 45–57.

Durant, J., W. Busby, A. Lafleur, B. Penman and C. Crespi. 1996. Human cell mutagenicity of oxygenated, nitrated and unsubstituted polycyclic aromatic hydrocarbons associated with urban aerosols. Mutat. Res. 371: 123–157.

Durant, J., A. Lafleur, W. Busby, L. Donhoffner, B. Penman and C. Crespi. 1999. Mutagenicity of C24H14 PAH in human cells expressing CYP1A1. Mutat. Res. 446: 1–14.

Duvall, R.M., B.J. Majestic, M.M. Shafer, P.Y. Chuang, B.R.T. Simoneit and J.J. Schauer. 2008. The water-soluble fraction of carbon, sulfur, and crustal elements in Asian aerosols and Asian soils. Atmos. Environ. 42: 5872–5884.

Dvorská, A., G. Lammel and J. Klánová. 2011. Use of diagnostic ratios for studying source apportionment and reactivity of ambient polycyclic aromatic hydrocarbons over Central Europe. Atmos. Environ. 45: 420–427.

EC. 2004. Council Directive 2004/107/EC relating to arsenic, cadmium, mercury, nickel and polycyclic aromatic hydrocarbons in ambient air.

Eiguren-Fernandez, A., A.H. Miguel, J.R. Froines, S. Thurairatnam and E.L. Avol. 2004. Seasonal and spatial variation of polycyclic aromatic hydrocarbons in vapor-phase and PM2.5 in Southern California urban and rural communities. Aerosol Sci. Technol. 38: 447–455.

Elovaara, E., J. Mikkola, H. Stockmann-Juvala, L. Luukkanen, H. Keski-Hynnilä, R. Kostiainen, M. Pasanen, O. Pelkonen and H. Vainio. 2007. Polycyclic aromatic hydrocarbon (PAH) metabolizing enzyme activities in human lung, and their inducibility by exposure to naphthalene, phenanthrene, pyrene, chrysene, and benzo(a)pyrene as shown in the rat lung and liver. Arch. Toxicol. 81: 169–182.

Englert, N. 2004. Fine particles and human health—a review of epidemiological studies. Toxicol. Lett. 149: 235–242.

Evagelopoulos, V., T.A. Albanis, A. Asvesta and S. Zoras. 2010. Polycyclic aromatic hydrocarbons (PAHs) in fine and coarse particles, Global Nest J. 12: 63–70.

Fang, G.-C., C.-N. Chang, Y.-S. Wu, P.P.-C. Fu, I.-L. Yang and M.-H. Chen. 2004. Characterization, identification of ambient air and road dust polycyclic aromatic hydrocarbons in central Taiwan, Taichung. Sci. Total Environ. 327: 135–146.

Fernandez, A.J., M. Ternero, F.J. Barragan and J.C. Jimenez. 2001. Size distribution of metals in urban aerosols in Seville (Spain). Atmos. Environ. 35: 2595–2601.

Fernández-Camacho, R., S. Rodríguez, J. de la Rosa, Sánchez, A.M. de la Campa, A. Alastuey, X. Querol, Y. González-Castanedo, I. Garcia-Orellana and S. Nava. 2012. Ultrafine particle and fine trace metal

(As, Cd, Cu, Pb and Zn) pollution episodes induced by industrial emissions in Huelva, SW Spain. Atmos. Environ. 61: 507–517.

Friend, A.J., G.A. Ayoko and H. Guo. 2011. Multi-criteria ranking and receptor modeling of airborne fine particles at three sites in the Pearl River Delta region of China. Sci. Total Environ. 409: 719–737.

Gaga, E.O. and A. Ari. 2011. Gas–particle partitioning of polycyclic aromatic hydrocarbons (PAHs) in an urban traffic site in Eskisehir, Turkey. Atmos. Res. 99: 207–216.

Galarneau, E. 2008. Source specificity and atmospheric processing of airborne PAHs: implications for source apportionment. Atmos. Environ. 42: 8139–8149.

Gianini, M.F.D., R. Gehrig, A. Fischer, A. Ulrich, A. Wichser and C. Hüeglin. 2012. Chemical composition of PM10 in Switzerland: an analysis for 2008/2009 and changes since 1998/1999. Atmos. Environ. 54: 97–106.

Gogou, A., N. Stratigakis, M. Kanakidou and E.G. Stephanou. 1996. Organic aerosols in Eastern Mediterranean: components source reconciliation by using molecular markers and atmospheric back trajectories. Org. Geochem. 25: 79–96.

Goyer, R.A. 1986. Toxic effects of metals. pp. 582–635. *In*: C.D. Klaassen, M.O. Amdur and J. Doul [eds.]. Casarett and Doul's Toxicology, 3rd ed. Macmillian, New York, USA.

Grimmer, G., J. Jacob and K.W. Naujack. 1983. Profile of the polycyclic aromatic compounds from crude oils—inventory by GC GC-MS. PAH in environmental materials, part 3. Fresen. J. Anal. Chem. 316: 29–36.

Gu, J., M. Pitz, J. Schnelle-Kreis, J. Diemer, A. Reller, R. Zimmermann, J. Soentgen, M. Stoelzel, H.-E. Wichmann, A. Peters and J. Cyrys. 2011. Source apportionment of ambient particles: Comparison of positive matrix factorization analysis applied to particle size distribution and chemical composition data. Atmos. Environ. 45: 1849–1857.

Guo, H., S.C. Lee, K.F. Ho, X.M. Wang and S.C. Zou. 2003. Particle-associated polycyclic aromatic hydrocarbons in urban air of Hong Kong. Atmos. Environ. 37: 5307–5317.

Halek, F., M. Kianpour-Rad, R.M. Darbani and A. Kavousirahim. 2010. Concentrations and source assessment of some atmospheric trace elements in northwestern region of Tehran, Iran. Bull. Environ. Contam. Toxicol. 84: 185–190.

Hanzalova, K., P. Rossner and R.J. Sram. 2010. Oxidative damage induced by carcinogenic polycyclic aromatic hydrocarbons and organic extracts from urban air particulate matter. Mutat. Res. 696: 114–121.

Harrison, R.M. and J. Yin. 2000. Particulate matter in the atmosphere: which particle properties are important for its effects on health? Atmos. Environ. 249: 85–101.

Harrison, R.M., D.J.T. Smith and L. Luhana. 1996. Source apportionment of atmospheric polycyclic aromatic hydrocarbons collected from an urban location in Birmingham. UK. Environ. Sci. Technol. 30: 825–832.

Harrison, R.M., R. Tilling, M.S.C. Romero, S. Harrad and K. Jarvis. 2003. A study of trace metals and polycyclic aromatic hydrocarbons in the roadside environment. Atmos. Environ. 37: 2391–2402.

Harrison, R.M., A. Jones, J. Gietl, J. Yin and D. Green. 2012. Estimation of the contribution of brake dust, tire wear and resuspension to nonexhaust traffic particles derived from atmospheric measurements. Environ. Sci. Technol. 46: 6523–6529.

Hassan, S.K. and M.I. Khoder. 2012. Gas-particle concentration, distribution, and health risk assessment of polycyclic aromatic hydrocarbons at a traffic area of Giza, Egypt. Environ. Monit. Assess. 184: 3593–3612.

Hernández, S., V. Mugica, M. Torres and R. García. 2010. Seasonal variation of polycyclic aromatic hydrocarbons exposure levels in Mexico City. J. Air Waste Manage. Assoc. 60: 548–555.

Hien, T.T., L.T. Thanh, T. Kameda, N. Takenaka and H. Bandow. 2007. Distribution characteristics of polycyclic aromatic hydrocarbons with particle size in urban aerosols at the roadside in Ho Chi Minh City, Vietnam. Atmos. Environ. 41: 1575–1586.

Hong, H.S., H.L. Yin, X.H. Wang and C.X. Ye. 2007. Seasonal variation of PM10-bound PAHs in the atmosphere of Xiamen. China. Atmos. Res. 85: 429–441.

Hüeglin, C., R. Gehrig, U. Baltensperger, M. Gysel, C. Monn and H. Vonmont. 2005. Chemical characterization of $PM_{2.5}$, PM_{10} and coarse particles at urban, near-city and rural sites in Switzerland. Atmos. Environ. 39: 637–651.

Jorquera, H. and F. Barraza. 2012. Source apportionment of ambient $PM_{2.5}$ in Santiago, Chile: 1999 and 2004 results. Sci. Total Environ. 435-436: 418–429.

Jorquera, H. and F. Barraza. 2013. Source apportionment of PM_{10} and $PM_{2.5}$ in a desert region in northern Chile. Sci. Total Environ. 444: 327–335.

Kamens, R.M., Z. Guo, J.N. Fulcher and D.A. Bell. 1988. Influence of humidity, sunlight and temperature in the daytime decay of polyaromatic hydrocabons on atmospheric soot particles. Environ. Sci. Technol. 22: 103–108.

Karlsson, H., J. Gustafsson, P. Cronholm and L. Möller. 2009. Size-dependent toxicity of metal oxide particles—A comparison between nano- and micrometer size. Toxicol. Lett. 188: 112–118.

Katsoyiannis, A., E. Terzi and Q.-Y. Cai. 2007. On the use of PAH molecular diagnostic ratios in sewage sludge for the understanding of the PAH sources. Is this use appropriate? Chemosphere 69: 1337–1339.

Kaur, S., K. Senthilkumar, V.K. Verma, B. Kumar, S. Kumar, J.K. Katnoria and C.S. Sharma. 2013. Preliminary analysis of polycyclic aromatic hydrocarbons in air particles (PM10) in Amritsar, India: Sources, apportionment, and possible risk implications to humans. Arch. Environ. Con. Tox. 65: 382–395.

Kavouras, I.G., P. Koutrakis, M. Tsapakis, E. Lagoudaki, E.G. Stephanou, D. Von Baer and P. Oyola. 2001. Source apportionment of urban particulate aliphatic and polynuclear aromatic hydrocarbons (PAHs) using multivariate methods. Environ. Sci. Technol. 35: 2288–2294.

Khalili, N.R., P.A. Scheff and T.M. Holsen. 1995. PAH source fingerprints for coke ovens, diesel and gasoline engines, highway tunnels, and wood combustion emissions. Atmos. Environ. 29: 533–542.

Khare, P. and B.P. Baruah. 2010. Elemental characterization and source identification of $PM_{2.5}$ using multivariate analysis at the suburban site of North-East India. Atmos. Res. 98: 148–162.

Kliucininkas, L., D. Martuzevicius, E. Krugly, T. Prasauskas, V. Kauneliene, P. Molnar and B. Strandberg. 2011. Indoor and outdoor concentrations of fine particles, particle-bound PAHs and volatile organic compounds in Kaunas, Lithuania. J. Environ. Monitor. 13: 182–191.

Kong, S.F., X. Ding, Z.P. Bai, B. Han, L. Chen, J. Shi and Z. Li. 2010. A seasonal study of polycyclic aromatic hydrocarbons in PM2.5 and PM2.5–10 in five typical cities of Liaoning Province, China. J. Hazard Mater. 183: 70–80.

Kopanakis, I., K. Eleftheriadis, N. Mihalopoulos, N. Lydakis-Simantiris, E. Katsivela, D. Pentari, P. Zarmpas and M. Lazaridis. 2012. Physico-chemical characteristics of particulate matter in the Eastern Mediterranean. Atmos. Res. 106: 93–107.

Kozielska, B. and J. Konieczyński. 2007. Polycyclic aromatic hydrocarbons in dust emitted from stoker—fired boilers. Environ. Technol. 28: 895–903.

Kozielska, B. and J. Konieczyński. 2008. Occurrence of polycyclic aromatic hydrocarbons in dust emitted from circulating fluidized bed boilers. Environ. Technol. 29: 1199–1207.

Kozielska, B., W. Rogula-Kozłowska and J.S. Pastuszka. 2013. Traffic emission effects on ambient air pollution by PM2.5-related polycyclic aromatic hydrocarbons in Upper Silesia, Poland. Int. J. Environ. Pollut. 53: 245–264.

Kristensson, A., Ch. Johansson, R. Westerholm, E. Swietlicki, L. Gidhagenb, U. Wideqvist and V. Vesely. 2004. Real-world traffic emission factors of gases and particles measured in a road tunnel in Stockholm, Sweden. Atmos. Environ. 38: 657–673.

Krudysz, M.A., S.J. Dutton, G.L. Brinkman, M.P. Hannigan, P.M. Fine, C. Sioutas and J.R. Froines. 2009. Intra-community spatial variation of size-fractionated organic compounds in Long Beach, California. Air Quality. Atmosphere and Health 2: 69–88.

Křůmal, K., P. MikuŠka and Z. Večeřa. 2013. Polycyclic aromatic hydrocarbons and hopanes in PM1 aerosols in urban areas. Atmos. Environ. 67: 27–37.

Kuo, C.-Y., P.-S. Chien, W.-C. Kuo, C.-T. Wei and J.-Y. Rau. 2013. Comparison of polycyclic aromatic hydrocarbon emissions on gasoline- and diesel-dominated routes. Environ. Monit. Assess. 185: 5749–5761.

Ladji, R., N. Yassaa, C. Balducci, A. Cecinato and B.Y. Meklati. 2009. Distribution of the solvent-extractable organic compounds in fine (PM1) and coarse (PM1–10) particles in urban, industrial and forest atmospheres of Northern Algeria. Sci. Total Environ. 408: 415–424.

Lakhani, A. 2012. Source apportionment of particle bound polycyclic aromatic hydrocarbons at an industrial location in Agra, India. The Scientific World Journal 2012. 781291.

Lammel, G., A.M. Sehili, T.C. Bond, J. Feichter and H. Grassl. 2009. Gas/particle partitioning and global distribution of polycyclic aromatic hydrocarbons—A modelling approach. Chemosphere 76: 98–106.

Lemos, A.T., M.V. Coronas, J.A.V. Rocha and V.M.F. Vargas. 2012. Mutagenicity of particulate matter fractions in areas under the impact of urban and industrial activities. Chemosphere 89: 1126–1134.

Lestari, P. and Y.D. Mauliadi. 2009. Source apportionment of particulate matter at urban mixed site in Indonesia using PMF. Atmos. Environ. 43: 1760–1770.

Lettino, A., R. Caggiano, S. Fiore, M. Macchiato, S. Sabia and S. Trippetta. 2012. Eyjafjallajökull volcanic ash in southern Italy. Atmos. Environ. 48: 97–103.

Li, C.K. and R.M. Kamens. 1993. The use of polycyclic aromatic hydrocarbons as sources signatures in receptor modeling. Atmos. Environ. 27A: 523–532.

Li, S., R. Olegario, J. Bandyasz and K. Shafer. 2003. Gas chromatography—mass spectrometry analysis of polycyclic aromatic hydrocarbons in single puff of cigarette smoke. J. Anal. Appl. Pyrol. 66: 155–163.

Li, Z., A. Sjodin, E.N. Porter, D.G. Patterson, Jr., L.L. Needham, S. Lee, A.G. Russell and J.A. Mulholland. 2009. Characterization of PM2.5-bound polycyclic aromatic hydrocarbons in Atlanta. Atmos. Environ. 43: 1043–1050.

Lim, J.-M., J.-H. Lee, J.-H. Moon, Y.-S. Chung and K.-H. Kim. 2010. Airborne PM10 and metals from multifarious sources in an industrial complex area. Atmos. Res. 96: 53–64.

Lippmann, M. 2008. Environmental toxicants: human exposures and their health effects, 3rd ed. John Wiley & Sons, Inc., Hoboken, New Jersey, USA.

Makkonen, U., H. Hellén, P. Anttila and M. Ferm. 2010. Size distribution and chemical composition of airborne particles in south-eastern Finland during different seasons and wildfire episodes in 2006. Sci. Total Environ. 408: 644–651.

Manoli, E., A. Kouras and C. Samara. 2004. Profile analysis of ambient and source emitted particle-bound polycyclic aromatic hydrocarbons from three sites in northern Greece. Chemosphere 56: 867–878.

Mansha, M., B. Ghauri, S. Rahman and A. Amman. 2012. Characterization and source apportionment of ambient air particulate matter ($PM_{2.5}$) in Karachi. Sci. Total Environ. 425: 176–183.

Mantis, J., A. Chaloulakou and C. Samara. 2005. PM10-bound polycyclic aromatic hydrocarbons (PAHs) in the greater area of Athens, Greece. Chemosphere 59: 593–604.

Martellini, T., M. Giannoni, L. Lepri, A. Katsoyiannis and A. Cincinelli. 2012. One year intensive PM2.5 bound polycyclic aromatic hydrocarbons monitoring in the area of Tuscany, Italy. Concentrations, source understanding and implications. Environ. Pollut. 164: 252–258.

Masiol, M., G. Formenton, A. Pasqualetto and B. Pavoni. 2013. Seasonal trends and spatial variations of PM10-bounded polycyclic aromatic hydrocarbons in Veneto Region, Northeast Italy, Atmos. Environ. 79: 811–821.

Massey, D.D., A. Kulshrestha and A. Taneja. 2013. Particulate matter concentrations and their related toxicity in rural residential environment of semi-arid region of India. Atmos. Environ. 67: 278–286.

McVeety, B.D. and R.A. Hites. 1988. Atmospheric deposition of polycyclic aromatic hydrocarbons to water surfaces: a mass balance approach. Atmos. Environ. 22: 511–536.

Miranda, R. and E. Tomaz. 2008. Characterization of urban aerosol in Campinas, São Paulo, Brazil. Atmos. Res. 87: 147–157.

Mirante, F., C. Alves, C. Pio, O. Pindado, R. Perez, M.A. Revuelta and B. Artiñano. 2013. Organic composition of size segregated atmospheric particulate matter, during summer and winter sampling campaigns at representative sites in Madrid, Spain. Atmos. Res. 132-133: 345–361.

Mohanraj, R., G. Solaraj and S. Dhanakumar. 2011a. PM2.5 and PAH concentrations in urban atmosphere of Tiruchirappalli, India. Bull. Environ. Contam. Toxicol. 87: 330–335.

Mohanraj, R., G. Solaraj and S. Dhanakumar. 2011b. Fine particulate phase PAHs in ambient atmosphere of Chennai metropolitan city, India. Environ. Sci. Pollut. R. 18: 764–771.

Monn, C. 2001. Exposure assessment of air pollutants: a review on spatial heterogeneity and indoor/outdoor/personal exposure to suspended particulate matter, nitrogen, dioxide and ozone. Atmos. Environ. 35: 1–32.

Moreno, T., T. Kojima, X. Querol, A. Alastuey, F. Amato and W. Gibbons. 2012. Natural versus anthropogenic inhalable aerosol chemistry of transboundary East Asian atmospheric outflows into western Japan. Sci. Total Environ. 424: 182–192.

Moroni, B., D. Cappelletti, F. Marmottini, F. Scardazza, L. Ferrero and E. Bolzacchini. 2012. Integrated single particle-bulk chemical approach for the characterization of local and long range sources of particulate pollutants. Atmos. Environ. 50: 267–277.

Murillo, J.H., A.C. Ramos, F.Á. García, S.B. Jiménez, B. Cárdenas and A. Mizohata. 2012. Chemical composition of $PM_{2.5}$ particles in Salamanca, Guanajuato Mexico: source apportionment with receptor models. Atmos. Res. 107: 31–41.

Na, K. and D.R. Cocker, III. 2009. Characterization and source identification of trace elements in PM2.5 from Mira Loma, Southern California. Atmos. Res. 93: 793–800.

Neilson, A.H. 1998. PAHs and Related Compounds. Springer, Berlin.

Nielsen, T., H. Jorgensen, J. Larsen and M. Poulsen. 1996. City air pollution of polycyclic aromatic hydrocarbons and other mutagens: occurrence, sources and health effects. Sci. Total Environ. 189-190: 41–49.

Nikolao, K., P.M. Masclet and G. Mouvier. 1984. Sources and chemical reactivity of polynuclear aromatic hydrocarbons in the atmosphere—a critical review. Sci. Total Environ. 32: 102–132.

Nisbet, I.C.T. and P.K. LaGoy. 1992. Toxic equivalency factors (TEFs) for polycyclic aromatic hydrocarbons (PAHs). Regul. Toxicol. Pharmacol. 16: 290–300.

Nriagu, J.O. 1979. Global inventory of natural and anthropogenic emissions of trace metals to the atmosphere. Nature 279: 409–411.

Nriagu, J.O. and J.M. Pacyna. 1988. Quantitative assessment of worldwide contamination of air, water and soils by trace metals. Nature 333: 134–139.

Oda, J., S. Nomura, A. Yasuhara and T. Shibamoto. 2001. Mobile sources of atmospheric polycyclic aromatic hydrocarbons in a roadway tunnel. Atmos. Environ. 35: 4819–4827.

Osornio-Vargas, A.R., J. Serrano, L. Rojas-Bracho, J. Miranda, C. García-Cuellar, M.A. Reyna, G. Flores, M. Zuk, M. Quintero, I. Vázquez, Y. Sánchez-Pérez, T. López and I. Rosas. 2011. *In vitro* biological effects of airborne $PM_{2.5}$ and PM_{10} from semi-desert city on the Mexico-US border. Chemosphere 83: 618–626.

Ostro, B., W.Y. Feng, R. Broadwin, S. Green and M. Lipsett. 2007. The effects of components of fine particulate air pollution on mortality in California: results from CALFINE. Environ. Health Perspect. 115: 13–19.

Pacyna, J.M. 1984. Estimation of the atmospheric emission of trace elements from anthropogenic sources in Europe. Atmos. Environ. 18: 41–50.

Pancras, J.P., M.S. Landis, G.A. Norris, R. Vedantham and J.T. Dvonch. 2013. Source apportionment of ambient fine particulate matter in Dearborn, Michigan, using hourly resolved PM chemical composition data. Sci. Total Environ. 448: 2–13.

Pandey, P.K., K.S. Patel and J. Lenicek. 1999. Polycyclic aromatic hydrocarbons: need for assessment of health risks in India. Study of an urban-industrial location in India. Environ. Monit. Assess. 59: 287–319.

Pant, P. and R.M. Harrison. 2013. Estimation of the contribution of road traffic emissions to particulate matter concentrations from field measurements: a review. Atmos. Environ. 77: 78–97.

Park, S.S., Y.J. Kim and C.H. Kang. 2002. Atmospheric polycyclic aromatic hydrocarbons in Seoul, Korea. Atmos. Environ. 36: 2917–2924.

Pastuszka, J.S., W. Rogula-Kozłowska and E. Zajusz-Zubek. 2010. Characterization of PM10 and PM2.5 and associated heavy metals at the crossroads and urban background site in Zabrze, Upper Silesia, Poland, during the smog episodes. Environ. Monit. Assess. 168: 613–627.

Patri, M., A. Padmini and P. Prakash Babu. 2009. Polycyclic aromatic hydrocarbons in air and their neurotoxic potency in association with oxidative stress: a brief perspective. Annals of Neurosciences 16: 22–30.

Pérez, N., J. Pey, X. Querol, A. Alstuey, J.M. Lopez and M. Viana. 2008. Partitioning of major and trace components in PM10-PM2.5-PM1 at an urban site in Southern Europe. Atmos. Environ. 42: 1677–1691.

Perrone, M.R., S. Becagli, J.A. Garcia Orza, R. Vecchi, A. Dinoi, R. Udisti and M. Cabello. 2013. The impact of long-range-transport on PM_1 and $PM_{2.5}$ at a Central Mediterranean site. Atmos. Environ. 71: 176–186.

Pey, J., N. Pérez, S. Castillo, M. Viana, T. Moreno, M. Pandolfi, J.M. López-Sebastián, A. Alastuey and X. Querol. 2009. Geochemistry of regional background aerosols in the Western Mediterranean. Atmos. Res. 94: 422–435.

Phuleria, H.C., R.J. Sheesley, J.J. Schauer, P.M. Fine and C. Sioutas. 2007. Roadside measurements of size-segregated particulate organic compounds near gasoline and diesel-dominated freeways in Los Angeles, CA. Atmos. Environ. 41: 4653–4671.

Pies, C., B. Hoffmann, J. Petrowsky, Y. Yang, T.A. Ternes and T. Hofmann. 2008. Characterization and source identification of polycyclic aromatic hydrocarbons (PAHs) in river bank soils. Chemosphere 72: 1594–1601.

Pietrogrande, M.C., G. Abbaszade, J. Schnelle-Kreis, D. Bacco, M. Mercuriali and R. Zimmermann. 2011. Seasonal variation and source estimation of organic compounds in urban aerosol of Augsburg, Germany. Environ. Pollut. 159: 1861–1868.

Pope, C.A. and D.W. Dockery. 2006. Health effects of fine particulate air pollution: lines that connect. J. Air Waste Manag. Assoc. 56: 709–742.

Prahl, F.G., E. Crecellus and R. Carpenter. 1984. Polycyclic aromatic hydrocarbons in Washington coastal sediments: an evaluation of atmospheric and riverine routes of introduction. Environ. Sci. Technol. 18: 687–693.

Puxbaum, H., B. Gomiscek, M. Kalina, H. Bauer, A. Salam, S. Stopper, O. Preining and H. Hauck. 2004. A dual site study of $PM_{2.5}$ and PM_{10} aerosol chemistry in the larger region of Vienna, Austria. Atmos. Environ. 38: 3949–3958.

Rajput, N. and A. Lakhani. 2012. Particle Associated Polycyclic Aromatic Hydrocarbons (PAHs) in Urban Air of Agra. Polycyclic Aromatic Compounds 32: 48–60.

Ravindra, K., L. Bencs, E. Wauters, J. de Hoog, F. Deutsch, E. Roekens, N. Bleux, P. Berghmans and R. Van Grieken. 2006a. Seasonal and site-specific variation in vapour and aerosol phase PAHs over Flanders (Belgium) and their relation with anthropogenic activities. Atmos. Environ. 40: 771–785.

Ravindra, K., E. Wauters, S.K. Taygi, S. Mor and R. Van Grieken. 2006b. Assessment of air quality after the implementation of CNG as fuel in public transport in Delhi, India. Environ. Monit. Assess. 115: 405–417.

Ravindra, K., S. Ranjeet and R. Van Grieken. 2008. Atmospheric polycyclic aromatic hydrocarbons: Source attribution, emission factors and regulation. Atmos. Environ. 42: 2895–2921.

Richter, P., P. Griño, I. Ahumada and A. Giordano. 2007. Total element concentration and chemical fractionation in airborne particulate matter from Santiago, Chile. Atmos. Environ. 41: 6729–6738.

Ringuet, J., A. Albinet, E. Leoz-Garziandia, H. Budzinski and E. Villenave. 2012. Diurnal/nocturnal concentrations and sources of particulate-bound PAHs, OPAHs and NPAHs at traffic and suburban sites in the region of Paris (France). Sci. Total Environ. 437: 297–305.

Rogula-Kozłowska, W. and K. Klejnowski. 2013. Submicrometer aerosol in rural and urban backgrounds in Southern Poland: primary and secondary components of PM_1. Bull. Environ. Contam. Toxicol. 90: 103–109.

Rogula-Kozłowska, W., B. Kozielska, B. Błaszczak and K. Klejnowski. 2012. The mass distribution of particle-bound PAH among aerosol fractions: a case-study of an urban area in Poland. pp. 163–190. *In*: T. Puzyn and A. Mostrag-Szlichtyng [eds.]. Organic Pollutants Ten Years after the Stockholm Convention—Environmental and Analytical Update. InTech.

Rogula-Kozłowska, W., B. Błaszczak, S. Szopa, K. Klejnowski, I. Sówka, A. Zwoździak, M. Jabłońska and B. Mathews. 2013a. PM2.5 in the central part of Upper Silesia, Poland: concentrations, elemental composition, and mobility of components. Environ Monitor and Assess. 185: 581–601.

Rogula-Kozłowska, W., B. Kozielska, K. Klejnowski and S. Szopa. 2013b. Hazardous compounds in urban PM in the central part of Upper Silesia (Poland) in winter. Arch. Environ. Prot. 39: 53–65.

Rogula-Kozłowska, W., B. Kozielska and K. Klejnowski. 2013c. Concentration, origin and health hazard from fine particle-bound PAH at three characteristic sites in Southern Poland. Bull. Environ. Contam. Toxicol. 91: 349–355.

Rogula-Kozłowska, W., K. Klejnowski, P. Rogula-Kopiec, L. Ośródka, E. Krajny, B. Błaszczak and B. Mathews. 2014. Spatial and seasonal variability of the mass concentration and chemical composition of $PM_{2.5}$ in Poland. Air Quality, Atmosphere & Health 7: 41–58.

Saarnio, K., M. Sillanpaä, R. Hillamo, E. Sandell, A.S. Pennanen and R.O. Salonen. 2008. Polycyclic aromatic hydrocarbons in size-segregated particulate matter from six urban sites in Europe. Atmos. Environ. 42: 9087–9097.

Safai, P.D., K.B. Budhavant, P.S.P. Rao, K. Ali and A. Sinha. 2010. Source characterization for aerosol constituents and changing roles of calcium and ammonium aerosols in the neutralization of aerosol acidity at a semi-urban site in SW India. Atmos. Res. 98: 78–88.

Saffari, A., N. Daher, M.M. Shafer, J.J. Schauer and C. Sioutas. 2013. Seasonal and spatial variation of trace elements and metals in quasi-ultrafine ($PM_{0.25}$) particles in the Los Angeles metropolitan area and characterization of their sources. Environ. Pollut. 181: 14–23.

Saliba, N.A., F. El Jam, G. El Tayar, W. Obeid and M. Roumie. 2010. Origin and variability of particulate matter (PM10 and PM2.5) mass concentrations over an Eastern Mediterranean city. Atmos. Res. 97: 106–114.

Salvador, P., B. Artíñano, X. Querol, A. Alastuey and M. Costoya. 2007. Characterisation of local and external contributions of atmospheric particulate matter at a background coastal site. Atmos. Environ. 41: 1–17.

Sánchez-Jiménez, A., M.R. Heal and I.J. Beverland. 2012. Correlations of particle number concentrations and metals with nitrogen oxides and other traffic-related air pollutants in Glasgow and London. Atmos. Environ. 54: 667–678.

Saunders, C.R., A. Ramesh and D.C. Shokley. 2002. Modulation of neurotic behavior in F-344 rats by temporal disposition of benzo(a)pyrene. Toxicol. Lett. 129: 33–45.

Schleicher, N.J., S. Norra, F. Chai, Y. Chen, S. Wang, K. Cen, Y. Yang and D. Stüben. 2011. Temporal variability of trace metal mobility of urban particulate matter from Beijing—A contribution to health impact assessments of aerosols. Atmos. Environ. 45: 7248–7265.

Schroeder, W.H., M. Dobson, D.M. Kane and N.D. Johnson. 1987. Toxic trace elements associated with airborne particulate matter: a review. J. Air Waste Manage. Assoc. 37: 1267–1285.

Shah, M.H. and N. Shaheen. 2007. Statistical analysis of atmospheric trace metals and particulate fractions in Islamabad, Pakistan. J. Hazard. Mater. 147: 759–767.

Shiraiwa, M., R.M. Garland and U. Pöschl. 2009. Kinetic double-layer model of aerosol surface chemistry and gas-particle interactions (K2-SURF): Degradation of polycyclic aromatic hydrocarbons exposed to O_3, NO_2, H_2O, OH and NO_3. Atmos. Chem. Phys. 9: 9571–9586.

Shridhar, V., P.S. Khillare, T. Agarwal and S. Ray. 2010. Metallic species in ambient particulate matter at rural and urban location of Delhi. J. Hazard. Mater. 175: 600–607.

Sicre, M.A., J.C. Marty, A. Saliot and X. Aparicio. 1987. Aliphatic and aromatic hydrocarbons in the Mediterranean aerosol. Int. J. Environ. An. Ch. 29: 73–94.

Sillanpää, M., R. Hillamo, S. Saarikoski, A. Frey, A. Pennanen, U. Makkonen, Z. Spolnik, R. Van Grieken, M. Braniš, B. Brunekreef, M.C. Chalbot, T. Kuhlbusch, J. Sunyer, V.M. Kerminen, M. Kulmala and R.O. Salonen. 2006. Chemical composition and mass closure of particulate matter at six urban sites in Europe. Atmos. Environ. 40: 212–223.

Singh, D.P., R. Gadi and T.K. Mandal. 2011. Characterization of particulate-bound polycyclic aromatic hydrocarbons and trace metals composition of urban air in Delhi, India. Atmos. Environ. 45(40): 7653–7663.

Singla, V., T. Pachauri, A. Satsangi, K.M. Kumari and A. Lakhani. 2012. Characterization and mutagenicity assessment of PM2.5 and PM10 PAH at Agra, India. Polycyclic Aromatic Compounds 32: 199–220.

Skupinska, K., I. Misiewicz-Krzeminska, K. Lubelska and T. Kasprzycka-Guttman. 2009. The effect of isothiocyanates on CYP1A1 and CYP1A2 activities induced by polycyclic aromatic hydrocarbons in Mcf7 cells. Toxicology *in Vitro* 23: 763–771.

Slezakova, K., D. Castro, A. Begonha, C. Delerue-Matos, M.D.C. Alvim-Ferraz, S. Morais and M.D.C. Pereira. 2011. Air pollution from traffic emissions in Oporto, Portugal: health and environmental implications. Microchemical Journal 99: 51–59.

Spindler, G., E. Brüggemann, T. Gnauk, A. Grüner and K. Müller. 2010. A four-year size-segregated characterization study of particles PM_{10}, $PM_{2.5}$ and PM_1 depending on air mass origin at Melpitz. Atmos. Environ. 44: 164–173.

Swaine, D.J. 2000. Why trace elements are important. Fuel Process Technol. 65-66: 21–33.

Szoboszlai, Z., Zs. Kertész, Z. Szikszai, A. Angyal, E. Furu, Zs. Török, L. Daróczi and Á.Z. Kiss. 2012. Identification and chemical characterization of particulate matter from wave soldering processes at a printed circuit board manufacturing company. J. Hazard. Mater. 203-204: 308–316.

Takada, H., T. Onda and N. Ogura. 1990. Determination of polycyclic aromatic hydrocarbons in urban street dusts and their source materials by capillary gas chromatography. Environ. Sci. Technol. 24: 1179–1186.

Tang, N., T. Hattori, R. Taga, K. Igarashi, X.Y. Yang, K. Tamura, H. Kakimoto, V.F. Mishukov, A. Toriba, R. Kizu and K. Hayakawa. 2005. Polycyclic aromatic hydrocarbons and nitropolycyclic aromatic hydrocarbons in urban air particulates and their relationship to emission sources in the PaneJapan Sea countries. Atmos. Environ. 39: 5817–5826.

Theodosi, C., U. Im, A. Bougiatioti, P. Zarmpas, O. Yenigun and N. Mihalopoulos. 2010. Aerosol chemical composition over Istanbul. Sci. Total Environ. 408: 2482–2491.

Tobiszewski, M. and J. Namieśnik. 2012. PAH diagnostic ratios for the identification of pollution emission sources. Environ. Pollut. 162: 110–119.

Tran, D.T., L.Y. Alleman, P. Coddeville and J.-C. Galloo. 2012. Elemental characterization and source identification of size resolved atmospheric particles in French classrooms. Atmos. Environ. 54: 250–259.

Valavanidis, A., K. Fiotakis and T. Vlachogianni. 2008. Airborne particulate matter and human health: toxicological assessment and importance of size and composition of particles for oxidative damage and carcinogenic mechanisms. J. Environ. Sci. Health C Environ. Carcinog. Ecotoxicol. Rev. 26: 339–62.

Valerio, F., P. Bottino and D. Ugolini. 1984. Chemical and photochemical degradation of polycyclic aromatic hydrocarbons in the atmosphere. Sci. Total Environ. 40: 169–188.

Van Jaarsveld, J.A., W.A.J. Van Pul and F.A. De Leeuw. 2011. Modelling transport and deposition of persistent organic pollutants in the European region. Atmos. Environ. 31: 1011–1024.

Vecchi, R., M. Chiari, A. D'Alessandro, P. Fermo, F. Lucarelli, F. Mazzei, S. Nava, A. Piazzalunga, P. Prati, F. Silvani and G. Valli. 2008. A mass closure and PMF source apportionment study on the sub-micron sized aerosol fraction at urban sites in Italy. Atmos. Environ. 42: 2240–2253.

Vercauteren, J., C. Matheeussen, E. Wauters, E. Roekens, R. van Grieken, A. Krata, Y. Makarovska, W. Maenhaut, X. Chi and B. Geypens. 2011. Chemkar PM10: an extensive look at the local differences in chemical composition of PM10 in Flanders, Belgium. Atmos. Environ. 45: 108–116.

Viana, M., T.A.J. Kuhlbusch, X. Querol, A. Alastuey, R.M. Harrison, P.K. Hopke, W. Winiwarter, M. Vallius, S. Szidat, A.S.H. Prévôt, C. Hueglin, H. Bloemen, P. Wåhlin, R. Vecchi, A.I. Miranda, A. Kasper-Giebl, W. Maenhaut and R. Hitzenberger. 2008a. Source apportionment of particulate matter in Europe: a review of methods and results. Aerosol Sci. 39: 827–849.

Viana, M., X. Querol, A. Alastuey, F. Ballester, S. Llop, A. Esplugues, R. Fernández-Patier, S.G. dos Santos and M.D. Herce. 2008b. Characterising exposure to PM aerosols for an epidemiological study. Atmos. Environ. 42: 1552–1568.

von Schneidemesser, E., E.A. Stone, T.A. Quraishi, M.M. Shafer and J.J. Schauer. 2010. Toxic metals in the atmosphere in Lahore, Pakistan. Sci. Total Environ. 408: 1640–1648.

Wang, H., Y. Zhou, Y. Zhuang, X. Wang and Z. Hao. 2009. Characterization of PM2.5/PM2.5–10 and source tracking in the juncture belt between urban and rural areas of Beijing, Chinese Sci. Bull. 54: 2506–2515.

Wang, J., Z. Hu, Y. Chen, Z. Chen and S. Xu. 2013. Contamination characteristics and possible sources of PM10 and PM2.5 in different functional areas of Shanghai, China. Atmos. Environ. 68: 221–229.

White, P.A. 2002. The genotoxicity of priority polycyclic aromatic hydrocarbons in complex mixtures. Mutat. Res. 515: 85–98.

WHO, Regional Office for Europe. 2000. Air Quality Guidelines for Europe, 2nd ed. Copenhagen.

Wingfors, H., L. Hagglund and R. Magnusson. 2011. Characterization of the size-distribution of aerosols and particle-bound content of oxygenated PAHs, PAHs, and n-alkanes in urban environments in Afghanistan. Atmos. Environ. 4360–4369.

Wiseman, C.L.S. and F. Zereini. 2009. Airborne particulate matter, platinum group elements and human health: A review of recent evidence. Sci. Total Environ. 407: 2493–2500.

Wong, L.S.N., H.H. Aung, M.W. Lamé, T.C. Wegesser and D.W. Wilson. 2011. Fine particulate matter from urban ambient and wildfire sources from California's San Joaquin Valley initiate differential inflammatory, oxidative stress, and xenobiotic responses in human bronchial epithelial cells. Toxicology *in Vitro* 25: 1895–1905.

Xu, L., X. Chen, J. Chen, F. Zhang, C. He, J. Zhao and L. Yin. 2012. Seasonal variations and chemical compositions of $PM_{2.5}$ aerosol in the urban area of Fuzhou, China. Atmos. Res. 104-105: 264–272.

Yang, L., X. Zhou, Z. Wang, Y. Zhou, S. Cheng, P. Xu, X. Gao, W. Nie, X. Wang and W. Wang. 2012. Airborne fine particulate pollution in Jinan, China: Concentrations, chemical compositions and influence on visibility impairment. Atmos. Environ. 55: 506–514.

Yin, J. and R.M. Harrison. 2008. Pragmatic mass closure study for $PM_{1.0}$, $PM_{2.5}$ and PM_{10} at roadside, urban background and rural sites. Atmos. Environ. 42: 980–988.

Yunker, M.B., R.W. Macdonald, R. Vingarzan, R.H. Mitchell, D. Goyette and S. Sylvestre. 2002. PAHs in the Fraser River basin: a critical appraisal of PAH ratios as indicators of PAH source and composition. Org. Geochem. 33: 489–515.

Zhang, W., S. Zhang, C. Wan, D. Yue, Y. Ye and X. Wang. 2008. Source diagnostics of polycyclic aromatic hydrocarbons in urban road runoff, dust, rain and canopy through fall. Environ. Pollut. 15: 594–601.

Zhang, W., Y. Sun, J. Guo, G. Zhuang, D. Xu, W. Wang and Z. Wu. 2010. Sources of aerosol as determined from elemental composition and size distributions in Beijing. Atmos. Res. 95: 197–209.

Zhang, W., T. Lei, Z.Q. Lin, H.S. Zhang, D.F. Yang, Z.G. Xi, J.H. Chen and W. Wang. 2011. Pulmonary toxicity study in rats with PM10 and PM2.5: differential responses related to scale and composition. Atmos. Environ. 45: 1034–1041.

Zhou, Ch., X. Zhu, Z. Wang, X. Ma, J. Chen, Y. Ni, W. Wang, J. Mu and X. Li. 2013. Gas-particle partitioning of PAHs in the urban air of Dalian, China: measurements and assessments. Polycyclic Aromatic Compounds 33: 31–51.

Zou, L.Y., W. Zhang and S. Atkiston. 2003. The characterizations of polycyclic aromatic hydrocarbons emissions from burning of different firewood species in Australia. Environ. Pollut. 124: 283–289.

Zwoździak, A., I. Sówka, B. Krupińska, J. Zwoździak and A. Nych. 2013. Infiltration or indoor sources as determinants of the elemental composition of particulate matter inside a school in Wrocław, Poland? Build and Environ. 66: 173–180.

4

Adverse Health Effects of the Exposure to the Spherical Aerosol Particles, Including Ultra-Fine Particles

Renata Zlotkowska

Anatomy and physiology of the respiratory system

Inhalable particles include a fraction of aerosol consisting of particles of a diameter below 10 μm, which could penetrate into the airways. The respiratory human airways are divided into upper and lower airways (Cotes 2009). The upper airways include the nasal cavity and pharynx. The lower airways include the trachea-bronchial region with the alveoli, where the gas exchange, which is the most important element of breathing, takes place. The main respiratory organ is the lung. The anatomy of the respiratory system is strictly connected to the respiratory physiology and function. The upper airways constitute the main barrier separating lower airways from the possible effect of particles, contained in the air. These particles, of diameter over 10 μm are "trapped" on the surface of the respiratory epithelium and removed from the airways by cough, or by the cilia movement. The main physical determinants of the particle's ability to penetrate the airways are the particle diameter, density, and shape (Donaldson et al. 2012). Most particles of a diameter over 5 μm are retained in the nasopharyngeal cavity and mechanically removed from the air, before they reach the lower part of the respiratory system. The respiratory epithelium is the "first line" of the pulmonary

School of Public Health in Bytom, Medical University of Silesia, Department of Social Medicine and Prevention, 18 Piekarska Str, 41-902 Bytom, Poland.
E-mail: rzlotkowska@sum.edu.pl

defense, since it assures the proper respiratory filtration. The respiratory epithelium lines the airways from the nasal cavity up to the terminal air sacks in the lungs. The main types of epithelial cells are ciliated, columnar, undifferentiated, secretory, and basal cells. The structure of the respiratory epithelium differs according to the part of the airways. In the large airways ciliated cells and secretory cells dominate, whereas in the small airways ciliated and Clara cells do. The alveolar epithelium is formed by type I and type II cells (Crystal et al. 2008). Regardless of the branch of the airways, the epithelium consists of seromucous and goblet cells. These cells produce mucin, which is the complex substance present on the surface of the respiratory epithelium. Mucin is the main component of the respiratory mucus, which also contains water, salts, and other soluble substances, free protein, and substances which chemically are carbohydrate-rich glycoproteids (Houtmeyers et al. 1999). The airway surface fluid (ASF) is a thin layer of fluid, covering the surface of the airways, which plays an important role in the defense of the respiratory system against the particles and chemical substances coming into human airways (Lillehoj and Chul 2002). ASF consists of periciliary fluid and an overlying mucus layer (Gray et al. 2004). The ASF consists of proteoglycans whose function is unknown. It is suggested that these substances play a role in the mucosal host defense mechanisms (Forteza et al. 2001). Lipids are also components of the ASF fluid. The results of numerous studies revealed that the amount of lipids in the purulent sputum is associated with the degree of bacterial infection and the metabolites of lipids present in ASF, like leukotrienes and prostaglandins, are involved in the airway inflammation (Nadziejko et al. 1993). It has also been suggested that lipids may be involved in the modification of rheological properties of mucus, modification of adhesiveness, and effect on mucocilliary clearance (Widdicombe 1987). The main function of mucin is to moisturize the air coming into the airways and to "trap" particles entering the airways. Then the particles are removed from the airways by the movement of the respiratory cilia. Mucus consistency is sticky and the cillia are in constant movement (Houtmeyers et al. 1999). The cilia move in a periciliary fluid, which determines the effective propulsion, and particles present in the upper part of the respiratory system are removed by the cilia with patches of mucous gel towards nares. The proper mechanism of mucociliary clearance depends on the size and composition of the particles. The mucociliary clearance (MCC) clears the airways through the mucin, and thus the substances which are trapped on the surface of the airways are removed. If the effectiveness of MCC fails, the cough is another defense mechanism which allows for the removing of particles with the fragments of mucin from the airways. In the lower airways, the alveolar clearance mechanism is responsible for the removing of the particles, deposited in the lower part of the respiratory tract. The MCC effectiveness is modified by age, sex, exercise, posture, and sleep (Houtmeyers et al. 1999). Some environmental agents may also affect the degree of MCC clearance. It is suggested by the results of studies that the effectiveness of MCC clearance decreases with age (Mortensen et al. 1994). The studies on the effect of gender on MCC clearance did not show the univocal effects (Hasani et al. 1994). Exercise is another factor, increasing the degree of MCC clearance in healthy, non-smoking males (Wolff et al. 1977). The postural factors are associated with the gravitational factors, especially in subjects with existing inflammation in the airways. The proper postural drainage of respiratory secretion ensures the proper purification of the airways. Sleep is the physiological factor which has a depressant effect on MCC clearance (Hasani et al. 1993). The osmotic agents, like hypertonic saline or mannitol enhance the mucus clearance in subjects with MCC dysfunction (Daviscas et al. 2010). Exposure to environmental factors, like tobacco smoke or gases like sulphur dioxide (SO_2), ozone (O_3), or nitrogen dioxide (NO_2) impair the proper MCC clearance (Houtmeyers et al. 1999). Smoking, both active and passive, affects the proper mucociliary clearance in the airways (Habesglou et al. 2012).

Components of respiratory fluid, like N-acetylneuraminic acid is responsible for the buffering function of the respiratory fluid. Chemical agents, like some iritative gases are hydrolyzed on the surface of the respiratory epithelium, which leads to their chemical neutralization. The ASF pH is another component of the defense mechanism. According to the numerous data, inflammation in the airways is associated with the acidic properties of ASF. Thus the alkalization of ASF is an element of proper defense mechanism.

Immune local mechanisms are another mechanism of defense in the upper airways (Watelet et al. 2002). Natural immunity is responsible for the host reaction. Important elements of the host reaction are the mucus membrane, inflammatory reaction, phagocytosis, complement synthesis, and mediator release. In the case of specific substances, the acquired immunity mechanisms are activated as the recognition of the specific antigens.

Deposition of particles in the airways

The deposition occurs when the particles enter the airways. The airflow in airways is the physiologic process which affects the type of particle deposition. Deposition depends on the size of the particles and their aerodynamics. The gravity affects the particle deposition (Haber et al. 2010). As the air is inhaled, it passes the airways towards the "blind end", which is the most distant part of the bronchial tree. Velocity of the particles depends on the segment of the bronchial tree, specifically on its diameter. The flow of particles passing the narrowest sections of the bronchi including the terminal bronchioli is extremely slow and reaches zero in the most distant parts of the bronchial tree (Donaldson et al. 2002). The flow of particles could be laminar or turbulent. Efficiency of deposition is defined as the difference between inlet and outlet particle concentration to the inlet concentration. The principal mechanisms of deposition include impaction, sedimentation, interception, and diffusion.

The impaction, sedimentation, and interception depend on aerodynamic characteristics of the particles, whereas the interception depends also on the particle geometry. The impaction occurs when the particle is more likely to impact the surface on the respiratory epithelium than to continue in the direction into the airstream. This phenomenon is observed often in the region of airways where the bifurcations occur. The impaction occurs when the airflow is high. When the airflow is low, the sedimentation occurs in the alveolar region. The size of the particles which are subject to sedimentation is estimated at > 0.5 μm. For the smaller particles, sedimentation is less likely. Because of their small size, these particle undergo the "Brownian motions" and come into discrete collisions. Thus, the kinetic energy of these particles is high and diffusion, which takes place in the centriacinar region is the way of impaction for particles with the size < 0.5 μm. Interception is the mechanism of deposition observed mainly for the fibers, which occurs when a particle's trajectory would escape contact with the adjacent surface, but some extremity of a particle would make contact (Donaldson et al. 2002). The efficiency of deposition depends on the breathing conditions, which are associated with the level of physical activity and gender.

Toxicological characteristics of particles

Particulate matter is a mixture of particles, different in terms of chemical and physical properties, origin, mass, number, and shape. Hence, the identification of specific toxicological effects of particulate matter is very difficult. From the point of view of health effects components of particulate matter have irritative, mutagenic, carcinogenic, and allergic effects (Feldman 1998).

The chemical properties affect the toxicological potential of particles causing health effects of exposure to the specific substance and they depend on the source of pollution. Particles present in the air come from natural and man-made processes. Natural particles come from soil dust, sea salts, volcanic ash, fungal spores, pollens, and products of forest fires (Kelly and Fussell 2012). Natural wind-blown particles are mainly the coarse fraction of particulate matter, and it is suggested that the potency of causing health effects is not high (Donaldson et al. 2003). The man-made sources of emission include the processes of combustion (including energy production and road transport), industry, construction works, mining, cigarette smoking, and wood-stove burning. The toxicological properties of particles depend also on the way of emission into the atmosphere. Primary particles come mainly from combustion and are emitted straight to the atmosphere. In urban areas, the road transport and domestic combustion sources are the main "producers" of primary particles, which are released directly from the source to the atmosphere. The primary particles emitted from the road transport from the chemical point of view are often composed from the carbon core of the organic chemical component or heavy metals (Kelly and Fussell 2012). Diesel exhaust particles (DEP) are the largest source of black carbon emission into the ambient air (Broday and Rozenzweig 2011). DEP also absorb organic particles (like allergens) and transition metals. Thus, the exposure to DEP components could have immune system, allergic and carcinogenic effects. The secondary particles are formed in the atmosphere as the product of chemical reactions (for example, ozone in the ambient air). These particles are more persistent in the environment since the chemical reactions which produce the secondary particles are relatively slow. The secondary particles have generally irritative properties. Sulfates and nitrates are produced in the atmosphere as the results of oxidation of sulfur dioxide (SO_2) and nitrogen dioxide (NO_2). The components of air pollution could interact with each other which potentially modify the toxicological effects. The important chemical component of particles coming from the combustion process is elemental and organic carbon (Chow et al. 2007). Carbon particles serve probably as the cores onto which other substances are present (like polyaromatic hydrocarbons or their derivates, endotoxins, and other biological particles). Metals, like lead (Pb), nickel (Ni), chromium (Cr), and iron (Fe) are also part of the particulate matter. Emission of metals comes mainly from the metallurgical industry.

The toxicological properties of particles depend on their size. Particles exceeding the size of 10 μm are removed from the upper part of respiratory system by the mucocilliary and other defense mechanisms, as described previously in this chapter. Particles of size below 10 μm penetrate into the respiratory system and form the respirable fraction of dust. Modern attitude in dust toxicology is based on the assumption that the fraction of dust with the diameter of particles over 2.5 μm (PM2.5) constitutes the coarse fraction of PM. This fraction of dust consists also of particles of a biological origin, like pollens, parts of plants, fungal spores, and mold. Particles of a size between 2.5 μm (PM2.5) and 0.1 (PM0.1) are the fraction of dust which penetrates the respiratory system. This fraction of dust comprises of primary and secondary particles. Fraction of dust below 0.1 μm constitutes the ultrafine fraction of PM. The ultrafine particles (UFP) come mainly from combustion and because of their small size penetrate into the smallest branches of the bronchial tree. Particles with the diameter below 100 nm, generated from the natural and industrial nanotechnological processes are called nanoparticles. In different sources of toxicological literature, the terms "ultrafine particles" and "nanoparticles" are used as synonyms, however the nanoparticles constitute the smallest fraction of dust and are characterizing by changed chemical and physical properties. The production of the nanoparticles is the consequence of processes when high pressure and high temperature are acting at the same time. The unique properties of nanoparticles are the consequence of a large surface area to volume ratio. Nanoparticles are divided into nanoparticles of a natural origin (mainly from volcano eruptions) and nanoparticles of anthropogenic origin (emitted from the road transport, or produced in the nanotechnological processes). Unique properties of nanoparticles affect their toxicological properties. The nanoparticles could enter the human organism not only by respiratory system but also transcutaneously. The exposure to nanoparticles is associated with the acute inflammation not only in the airways but also in other organs and cells (Cetta 2009). Nanotoxicology is a new branch of toxicology which deals with the toxicological properties and effects of exposure to nanoparticles.

Health effects of exposure to air pollution

Health effects of exposure to air pollution could be short-term and long-term depending on the exposure. Historically, the health effects resulting from the short-term exposure to air pollution were reported in 1952, during the smog episode in London (Bruneekreef and Holgate 2002). As shown by a number of epidemiological studies, short-term effects of exposure to air pollution are associated with increased respiratory and cardiovascular mortality and increased hospital admissions in Europe and North America (Dockery 2009). According to findings from the European APHEA study, the all-cause mortality as well as hospital admissions due to asthma and chronic obstructive pulmonary disease (COPD) among people aged 65 years and due to cardiovascular diseases were more noticeable during days characterized with elevated levels of PM10 and diesel exhaust (Atkinson et al. 2001). During the smog episodes, an increased number of exacerbations of chronic respiratory diseases and increased number of acute myocardial events were also noticed (Peters et al. 2001).

The main long-term effects of population exposure to air pollution come from cohort and cross-sectional studies performed in USA, Europe, and Asia (Ostro et al. 1999; Gauderman et al. 2000). Exposure to particulate air pollution is associated with the reduced level of lung function in adults and children, increased occurrence of chronic respiratory symptoms and reduced lung growth in children. Chronic exposure to particulate air pollution is responsible for the diminished life years in the population living in large urban environments. According to WHO, air pollution is responsible for an increase of 1.4% in total mortality, 0.4% of all disability-adjusted life years (DALY's) and 2.0% of cardiovascular morbidity worldwide (Ostro 2004). The relation between increased asthma and allergy prevalence and exposure to air pollution has not been univocally explained by the epidemiological data and probably this association is complicated.

Factors modifying the health effects of exposure to air pollution

The health effects of exposure to certain components of particulate matter could be modified by the physiological and socioeconomic factors. The important factor which affects the health effects attributable to air pollution is physical exercise. During exercise, the frequency of breathing and minute ventilation increases. Thus the deposition of inhaled particles is enhanced. Diesel exhaust particles (DEP) are the example of the substance which contains small soot particles that potentially easily penetrate into small airways. It was revealed that for an adult person who is performing heavy exercise, the exposure to DEP increases more than for sedentary persons (Broday 2011). Another important physiological factor modifying the health effects of exposure is age. Children in the population are more susceptible to the effects of air pollution than adults which has a physiological background. Development of respiratory system begins in the fourth week of fetal life and maturity is accomplished at about 16 years of age. The chest and lungs are smaller in children than in adult persons and indices of the lung function expressed in absolute values are lower. Respiratory resistance in the airways is higher in children than in adults and chest compliance is lower. Another physiological factor is associated with the immaturity of the immune system in children. The local immune mechanisms of defense in the airways are not sufficiently developed in children. Thus, special attention is given to the fetal period of life in children. This period of life is an age of increased susceptibility to the effect of exposure to xenobiotics penetrating through placenta since this is a period of organogenesis and rapid development of fetus (Moya et al. 2014).

Health effects of exposure to air pollution are more prominent in older persons (Colais et al. 2012). The socioeconomic status remains another factor which is attributed to the modification of health effects of exposure to air pollution. Gender also modifies the health effects of exposure to air pollution, especially indoor air pollution. At school age boys are more susceptible to the effects of air pollution than girls, as proven for passive smoking exposure (Zlotkowska and Zejda 2005). Results of some epidemiological studies suggest that young women are more susceptible to the effects of pollutants, present in the indoor air, like nitrogen dioxide. These differences in susceptibility could stem from the habitual factors. Another factor which could potentially modify the health effects of exposure to air pollution is genetic polymorphism.

Role of oxidative stress

Oxidative stress is the most important cellular and biochemical mechanism underlying the health effects of exposure to particulate matter. This phenomenon is defined as the imbalance between oxidants and antioxidants in the cells, in favor of the oxidants, which is caused by the excessive production of reactive oxygen species (ROS). The presence of oxidative stress leads to the damage of cells. The consequence of oxidative stress is the process of inflammation, which is in turn responsible for the pathophysiological effects of air pollution. The inhalation is the main route of exposure to particles. The inflammation primarily takes place in the respiratory system but this process could occur in other organs and systems in the human body. The components of particulate matter with the great potential for the generation of oxidative stress are ultrafine particles, organic particles, and transition metals (Donaldson et al. 2002). The ability of particles to induce oxidative stress increases with the decreased particle size and thus, the ultrafine particles are considered as having the more prominent toxicity as compared to the larger particles. In the case of ultrafine particles, the ability to cause oxidative stress is a consequence of the large surface area and the substances released from the surface like transition metals. Metals like lead or nickel present on the surface of nanoparticles could generate free radicals and the larger is surface area of the particles, the higher is the transition metal concentration. Oxidative stress itself is a process of formation of free radicals and imbalance in the antioxidative defense (Shi et al. 2004). During the activation of oxidative stress, reactive oxygen species (ROS) and other ions, like hydrogen peroxide (H_2O_2), nitric oxide (NO), and superoxide (O_2^-) are produced. In normal conditions, the generation of ROS and other substances and their possible harmful effects are eliminated by the natural antioxidant mechanisms. In the culmination of oxidative stress when the antioxidant capacity is diminished, the ROS which have high potential for chemical reactivity cause tissue damage by attacking and denaturizing the structural molecules which are responsible for their mutagenic and carcinogenic effect (Sanchez-Perez et al. 2009). ROS activate the redox-sensitive transcription factors, like AP-1 and nuclear factor kappa B and signal transduction pathways (Vaziri 2008). This results in chromatin change, and expression of genes responsible for the inflammation, apoptosis, proliferation, transformation, and differentiation (Donaldson 2003). The hydroxyl radicals are responsible for the DNA damage and carcinogenic effect, especially when considering the effect of PM on lung cancer (Donaldson 2003). Some studies indicate that inflammatory lung diseases like COPD or pneumoconiosis are attributed to the higher lung cancer risk. It was demonstrated in the *in vivo* studies that neutrophils are responsible for the induction of DNA damage and tumor formation in response to silica (Donaldson 2003). The DNA damage could be responsible for the carcinogenic effect but also could result in accelerated cell and tissue aging (Mena et al. 2009). Resident macrophages are activated as the result of air pollution exposure and number of neutrophils in the airways is also increased as the result of exposure to particles (Aurebrach and Hernandez 2012). However, the exact mechanism of inflammation as the result of exposure is not completely investigated since the role of cytokines involved in this process is not sufficiently explained. The pro-inflammatory cytokines could also be involved in the anti-inflammatory process (Mena et al. 2009). Exposure to diesel exhaust particles could interfere with the cell—signaling pathways. Presence of oxidative stress is also associated with the decreased levels of expression of antioxidant genes (Yang et al. 2009). Glutathione (GSH) is the substance present in the respiratory epithelial lining fluid regarded as the physiological antioxidant responsible for the protection of thiol (SH) groups in protein. Some observations from the literature suggest that oxidative stress could be associated with the decreased expression of GST (Glutathione S-transferase) (Li et al. 2013). The vascular endothelium plays a role in the inflammatory process and production of inflammatory mediators. Results of several studies revealed that exposure to ultrafine particles could result in an altered endothelial function (Montiel-Davalos et al. 2010). Depletion of reduced GSH as the result of oxidative stress could result in increased permeability of epithelial cells in lungs which in turn allows the particles to reach the systemic circulation. Another effect of oxidative stress is increased lipid oxidation, including LDL cholesterol particles as the result of exposure to transition metals. Results of experimental studies in animals revealed that mice exposed to a fine fraction of dust (PM2.5), were more likely to develop atherosclerotic plaques through a process independent of dyslipidemia (Sun 2005). Another suggested mechanism of the effect of inflammation on atherosclerosis plaque disruption is based on the literature data which reveals that inflammation as the result of neutrophil activation could result in plaque rupture (Van Eeden et al. 2012). This potentially explains the effect of exposure to high levels of air pollution on the increased cardiovascular morbidity and mortality including the effect on the increased occurrence of acute myocardial incidents. The inflammation resulting from the exposure to air pollution has the systemic dimension, not restricted to the respiratory system. It is also suggested that presence of oxidative stress in kidneys co-existing with the vascular changes as the result of lead exposure could be the potential mechanism for the hypertensive effect of exposure to this metal (Vaziri 2008). The presence of oxidative stress in the central nervous system was also reported as the result of ultrafine particles.

The effect of oxidative stress could be inhibited by antioxidants. The natural antioxidant system consists of enzymes, like catalase, glutathione peroxidase, and substances present in the diet, like carotenoids, vitamin C, flavonoids (Maulik et al. 2013). These substances could be beneficial and have a protective effect for the human organism exposed to different xenobiotics.

Nanoparticles: exposure and health effects

The toxicity of particles depends mainly on particle size and increases with the decreased size (Donaldson and Seaton 2012). This observation is not related to the nanoparticles. From the toxicological point of view, nanoparticles are ultrafine particles having the dimension of 1–100 nm, characterized by a changed chemical structure. Nanoparticles are produced naturally as the consequence of processes where the increased temperature and increase pressure act synergistically like volcano eruptions or are the product of nanotechnological man-made processes (Andujar et al. 2011). Nanoparticles are also emitted to the environment as the by-product of air pollution, combustion processes or emitted from diesel engines. Palladium (Pd), ruten (Ru), platinum (Pt) and irydium (Ir) are group of metals emitted from the automobiles (catalytic converter) into the ambient air in a form of nanoparticles which have not been observed before. Another important source of nanoparticles emission to the environment is the automotive brake (Kukutschkova et al. 2011). Different chemical substances may occur in a form of nanoparticles, for example, minerals (graphite, silica), transition metals, like silicon dioxide and titanium dioxide, or organic compounds (single-walled carbon nanotubes and multi-walled carbon nanotubes, carbon fullerens) (Andujar et al. 2011).

The main factor characterizing nanoparticles and affecting their potential toxicity is the high ratio of area size to mass which is a result of reduced size. These special properties of nanoparticles may increase their ability to catalyze the chemical reactions. Thus due to their changed structure and small size, nanoparticles may have greater reactivity than other particles with the identical chemical composition. Another factor affecting toxicity of nanoparticles is their chemical composition. Transition metals occurring in a form of nanoparticles induce the oxidative stress process and the free radicals which could cause cell damage (Andujar et al. 2011). There are three pathways for the nanoparticles to enter the human body: inhalation, skin absorption, and gastrointestinal tract absorption (Boczkowski and Hoet 2010). Nanoparticles tend to form agglomerates since they are subjected to electrostatic forces and Van der Waals forces. The agglomeration of nanoparticles probably affects their deposition in the airways. According to the results of the studies performed on mice and rats, single-walled carbon nanotubes exposure is associated with the granuloma formation and presence of inflammation in lungs (Sharma et al. 2007). This observation confirms the hypothesis on nanoparticles toxicity. The inhalation is the main route of exposure to nanoparticles. Nanoparticles deposit in the lungs and the probability of reaching the alveoli is highest for the particles of a size of 20 nm (Schins 2007). Inhaled nanoparticles have greater lung retention when compared to other particles with comparable chemical composition. According to literature data, the retention of particles is greater in patients with chronic respiratory diseases like asthma. The nanoparticle clearance from the airways is comparable to the process undergoing in case of exposure to the conventionally–structured particles. Nanoparticle clearance from the respiratory system depends mainly on the surface area, not the mass and it is suggested that this process would also depend on the length of a specific fiber (Boczkowski and Hoet 2010). The inhaled nanoparticles show differences when compared to other particles in terms of their translocation; however, the results of the studies are conflicting (Andujar et al. 2011). The hypothesis that nanoparticles migrate to the interstitium, epithelium, and neurons is supported by a number of the studies (Andujar et al. 2011). The nanoparticle translocation occurs in the systemic circulation but the literature data on this issue is very limited. It is suggested that the translocation from the respiratory system to the bloodstream may occur via endocytosis or nanoparticles could cross the alveolar barrier and enter the blood capillaries (Nemmar et al. 2003). The process of phagocytosis is also involved in clearance of nanoparticles from the respiratory system. As suggested by the data from the experimental studies, long nanoparticle fibers do not undergo the physiological clearance mechanisms. The carbon nanotubes probably show the bio-durability which supports the hypothesis that they pose a risk for human health (Donaldson and Seaton 2012). As shown in experimental studies, nanoparticles could translocate along the neurons, and through the olfactory nerves reach the central nervous system. The nanoparticles also reach the autonomic nervous system and according to the experimental data, activate the sympathetic neurons (Nemmar et al. 2013). Through the activation of the sympathetic nerves, the nanoparticles affect the breathing pattern and heart rate. Another suggested mechanism which explains neurotoxicity of nanoparticles involves the transport via blood-brain barrier by passive diffusion or endocytosis. The transport of nanoparticles through the blood-brain barriers would have the implications in neuropharmacology. Another route of exposure to nanoparticles is the skin. Data from the literature indicates that titanium dioxide (TiO_2) particles, solid lipid particles, and polymer particles could enter the body via skin. TiO_2 nanoparticles are used in sunscreens and they enter the skin thorough the stratum corneum and hair follicles. As postulated by results of some observations, TiO_2 nanoparticles could interact with the immune system. The glass fibers and Rockwool fibers could probably induce skin irritation and cause dermatitis. According to the case report, the dendrimers could be the etiologic factor for allergic dermatitis (Toyama et al. 2008). The gastrointestinal tract could be also the route of exposure to nanoparticles. The aggregates of nanoparticles could migrate to the lumen of the intestine and they enter the intestine through the intestinal lymphatic tissue. The gastrointestinal route of exposure is postulated in the case of exposure to carboxylated polystyrene nanoparticles. The toxicological data concerning this route of exposure is very limited.

The literature data on health effects of nanoparticles is very limited and mainly comes from *in vitro* and *in vivo* animal studies. In general exposure to nanoparticles is associated with the presence of oxidative stress and inflammation in different organs, not only in the respiratory system (Nemmar et al. 2013). The manufactured nanoparticles could induce oxidative stress in the respiratory system, central nervous system, and other organs like liver (Andujar et al. 2011). The exposure to nanotubes is associated with weakened antioxidant response, resulting in decreased levels of glutathione and other antioxidant enzymes (Sharma et al. 2007). The exposure to TiO_2 was proven to produce the oxidative stress according to the results of experimental studies. The exposure to carbon nanotubes and to metallic nanoparticles is associated with the DNA damage and genotoxicity (Schins et al. 2007). Experimental studies confirmed that the exposure to nanoparticles mentioned above induced the inflammation process in the respiratory system (Andujar et al. 2011). After the administration of titanium oxide nanoparticles, the bronchoalveolar lavage total cells count was increased, especially macrophages and neutrophils (Grassian et al. 2007). Pro-inflammatory effect of exposure to nanoparticles is the result of the expression of pro-inflammatory gene transcription which leads to the production of pro-inflammatory eicosanoids (Donaldson and Seaton 2012).

As suggested by the results of epidemiological studies, nanoparticles are responsible for the cardiopulmonary effects of exposure to air pollution. Nanoparticles have the strong pro-inflammatory properties due to their ability to cause oxidative stress and inflammation systemically. This is the consequence of their facilitated translocation to systemic circulation compared to other ultrafine particles. The systemic toxicity of nanoparticles is related to the cardiovascular and central nervous systems (Andujar et al. 2011).

Polystyrene particles as shown in an experimental study have the effect on homeostasis, specifically on the process of thrombus formation (Nemmar et al. 2003). This data confirms the possible effect of nanoparticles on thrombus growth, resulting in the atherosclerotic plaque rupture which may show clinically as an acute myocardial incident. Also, accelerated atherosclerosis and effect on cardiac autonomic function could also explain the cardiotoxicity of nanoparticles. Another factor could be the effect of exposure to nanoparticles on increased plasma viscosity (Peters et al. 1997).

The central nervous system is also the target organ for health effects of exposure to nanoparticles. Neurotoxicity of nanoparticles stems from the inflammatory systemic pathway. The release of inflammatory mediators results in neuroinflammation or neurodegeneration in the central nervous system. Nanoparticles due to their small size could also directly cross the blood-brain barrier. Another suggested route of exposure is the nasal pathway and further axonal transport through the olfactory nerves to the olfactory bulb and then to the central nervous system. The exact mechanism is not explained yet. Literature data suggests the possible link of exposure to air pollution to stroke incidents and the effect of ultrafine and nanoparticles is considered. Another deliberated effect of nanoparticles exposure is their effect on neurodegenerative diseases, like Alzheimer disease but literature data is scarce (Calderon–Garciduenas et al. 2008). The exposure to TiO_2 and silver nanoparticles is also associated with the negative effect on neurodevelopment as suggested in experimental studies. The literature data also suggests the effect of exposure to nanoparticles on the cognitive function in children. This phenomenon is probably caused by the presence of lead but black carbon also has the same effect (Chiu et al. 2013).

The carcinogenic effect of exposure to nanoparticles is also considered. It is suggested by some authors that the carbon nanotubes might cause asbestos-like pathology in pleural mesothelium (Donaldson and Seaton 2012). As revealed in the experimental studies, especially long carbon nanotubes, longer than 15 μm, could induce the process of inflammation in the parietal pleura. The chronic inflammation is regarded as the driver for cell proliferation and genotoxicity suggesting that carbon nanotubes could be involved in asbestos like pathologies such as fibrosis, pleural effusions, and mesothelioma. The underlying immunological process is associated with the process of "frustrated" phagocytosis.

The data on health effects of exposure to nanoparticles is still very limited and this issue is under intensive study in many research centers.

Summary

Health effects of exposure to spherical aerosol particles are the consequence of circumstances of exposure, toxicological characteristics of particles, physiological mechanisms of airway defense, and modifying factors. The pathological process underlying the health effects is inflammation, which in turn is caused by oxidative stress in cells and organs stemming from the exposure to spherical particles. Historically, the first mentions of health effects from exposure to dust came from occupational populations and date back to the fifteenth century. As new particles are introduced to the environment, the issue of health effects of exposure still remain a topic under study. The nanoparticles are the products of natural processes occurring in the environment whose role have not been deliberated so far and they were simultaneously introduced to the environment as the result of nanotechnological man-made processes. The health effects of exposure to nanoparticles are not sufficiently examined yet. Thus, the issue of toxicity of nanoparticles and their health effects begins a new era in research on occupational and environmental toxicology.

References

Andujar, P., S. Lanone, P. Brochard and J. Boczkowski. 2011. Respiratory effects of exposure to manufactured nanoparticles. Rev. Mal. Respir. 28: 66–75.

Atkinson, R.W., H.R. Anderson, J. Sunyer, J. Ayres, M. Baccini, J.M. Vonk, A. Boumghar, F. Forastiere, B. Forsberg, G. Touloumi, J. Schwartz and K. Katsouyanni. 2001. Acute effects of particulate air pollution on respiratory admissions: results from APHEA2 project. Air pollution and health: a European approach. Am. J. Respir. Crit. Care Med. 164: 1860–1866.

Aurebrach, A. and M.L. Hernandez. 2012. The effect of environmental oxidative stress on airway inflammation. Curr. Opin. Allergy Clin. Immunol. 12: 133–139.

Boczkowski, J. and P. Hoet. 2010. What's new in nanotoxicology: implication for public health from a brief review of the 2008 literature. Nanotoxicology 4: 1–14.

Broday, D.M. and R. Rozenzweig. 2011. Deposition of fractal-like soot aggregates in the human respiratory tract. J. Aerosol Sci. 42: 372–386.

Bruneekreef, B. and T. Holgate. 2002. Air pollution and health. Lancet 360: 1233–1242.

Calderón-Garcidueñas, L., W. Reed, R.R. Maronpot, C. Henríquez-Roldán, R. Delgado-Chavez, A. Calderón-Garcidueñas, I. Dragustinovis, M. Franco-Lira, M. Aragón-Flores, A.C. Solt, M. Altenburg, R. Torres-Jardón and J.A. Swenberg. 2004. Brain inflammation and Alzheimer's like pathology in individuals exposed to air pollution. Toxicol. Pathol. 32: 650–8.

Cetta, F., A. Dhamo, L. Moltoni and E. Bolzacchini. 2009. Adverse health effects from combustion-derived nanoparticles: the relative role of intrinsic particle toxicity and host response. Environ. Health Perspect. 117: A190. doi: 10.1289/ehp.0800218.

Chiu, Y.H., D.C. Bellinger, B.A. Coull, S. Anderson, R. Barber, R.O. Wright and R.J. Wright. 2013. Associations between traffic-related black carbon exposure and attention in a prospective birth cohort of urban children. Environ. Health Perspect. 121: 859–864.

Chow, J.C., J.Z. Yu, J.G. Watson, S.S. Ho, T.L. Bohannan, M.D. Hays and K.K. Fung. 2007. The application of thermal methods for determining chemical composition of carbonaceous aerosols: a review. J. Environ. Sci. Health (A) Tox. Hazard Subst. Environ. Eng. 7(42): 1521–1541.

Colais, P., A. Faustini, M. Stafoggia, G. Berti, L. Bisanti, E. Cadum, A. Cernigliaro, S. Mallone, B. Pacelli, M. Serinelli, L. Simonato, M.A. Vigotti and F. Forastiere. 2012. Particulate air pollution and hospital admissions for cardiac diseases in potential sensitive groups. Epidemiology 23: 473–481.

Cotes, J.E. 2009. Lung Function: Assessment and Application in Medicine. Blackwell Scientific Publications, London, UK.

Crystal, R.G., S.H. Randell and J.F. Engelhardt. 2008. Airway epithelial cells. Current concepts and challenges. Proc. Am. Thor. Soc. 5: 772–777.

Daviscas, E., M. Robinson and S.D. Anderson. 2010. Osmotic stimuli increase clearance of mucus in patients with mucociliary dysfunction. J. Aerosol Med. 15: 331–341.

Dockery, D.W. 2009. Health effects of air pollution. Ann. Epidemiol. 19: 257–263. doi: 10.1016/j.

Donaldson, K. and A. Seaton. 2012. A short history of the toxicology of inhaled particles. Particle and Fibre Toxicology 9: 13. doi: 10.1186/1743-8977-9-13.

Donaldson, K., C. Lang and W. MacNee. 2002. Deposition and effects of fine and ultrafine particles in the respiratory tract. pp. 77–92. *In*: G. D'Amato and S.T. Holgate [eds.]. The Impact of Air Pollution on Respiratory Health. Eur. Respir. Mon., European Respiratory Society, Sheffield, UK.

Feldman, R.G. 1998. Occupational and Environmental Neurotoxicology. Lippincott, Williams and Wilkins Publishers. Philadelphia—New York.

Forteza, R., T. Lieb, T. Aoki, R.C. Savani, G.E. Conner and R. Salathe. 2001. Hyaluronian serves a novel role in airway mucosal host defence. FASEB J. 15: 2179–2186.

Gauderman, W.J., G.F. Gilliland, H. Vora, E. Avol, D. Stram, R. McConnell, D. Thomas, F. Lurmann, H.G. Margolis, E.B. Rappaport, K. Berhane and J.M. Peters. 2000. Association between air pollution and lung function growth in southern California children. Am. J. Respir. Crit. Care Med. 162: 1383–1390.

Grassian, V.H., P.T. O'shaughnessy, A. Adamcakova-Dodd, J.M. Pettibone and P.S. Thorne. 2007. Inhalation exposure study of titanium dioxide nanoparticles with a primary particle size of 2–5 nm. Environ. Health Perspect. 115: 397–402.

Gray, T., R. Coakley, A. Hirsh, D. Thornton, S. Kirkham, J.S. Koo, L. Burch, R. Boucher and P. Nettesheim. 2004. Regulation of MUC5AC mucin secretion and airway surface liquid metabolism by IL-1B in human bronchial epithelia. Am. J. Physiol. Lung Cell Mol. Physiol. 286: L320–L330.

Haber, S., D. Yitzhak and A. Tsuda. 2010. Trajectories and deposition sites of spherical particles moving inside rhythmically expanding alveoli under gravity-free conditions. J. Aerosol Med. Pulm. Drug Deliv. 23: 405–413.

Habesglou, M., K. Demir and A.C. Yumusakhuylu. 2012. Does passive smoking has an effect on nasal mucocilliary clearance? Otol. Head Neck Surg. 147: 152–156.

Hasani, A., J.E. Agnew and D. Pavia. 1993. Effect of oral bronchodilators on lung mucociliary clearance during sleep in patients with asthma. Thorax 48: 287–289.

Hasani, A., H. Vora and D. Pavia. 1994. No effect of gender on lung mucocilliary clearance in young healthy adults. Respir. Med. 88: 697–700.

Houtmeyers, E., R. Gosselink, G. Gayan-Ramirez and M. Decramer. 1999. Regulation of mucocilliary clearance in health and disease. Eur. Respir. J. 13: 1177–1188.

Kelly, F.J. and J. Fussell. 2012. Size, source and chemical composition as determinants of toxicity attributable to ambient particulate matter. Atmos. Environ. 60: 504–526.

Kukutschová, J., P. Moravec, V. Tomášek, J. Matějka, J. Smolík, J. Schwarz, K. Seidlerová, K. Šafářová and P. Filip. 2011. On airborne nano/micro-sized wear particles released from low-metallic automotive brakes. Environ. Pollut. 159: 998–1006.

Li, T., X.P. Zhao, L.Y. Wang, S. Gao, J. Zhao, Y.C. Fan and K. Wang. 2013. Glutathione S-Transferase P1 correlated with oxidative stress in hepatocellular carcinoma. Int. J. Med. Sci. 10: 683–690.

Lillehoj, E.P. and K. Chul Kim. 2002. Airway mucus: its components and function. Arch. Pharm. Res. 25: 770–780.

Maulik, N.D., H. McFadden, M. Otani, M. Thirunavukkarasu and N.L. Parinandi. 2013. Antioxidants in longevity and medicine. Oxid. Med. Cell Longev. doi: 10.1155/2013/820679.

Mena, S., A. Ortega and J.M. Estrela. 2009. Oxidative stress in environmental-induced carcinogenesis. Mut. Res. 674: 36–44.

Montiel-Davalos, A., I. Ibarra-Sanchez, J. Mde, J.L. Ventura–Gallegos, E. Alfaro-Moreno and R. Lopez-Mallure. 2010. Oxidative stress and apoptosis are induced in human endothelial cells exposed to urban particulate matter. Toxicol. *in Vitro* 24: 135–141.

Mortensen, J., P. Lange and N. Jorgen. 1994. Lung mucociliary clearance. Eur. J. Nucl. Med. 21: 953–961.

Moya, J., L. Phillips, J. Sanford, W. Wooton, A. Gregg and L. Schuda. 2014. A review of physiological and behavioral changes during pregnancy and lactation: potential exposure factors and data gaps. J. Expo. Sci. Environ. Epidemiol. doi: 10.1038/jes.2013.92.

Nadziejko, C.E., B.L. Slomiany and A. Slomiany. 1993. Most of the lipid in purulent sputum is bound to mucus glycoprotein. Exp. Lung Res. 19: 671–684.

Nemmar, A., M.F. Hoylaerts, P.H. Hoet, J. Vermylen and B. Nemery. 2003. Size effect of intratracheally instilled particles on pulmonary inflammation and vascular thrombosis. Toxicol. Appl. Pharmacol. 186: 38–45.

Nemmar, A., J.A. Holme, I. Rosas, P.E. Schwarze and E. Alfaro-Moreno. 2013. Recent advances in particulate matter and nanoparticle toxicology: a review of the *in vitro* and *in vivo* studies. Biomed. Res. Int. doi: 10.1155/2013/2771.

Ostro, B. 2004. Outdoor Air Pollution. Assessing the Environmental Burden of Disease at National and Local Levels. WHO, Geneva, Switzerland.

Ostro, B., L. Chestnut, N. Vichit-Vadakan and A. Laixuthai. 1999. The impact of particulate matter on daily mortality in Bangkok, Thailand. J. Air Waste Manag. Assoc. 49: 100–107.
Peters, A., A. Doring and H.E. Wichmann. 1997. Increased plasma viscosity during an air pollution episode: a link to mortality? Lancet 349: 1582–1587.
Peters, A., D.W. Dockery, J.E. Mueller and M.A. Mittelmann. 2001. Increased particulate air pollution and the triggering of myocardial infarction. Circulation 103: 2810–2815.
Sánchez-Pérez, Y., Y.I. Chirino, A.R. Osornio-Vargas, R. Morales-Bárcenas, C. Gutiérrez-Ruíz, I. Vázquez-López and C.M. García-Cuellar. 2009. DNA damage response of A549 cells treated with particulate matter (PM_{10}) of urban air pollutants. Cancer Letters 278: 192–200.
Schins, R.P. and A.M. Knaapen. 2007. Genotoxicity of poorly soluble particles. Inhal. Toxicol. 19: 189–198.
Sharma, C.S., S. Sarkar, A. Periyakarunppan, J. Barr, K. Wise, R. Thomas, B.L. Wilson and G.T. Ramesh. 2007. Single-walled carbon nanotubes induce oxidative stress in rat lung epithelial cells. J. Nanosci. Nanotechnol. 7: 2466–2472.
Shi, H., L.G. Hudson and K.J. Liu. 2004. Oxidative stress and apoptosis in metal ion-induced carcinogenesis. Free Radical Biology and Medicine 37: 582–593.
Sun, Q., A. Wang, X. Jin, A. Natanzon, D. Duquaine, R.D. Brook, J.G. Aguinaldo, Z.A. Fayad, V. Fuster, M. Lippmann, L.C. Chen and S. Rajagopalan. 2005. Long-term air pollution exposure and acceleration of atherosclerosis and vascular inflammation in an animal model. JAMA 294: 3003–3010.
Toyama, T., H. Matsuda, I. Ishida, M. Tani, S. Kitaba, S. Sano and I. Katayama. 2008. A case of toxic epidermal necrolysis-like dermatitis evolving from contact dermatitis on the hands associated with exposure to dendrimers. Contact Dermatitis 59: 122–123.
Van Eeden, S., J. Leipsic, S.F. Paul Man and D.D. Sin. 2012. The relationship between lung inflammation and cardiovascular disease. Am. J. Respir. Crit. Care Med. 186: 11–16.
Vaziri, N.D. 2008. Mechanism of lead—induced hypertension and cardiovascular disease. Am. J. Physiol. Heart Circ. Physiol. 295: H454–H456.
Watelet, J.B., Van Cauwenberger and C. Bachert. 2002. The role of the upper airways in the protection against air pollution. pp. 66–76. *In*: G. D'Amato and S.T. Holgate [eds.]. The Impact of Air Pollution on Respiratory Health. Eur. Respir. Mon., European Respiratory Society, Sheffield, UK.
Widdicombe, J.G. 1987. Role of lipids in airway function. Eur. J. Respir. Dis. Suppl. 153: 197–204.
Wolff, R.K., M.B. Dolovitch and G. Obminsky. 1977. Effects of exercise and eucapnic hyperventilation on bronchial clearance in men. J. Appl. Physiol. 43: 46–50.
Yang, W. and S.T. Omaye. 2009. Air pollutants, oxidative stress and human health. Mut. Res. 674: 45–54.
Zlotkowska, R. and J.E. Zejda. 2005. Fetal and postnatal exposure to tobacco smoke and respiratory health in children. Eur. J. Epidemiol. 20: 719–721.

5

Fibers: Their Characteristics and Health Implications

Jozef S. Pastuszka

Introduction

Although aerosol particles have differentiated sizes and various shapes, some of them seem to be especially interesting and important. These particles with an aspect ratio of at least three are called "fibers". However, in addition to isolated fibers, assemblages of spherical particles and fibers frequently occur in air samples. Groupings of fibers and spherical particles, referred to as "structures", for example, "asbestos structures", are defined as fibrous bundles, clusters, and matrices. Examples of the various types of morphological structure, and the manner in which these shall be recorded, are shown in Fig. 5-1.

Since the aerodynamic behavior of airborne fibers depends on the orientation of these fibers in the flow, theories and data valid for spherical particles are not applicable for fibers (Podgórski and Gradoń 1998). Therefore, knowledge about the emission of fibers, their atmospheric and indoor transport, and penetration from outdoor into indoor environment, as well as, deposition and resuspension is still very poor. This has been largely due to the enormous numbers of variables present in the outdoor environment and the nearly impossible feasibility to reproduce them all in a laboratory. Another reason is the lack of a method to inexpensively produce length and diameter mono-dispersive fibers. Only recently Gilbertson et al. (2005) have developed such a method, which produces straight fibers with controllable lengths, using thin film grown by physical vapor deposition.

Silesian University of Technology, Division of Energy and Environmental Engineering, Department of Air Protection, 22B Konarskiego St., 44-100 Gliwice, Poland.
E-mail: Jozef.Pastuszka@polsl.pl

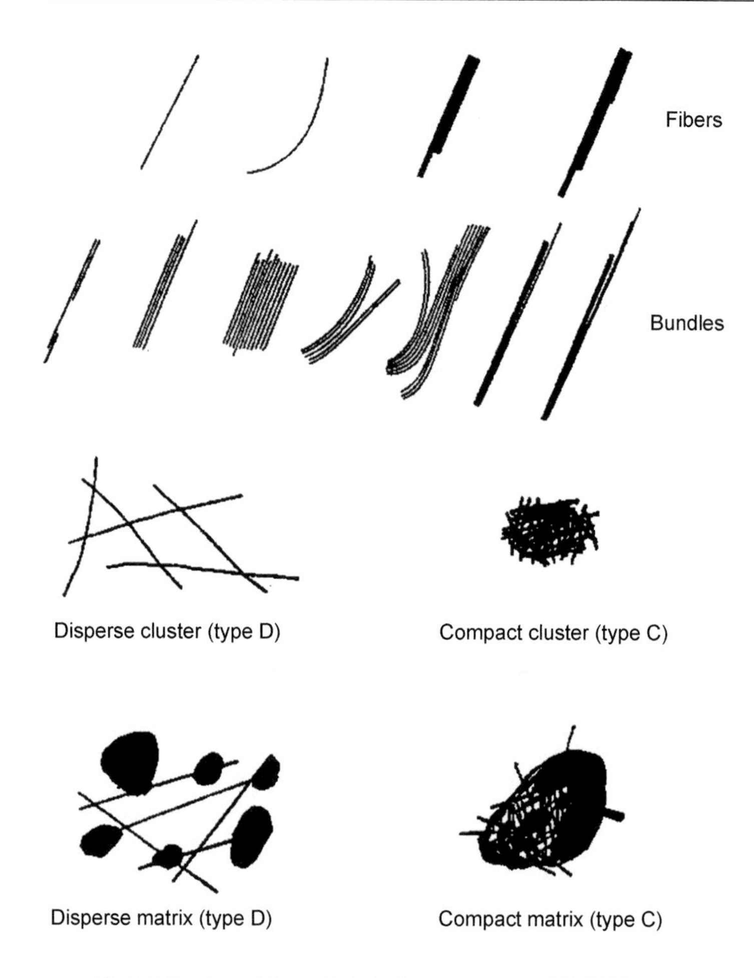

Fig. 5-1. Fundamental morphological structure types (ISO 1999).

Asbestos fibers

Most of the fibers studies were carried out on the asbestos because the asbestos fibers have been precisely shown to induce fibrosis, lung cancer, mesothelioma, and probably other kinds of intestinal cancer (LaDou 2004; Kazan-Allen 2005; Topinka et al. 2004; Xu et al. 2007). There are natural and anthropogenic sources of asbestos. For example, chrysotile is present in most serpentine rock formations and can be emitted due to natural weathering. Although very little is known about the amounts emitted from natural sources they are important, because asbestos minerals are widely spread through the earth's crust. Man-made emissions originate from activities in the following categories (WHO 1987): mining and milling, manufacture of products, construction activities, transport and use of asbestos-containing products, and disposal.

Asbestos is a common name for several naturally occurring silicate minerals with a fibrous structure. These are divided into two principal groups: serpentine asbestos and the amphibolic asbestos types. Chrysotile (white asbestos), presented in Fig. 5-2, is the sole member of the serpentine group whereas asbestos types as, crocidolite (blue asbestos—Fig. 5-3), amosite (brown asbestos—Fig. 5-4), antophyllite (Fig. 5-5), actinolite, and tremolite (Fig. 5-6) belong to the amphibole group.

It should be noted, however, that 95% of the former world production of asbestos was chrysotile.

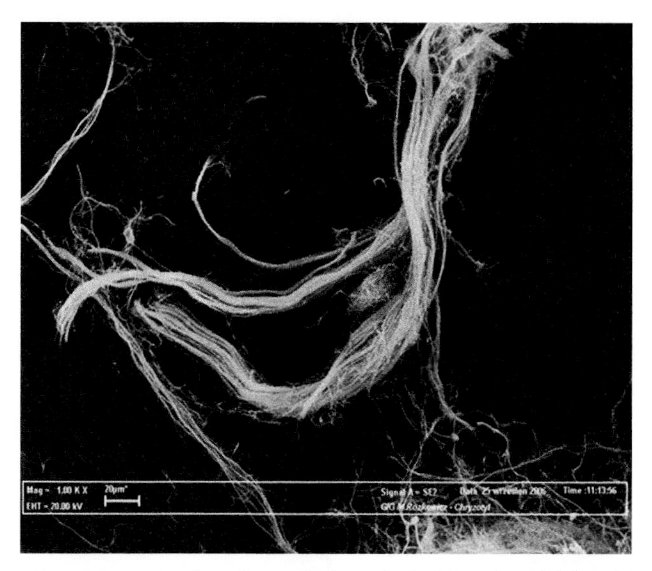

Fig. 5-2. Chrysotile structures taken from the standard sample (SPI Supplies Division of Structure Probe, Inc.). Picture made by Różkowicz (2007).

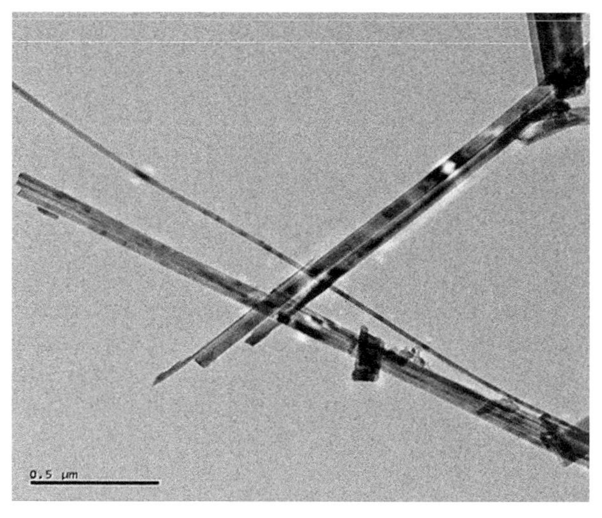

Fig. 5-3. Micrograph of crocidolite asbestos fibers (Różkowicz 2007).

Fig. 5-4. Micrograph of amosite asbestos fibers (Różkowicz 2007).

Fig. 5-5. Micrograph of antophyllite asbestos fibers (Różkowicz 2007).

The crystalline structures of the two main groups of asbestos differ greatly. Chrysotile consists of large composite sheets of one layer of silicate and one layer of magnesium hydroxide joined together. Due to a misfit between these layers, the sheets are scrolled into a tabular structure with the magnesium hydroxide layer facing outwards (Yada 1971). These tubes or fibrils have a central pore with a diameter of just under 100 Å and an outside diameter of about 400 Å (Gerde 1987). A chrysotile fiber is built up of bundles of roughly parallel fibrils, which gives the fiber a porous but curly and flexible structure (Gerde 1987). In contrast, the amphibole fibers consist of much more solid stacks built up of rod like fibrils with a trapezoidal cross section (Mossman and Craighead 1981). The

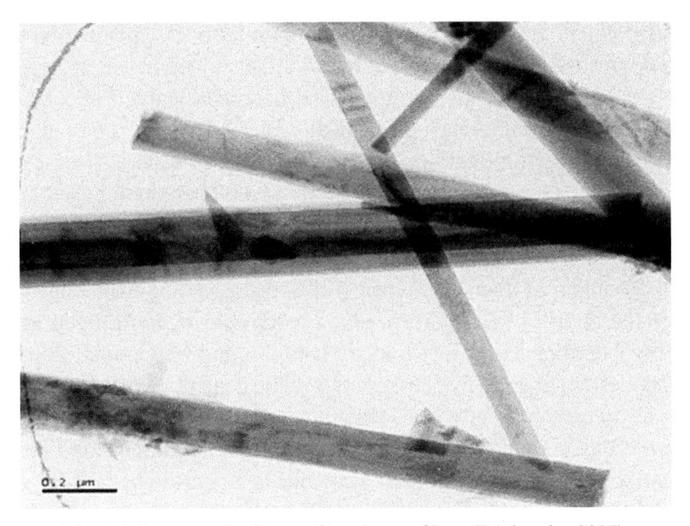

Fig. 5-6. Micrograph of tremolite asbestos fibers (Różkowicz 2007).

fibril consists of cross-linked double chains of silicon-oxygen tetrahedral. This results in a stiff and harsh fiber. The various amphibolic asbestos types have different compositions of the metal ions located along the parallel sides of the fibrils. This in turn gives rise to a great variation in the length and diameter distributions of amphibolic fibers, not only between the various types but also within the same species of asbestos.

The macroscopic asbestos fibers are actually bundles of thinner fibers made up of fibrils which, in the case of chrysotile, have a diameter of 20–25 nm. Once emitted into the atmosphere, asbestos fibers may travel long distances owing to their aerodynamic properties. Because no chemical breakdown of these fibers occurs, washout by rain or snow is only a cleaning mechanism (WHO 1987).

In the occupational environment, asbestos fibers can be detected and monitored by using phase-contrast optical microscopy (the NIOSH 7400 method); however their assessment in the atmospheric and in the indoor air needs an integrated method capable of micro-chemical analysis of single fibers, measurement of fiber length and diameter, and count of fiber numbers in given air samples. The reason is the fact that asbestos fibers normally constitute only a relatively small fraction of the total fibrous aerosols in the lower atmosphere. Electron microscopy with an X-ray diffraction analysis is the only method which can detect and identify asbestos fibers. Unfortunately, such instrumentation is costly and highly-trained personnel are required in order to obtain reliable results. Therefore, the combinations of some different sampling and analytical methods are still used.

As mentioned above, there are various diseases, including lung cancer, which can be caused by the inhalation of airborne fibrous dust clouds (Walton 1991; Peters and Peters 1997a,b; Moolgavkar et al. 2001; Mossman et al. 2007). The so-called "asbestos panic", which appeared in US in the eighties of the last century (Zurer 1985), and then in many other countries in the world, is now over. However, the mutagenic and carcinogenic properties of airborne asbestos are still the object of scientific studies. For example, quite recently Ben-Shlomo and Shans (2011) tested the genetic diversity in wild mice inhabiting an asbestos-polluted area as a model for the long-term mutagenic effect of asbestos. Their data confirmed elevated health risk for humans living in asbestos-polluted areas.

Over the past two decades, a great deal has been learned about how and why asbestos fibers cause pulmonary diseases (example: Burmeister et al. 2004). There are three essential factors that are required to develop such a disease (Maxim and McConnell 2001): adequate dose, dimensions of the fibers in the alveolar region, and fiber biopersistence. Other fiber properties, such as presence of iron or other transition metals on fibers, ability of fibers to generate free radicals (Gold et al. 1997; Kadiiska et al. 2004), and the ability of fibers to interact with and alter biologically relevant molecules, as well as, the ability of fibers to produce reactive oxygen/nitrogen species (ROS) may also be determinants of fiber toxicity (Maxim and McConnell 2001), especially among biopersistent fibers. Fibers may catalyze the generation of oxidants directly from molecular oxygen or indirectly from reactive oxygen and nitrogen species released by inflammatory cells recruited to the lungs or pleura (Kane 2003). The fundamental property of the fiber toxicity is that, in contrast to chemicals, fibers are believed to cause disease through a physical/chemical interaction (WHO 1987), which means that the health effects depend not only on the type of fiber but also upon its diameter and length. It is also possible that the physical form of a fiber is even more important than its chemical composition (Maxim and McConnell 2001; WHO 2000).

Knowing that the lung cancer can be caused by inhaled fibers, especially asbestos fibers, it is clear that these fibers must first be deposited in the lungs and then should remain there for a long time because of the so called "latency time" (needed for lung cancer to develop) which is equal to about 20 years. To reach a lung, the aerodynamic diameter of the fiber should be less than 3.5 μm (such particles are respirable) what means that the fiber should be thin. On the other hand, such carcinogenic fibers must be long. The reason is that the specific cells—macrophages, living in the lung, biochemically destroy various particles, including fibers, present in the lung and only large/long particles can remain (partially) there for a period of time long enough to generate the cascade of processes leading to cancer. Typically, it is assumed that the most hazardous fibers are those longer than 5 μm and of diameter up to 3-4 μm with an aspect ratio equal to or greater than 3:1 (WHO 1987). The US EPA uses the IRIS model (Integrated Risk Information System) to calculate risk based on the concentration of optically equivalent fibers, usually described as 5 μm and longer, a minimum width of 0.25 μm, and a minimum aspect ratio of 3:1 (EPA 1988). Although the IRIS model suggests that optical microscopy of asbestos fibers counts fibers longer than 5 μm and wider than 0.4 μm, the analytical procedure (NIOSH 1994) indicates the observable width of asbestos fibers is 0.25 μm. The effect of including these thinner fibers is to slightly increase the risk estimates (Lee and Van Orden 2008).

Actual atmospheric and indoor concentrations of asbestos range from below one hundred to several thousand fibers longer than 5 micrometers per cubic meter. In rural areas, remote from asbestos emission sources, the concentration of airborne asbestos is below 100 F m^{-3} (where F means respirable fiber, longer than 5 μm). In urban areas, general levels may vary from below 100 to 1,000 F m^{-3} (WHO 2000). In buildings without specific asbestos sources, concentrations are generally below 1,000 F m^{-3} but in buildings with friable asbestos, concentrations vary irregularly; usually less than 1000 F m^{-3} are found, however in some cases, exposure reaches 10,000 F m^{-3} (WHO 2000).

Till now, it is not fully understood whether the health impact is due to toxic/carcinogenic compounds absorbed on the fibers. Such compounds could be, for example, PAHs. Figures 5-7 and 5-8 seem to confirm such hypothesis. It can be seen that a number of fine particles are closely connected with a fiber, forming the enlarged and certainly chemically interesting structure.

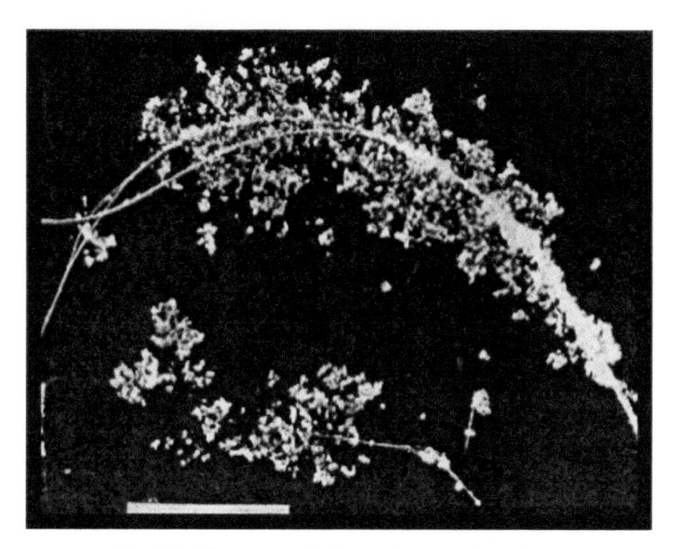

Fig. 5-7. Micrograph of asbestos fiber (Walton 1991).

Twenty five years ago, Gerde (1987) tried to explain how physicochemical mechanisms such as molecular diffusion, convection, adsorption onto solids, and partition between various phases can influence the uptake of particle-associated Polycyclic Aromatic Hydrocarbons (PAHs) by the human lung. He especially discussed the possible effects of inhaled mineral fibers on this transport process and showed that the uptake of PAHs by the human lung is probably governed much more by properties of the surface lining layer that covers the airways of the lung than by properties of the carrier of PAHs. He stated that mineral fibers, particularly asbestos fibers, adsorb layers of phospholipids from lung surfactant, and if such a fiber transverses the bronchial lining layer this can create a lipid connection between contaminated surfactant lipids at the air interface and epithelial cellular membranes. However, this hypothesis still needs additional experimental confirmation. On the other hand, it should be mentioned that generally, surface reactivity is related to biological reactivity. Surface reactivity depends on the bioavailability of iron on the surface of the fiber. The source of this iron could be from asbestos or a trace contaminant, or endogenous deposition in the lungs (Kane 2003).

Although some extremely important questions concerning the relationships between exposure to asbestos and adverse health effects still need precise answers, the World Health Organization officially recognized that asbestos is a proven human carcinogen (IARC Group 1). No safe level can be proposed for asbestos because a threshold is not known to exist (WHO 2000). Exposure should therefore be kept as low as possible.

a

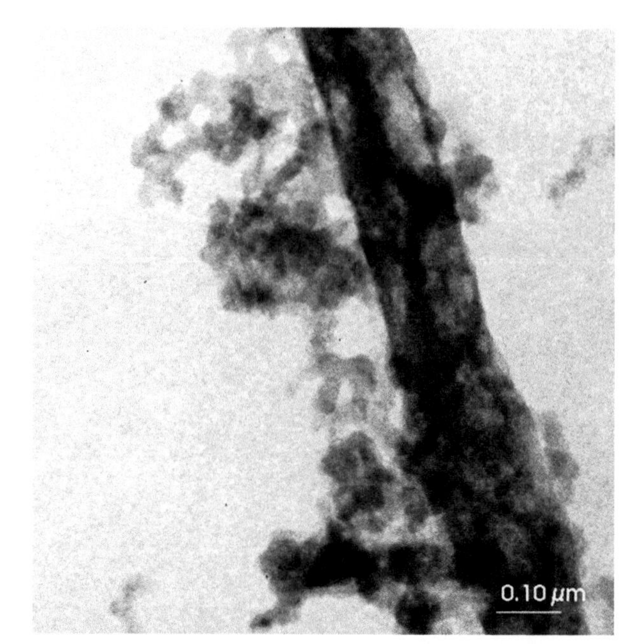

b

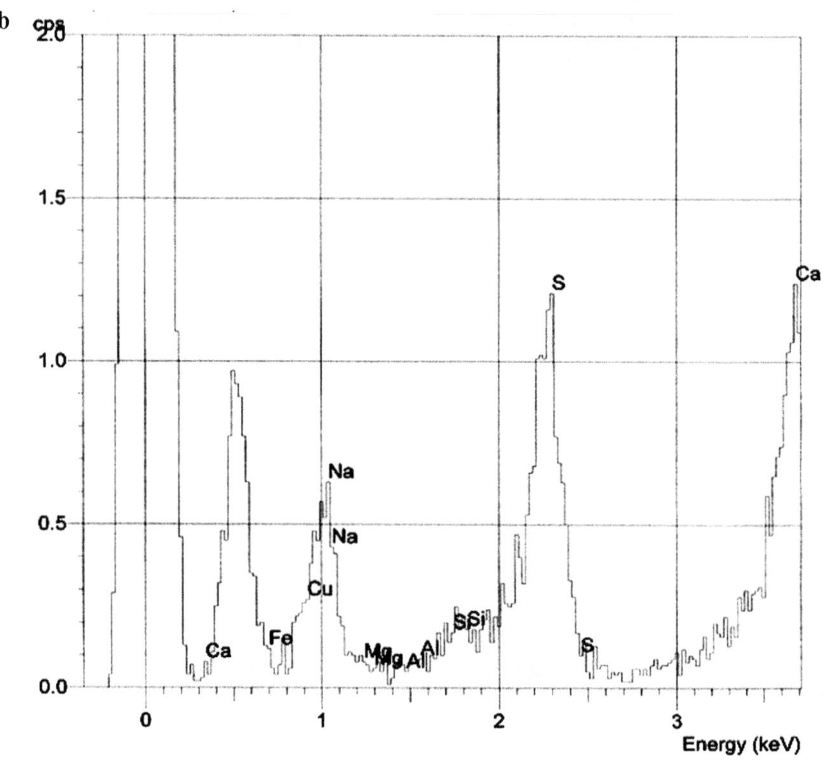

Fig. 5-8. Micrograph of the crocidolite fiber (a) and the X-Ray spectrum of the whole structure (b) (Różkowicz 2007).

Emission of fibers from asbestos-containing materials

The health effects of inhalation of fibrous aerosol can concern huge groups of people becoming very important for some local populations. It should be noted that according to a WHO report (1987), in many industrialized countries, most of the asbestos used was in the building industry (up to 70–90%). Because of its specific technical properties, asbestos has found an extremely large variety of applications. The major uses in buildings include thermal system insulation, surface treatments, and other materials. For example, friable asbestos is commonly found in buildings where the walls and ceilings have been sprayed or trowelled with asbestos-containing materials. This technique was widely used for thermal and acoustic insulation (or even for decorative purposes) between 1950 and 1973 in the USA, and later in Europe. In most countries, the spraying of asbestos-containing materials in buildings is now prohibited. However, many school buildings still contain friable asbestos materials and these constitute a major public health concern, particularly since large numbers of children are potentially exposed (Bignon et al. 1989). It should be noted that even though asbestos use is currently prohibited in most countries, asbestos containing-products are expected to remain in the environment for several decades.

The asbestos fibers found in the buildings may be of indoor or outdoor origin. The first group of fibers may have been incorporated directly and deliberately into buildings materials, be accidental contaminants of building materials, and have developed during aging of originally nonfibrous materials (Brown and Hoskins 1992). Analyzing the literature data it can be concluded that the emission of mineral fibers from materials used for building construction is the main source of indoor fiber pollution at the present time (Bignon et al. 1989; Pastuszka 1997; Pastuszka et al. 1999; Pastuszka 2009; Kim et al. 2015). The daily use of domestic equipment (such as hair dryers, ovens, etc.), hobbies, and the use of pet litters (attapulgite, sepiolite) at home are also potential sources of indoor fiber release. Additionally, the deposits of fiber on floors and different surfaces may become new sources of indoor air pollution due to their resuspension from the deposition surface into the air. The preliminary study of the resuspension of fibrous particles (mainly asbestos) from the carpet, generated by the physical activity of children, was carried out in 1993 (Pastuszka et al. 1993), and some years later USEPA prepared the memorandum on the testing carpet being the asbestos reservoir (USEPA 2002). Crossman et al. (1996) reported the quantification of fiber releases for various floor tile removal methods. They documented that fibers are released when a floor tile is broken and/or abraded during removal procedures. They established that fiber levels vary with the aggressiveness of the procedures but they did not study the emission rate.

It is certainly possible to remove the friable materials efficiently with proper working methods and proper protection. Unfortunately, inappropriate removal methods increase the fiber dust levels inside buildings. On the other hand, non-friable materials containing fibers do not cause major problems because the fibers are sealed inside the compact materials (cement, plastic). A problem will develop only when the building is demolished or renovated. Consequently, the indoor air fiber concentrations depend strongly on human activities which may damage the surface of the walls and redisperse fibers deposited on the floor.

Another problem appears in Central and Eastern Europe where many buildings, mostly apartment buildings are covered with thermal insulation containing asbestos-cement sheets, which weather and corrode. The example of the section of such asbestos-cement sheets from Upper Silesia is presented in Fig. 5-9. The micrograph was taken by using the camera mounted on the electron scanning microscope (ESM).

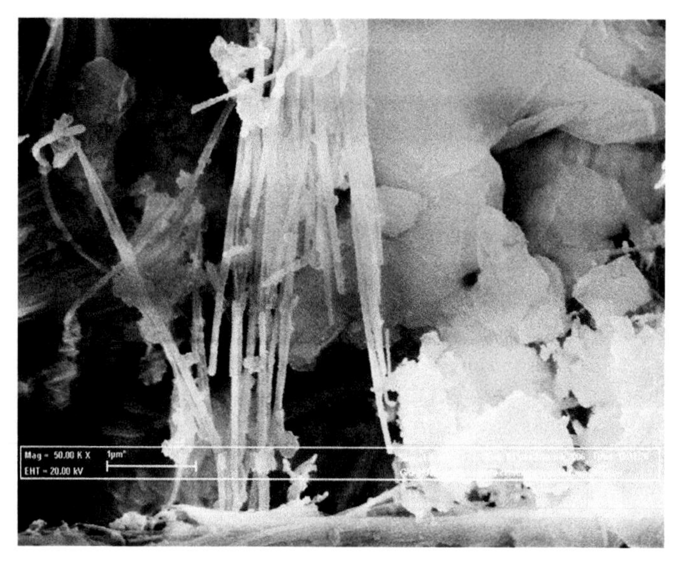

Fig. 5-9. The micrograph of the section of asbestos-cement sheet (Różkowicz 2007). The picture was taken using ESM at the Silesian University of Technology in Gliwice.

Cement particles, asbestos fibers, and agglomerates of both particles and fibers are released from the plate surface and become dispersed in the air (Bornemann and Hildebrandt 1986; Spurny 1989). These airborne asbestos fibers can create a health risk, including a lung cancer risk. The problem of exposure to asbestos is, therefore, still actual. Asbestos will become airborne also from coatings, mastics, and adhesives used in the insulation industry from about 1930 to the present (Paustenbach et al. 2004). Asbestos fibers can also be released from brake pads of overhead industrial cranes (Spencer et al. 1999). This fact creates exposure to the asbestos in these products during the application, spill, cleanup, sanding, cutting, and removal. Also vermiculite, a naturally occurring material, from the mine that operated near Libby, Montana, USA, from the early 1920s until 1990 was contaminated with asbestos and other amphibole minerals. The historical cohort mortality study carried out recently by Sullivan (2007) indicated significant elevations in standardized mortality ratios for asbestosis, lung cancer, and cancer of the pleura among Libby vermiculite workers. Unfortunately, the exposed population is much larger. Since vermiculite from Libby mine was used to make loose-fill attic insulation that remains in millions of homes, these findings highlight the need for control of exposures that currently occur when homeowners or such workers as cable installers, electricians, insulators, and so on, disturb these insulation materials made with asbestos-contaminated vermiculite (Sullivan 2007). Generally, on the world scale, a correlation between national asbestos consumption and the incidence of asbestos diseases has been observed (Kazan-Allen 2005).

It should be noted that asbestos fiber-containing materials outside and even inside a building become dangerous only when the microscopic fibers are emitted from these materials into air. Such phenomenon appears if the asbestos materials become friable (crumbling), for example, due to the atmospheric weather, or if building maintenance, repair, renovation, or other activities (vibration and vandalism) disturb these materials. Therefore, it can be concluded that the important factor that generates, or considerably increases the emission of fibers from these fibers-containing materials used in buildings is the mechanical impact. The risk of human exposure to asbestos-containing material is also directly related with the condition of this material. Unfortunately, till now the phenomenon of the mechanically generated emission of fibers from fiber-containing slabs is only partially recognized. After the pioneering work of Spurny (1989) on the release of asbestos fibers from weathered and corroded asbestos-cement products, no similar paper appeared during twenty years in the available literature although some works describing in qualitative way the process of fibers emission have been published. Only in 2009 the emission rates of fibers released from the asbestos-cement plates due to the mechanical impact were determined (Pastuszka 2009). The obtained values of the surface emission rate (ϵ_S), defined as the number of fibers released from the area of 1 m^2 of the asbestos-containing surface under the impact energy of 1 J (1 Joul) are presented in Table 5-1. The basic principle of the experiments was as follows: the falling weight generated the emission of fibers from the samples of asbestos-containing sheets. The increasing concentration of fibers in the cabinet volume (adopted for these experiments) was measured by the Laser Fiber Monitor (MIE, Inc., Billerica, MA). The method used in this monitor is based on the electric field-induced alignment and oscillation of particles, combined with light scattering, resulting in the highly selective detection of individual fibers, even in the presence of a population of predominantly non-fibrous particles. After every experiment, the cabinet was cleaned using the ventilation system with HEPA filter. During all experiments, the air temperature and humidity inside the cabinet were almost constant, keeping the level 24–25°C and 30–31%, respectively.

Counting the highest increase in the measured concentration of fibers inside the cabinet, with known volume after the impaction of the falling weight, and knowing the surface of the sample, as well as the impact energy, it was possible to calculate the factor ϵ_S for all experiments.

Table 5-1. The averaged surface emission rate of the long (L > 5 μm) and short (L ≤ 5 μm) fibers emitted from the asbestos-cement sheets (Pastuszka 2009).

	ϵ_S [$m^{-2}J^{-1}$]*	
	fibers > 5 μm	fibers ≤ 5 μm
Mean	5.1 x 10^3	5.7 x 10^3
Standard deviation	1.9 x 10^3	3.1 x 10^3
Median	5.2 x 10^3	5.1 x 10^3
N (samples)	6	6

*In traditional units [F/(m^2J)].

It can be seen that the averaged emission rate for all studied samples was about 5,000 and 6,000 of long and short fibers, respectively, emitted per one square meter because of the impact energy equal to 1 J. It was also documented that ϵ_S slightly increased (more than two times) with deteriorating surface.

The results of the field studies of airborne fibers carried out in Upper Silesia, Poland, also seem to be very interesting. Upper Silesia Industrial Zone is an extensively urbanized and industrialized province in southern Poland where coal mining and metallurgy, as well as chemical industry are still the major industrial activities. These kinds of activities, especially the underground coal mining, significantly influence the quality of the building facades. The reason is the location of these coal mines—exactly under the cities, which causes frequent vibration of the surface of the land area in Upper Silesia resulting in damages of the buildings. Hence, many of the asbestos-cement buildings' facades show breakdown likely to cause release of asbestos fibers into the environment. Therefore, asbestos can enter the homes as fibers suspended in outdoor air. Once such fibers are indoors, they can be resuspended by normal household activities, including even vacuuming (as many fibers will simply pass through vacuum cleaner bags).

The levels of airborne fibers found in the studied dwellings are shown in Table 5-2.

It can be seen that the mean concentration of fibers in buildings covered with asbestos-cement sheets (Group I) was 850 fibers per cubic meter, which was three times higher than in Group II of control dwellings (280 fibers/m^3). These data indicate, especially in the context of the laboratory studies (Pastuszka 2009), that significant sources of fibrous aerosol in the indoor environment in Upper Silesia are outdoor asbestos materials (mainly asbestos-cement buildings facades). These results agree with the previous, preliminary studies on concentration of airborne asbestos in Polish dwellings obtained using the Laser Fiber Monitor (Pastuszka 1997; Pastuszka et al. 1993, 1999, 2000). Especially, the background level indoors (Group II of dwellings) was the same as the concentration of airborne fibers longer than 5 μm obtained in the family house (280 fibers/m^3) in Davis, California, while the concentration level measured in the offices with HVAC-in was 110 fibers/m^3 (Pastuszka et al. 2000). Also Lee and Van Orden (2008) reported the level 120 fibers/m^3 as the average concentration of airborne asbestos ≥ 5 μm long in buildings nationwide in

Table 5-2. Concentration of the airborne fibers in two groups of dwellings in four towns in Upper Silesia, Poland; Katowice, Chorzów, Sosnowiec, Bytom (Pastuszka 2009).

Group of dwellings	Concentration level [m^{-3}]* of airborne fibers			
	longer than 5 μm		shorter and equal to 5 μm	
	Mean	Range	Mean	Range
I (Buildings covered with asbestos-cement sheets) n = 25	850	300–1800	1290	600–3400
II (Reference buildings) n = 20	280	< 70–700	580	< 70–1500

*In traditional units [F/m^3].

USA. It should be noted that they used transmission electron microscopy (TEM) in the measurement of asbestos concentration. Only Woźniak et al. (1994) who used the Laser Fiber Monitor FM-7400 (MIE, Inc., USA) in their indoor study in Wroclaw reported that the concentration of fibers > 5 μm reached the level of 11,100 m^{-3} but their measurements were carried out in the flat located in the vicinity of the playground covered with serpentinite.

Since the length of the asbestos fibers is an important factor of its toxicity, the average length distribution of fibers for the two groups of dwellings is shown in Figs. 5-10 and 5-11. The fiber length distributions are seen to have a peak concentration level in the 3–5 μm length range, representing most of the total concentration in studied

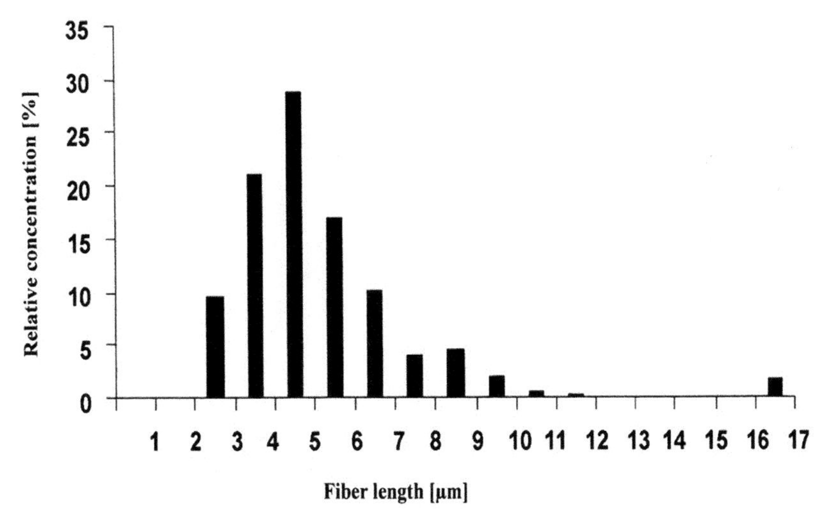

Fig. 5-10. Length distribution of the fibrous aerosol in buildings with asbestos-cement facades (Pastuszka 2009).

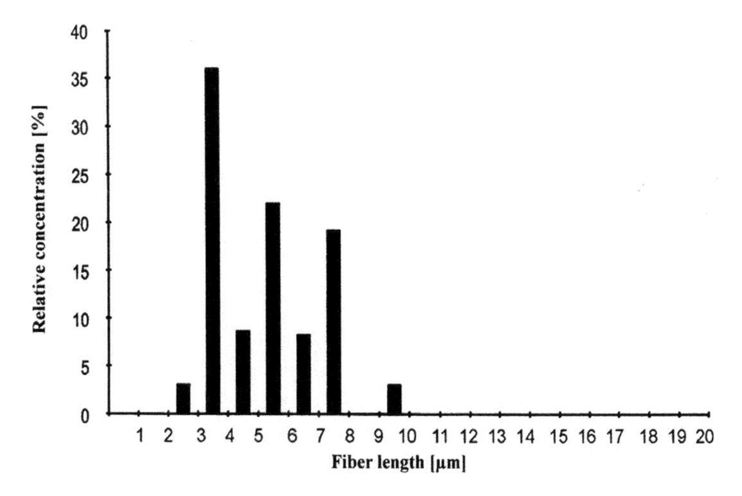

Fig. 5-11. Length distribution of the fibrous aerosol in the reference buildings (Pastuszka 2009).

dwellings. In homes covered with asbestos-cement sheets, a peak was observed in the 4–5 μm length range (Fig. 5-10) while in the reference homes (Group II), a peak appeared in the 3–4 μm length range (Fig. 5-11). It seems to be very important to note, once again, the significant contribution of short fibers to the total concentration of fibrous aerosol in the dwellings in Upper Silesia, because some papers have recently appeared in literature, indicating that short asbestos fibers could also be carcinogenic (example: Suzuki et al. 2005).

It should also be noted that the length distributions of airborne fibers obtained in the buildings with asbestos-cement facades (Fig. 5-10) and in the laboratory experiments where fibers were emitted from asbestos-cement sheets due to the mechanical impact (Fig. 5-12) are very similar.

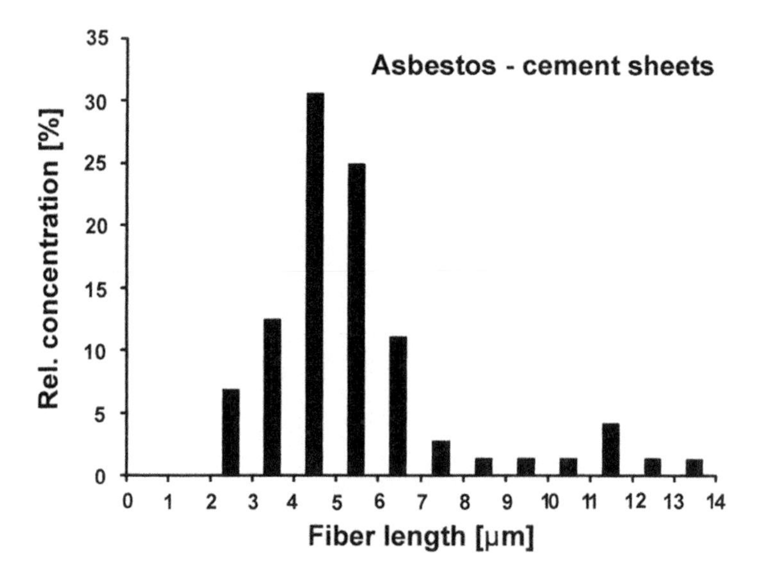

Fig. 5-12. Length distribution of the fibrous aerosol generated from the asbestos-cement sheets by the mechanical impaction (Pastuszka 2009).

These indoor data seem to be in good agreement with the results obtained by Janeczek et al. (2001) who measured the concentration of asbestos fibers in ambient air in Sosnowiec, Upper Silesia, Poland, using the Phase Contrast Microscope. The lowest concentrations of respirable asbestos fibers were found in the Sosnowiec suburb ranging from 0 to 533 fibers per cubic meter while the highest concentrations were recorded near the crossroads (768–1,646 fibers/m^3). In the residential districts of Sosnowiec, the asbestos concentration was reported to be 768–1,067 fibers/m^3. All these microscopic data were obtained for the samples collected during 70–12 minutes only, what means that the 24-hours levels would probably be slightly lower. It should be noted, however, that PCM cannot distinguish asbestos fibers. Therefore, further studies using TEM method are needed to confirm the results obtained in Upper Silesia flats.

In fact, some studies of asbestos structures suspended in ambient air in Katowice, the capital city of Upper Silesia, were performed in this decade by the use of an electron transmission microscope (ETM) with an X-ray analyzer (Różkowicz 2007). These results indicate that the mean level of asbestos in outdoor air in the residential districts of Katowice (where a great number of buildings with asbestos-cement facades are located) is between 510 and 2,532 asbestos structures, longer than 5 μm, per cubic meter. Unfortunately, these data are not fully comparable with the concentration of individual, clearly separate fibers indoors.

It is also rather difficult to compare Pastuszka's results with data obtained in other countries due to the difference in the kind of asbestos emitters influencing the airborne fibers indoors. This is especially since the asbestos-containing materials (ACM) used in buildings in different countries are different. For this reason and because the reported measurements were carried out, as a rule, under different conditions, the range of published data is from almost zero up to 10^6 m^{-3}. For example, Crump and Ferrar (1989) made the statistical analysis of the data from the EPA study of airborne asbestos levels in 49 buildings occupied by the General Service Administrations. They detected no statistically significant differences in asbestos levels among three categories of buildings: (1) with no ACM, (2) with ACM in good condition, and (3) in buildings containing damaged ACM. The average indoor concentration of asbestos was 70 m^{-3} for fibers 5 micrometers or longer. Corn (1994) determined the concentration of asbestos in air from analysis of samples collected in over 300 buildings involved in litigation and found that concentrations of fibers > 5 μm were generally less than 500 m^{-3}. Much higher levels were obtained by Ganor et al. (1992) who measured the concentrations of asbestos in the air of a communal dining room in Israel in which the damaged ceiling had a sprayed on coating of insulation containing asbestos. They obtained the concentration 4 x 10^6 m^{-3}. It should be noted that the literature data presented above were obtained in the 1990s. Towards the end of the 20th century, governments in many developed countries banned or seriously restricted the use of asbestos. Unfortunately, consumption of asbestos is increasing in Asia, Latin America, and the Commonwealth of Independent States (Kazan-Allen 2005).

In most of the countries there is little, if any, control of hazardous asbestos exposures from occupational, environmental, and domestic sources (Kazan-Allen 2005). The illustration of this thesis can be the result of the study recently carried out by Ansari et al. (2007) in India where around 60% of the total production of asbestos is processed in unorganized sectors including milling and manufacturing of asbestos-based products. Asbestos fibers levels in the range of (2–16) x 10^6 m^{-3} were found in air at work zone areas of asbestos milling units (Ansari et al. 2007) but due to the specific situation in India (people living close to the industrial sectors of the towns), part of the general population is also exposed to the similar concentration of asbestos.

Carcinogenic risk

It is well known that the health effects of the exposure to a toxic substance depend on the received dose of this substance but the dose must exceed the so called "threshold dose", usually more or less precisely determined. For air pollutants, the received dose D depends on the concentration C_k of the pollutant in the air in the microenvironment k, time of exposure to this pollutant in the k-microenvironment (t_k) and on the rate of inhalation w.

In case of carcinogenic substances, the most important effect is the occurrence of cancer; therefore the health prognosis for the selected human population exposed to carcinogenic pollutants concerns the calculation of the probability of cancer (i.e., "carcinogenic risk") among the people who received the real dose. Hence, instead of the dose-effect relationship, the dose-risk relationship is as a rule analyzed for the carcinogenic substances.

In contrast to toxic substances, carcinogenic substances have no threshold level which means that only lack of exposure to these substances generates zero-level risk. Generally, risk is a function of dose

$$R = f(D) \tag{5-1}$$

It should be noted that the cancer risk classification for the so called general population is till now not fully completed. The only general agreement is that the risk equal to 10^{-6} or less is classified as acceptable while the risk level 10^{-3} or higher is unacceptable. The recommendation for people living in the areas where the environmentally-generated cancer risk ranges between 10^{-4} and 10^{-5} is rather general, and varies in the literature, still being the subject of emotional discussions among the experts.

Assuming that the life exposure will be unchanged for 70 years, i.e., the measured concentration C of airborne, respirable asbestos fibers, longer than 5 μm, will remain at the same level during the life period it is possible to calculate so called "life risk" as follows (see Chapter 11 for the details):

$$R_{LIFE} = K_{LIFE}C \tag{5-2}$$

Equations (5-1) and (5-2) indicates that, generally, it is possible to provide predictions of asbestos-related mortality even in a selected cohort of exposed subjects, using previous knowledge about the exposure-response relationship. Unfortunately, precision of such estimation is usually not so high. However, Gasparrini et al. (2008) showed that the inclusion of individual information in the projection model helps reduce misclassification and improves the results.

On the basis of literature review, especially the WHO guidelines (2000), it is possible to write the typical value of K_{LIFE} [m³/No. of fibers] as follows:

- Nonsmokers: K_{LC} [F/m³] $= 2 \cdot 10^{-8}$
- Smokers K_{LC} [F/m³] $= 2 \cdot 10^{-7}$

where K_{LC} [F/m³] is a factor describing the carcinogenic potency of asbestos developing a lung cancer.

For the same lifetime exposure, the mesothelioma risk for the general population would be about ten times higher.

It should be noted that, according to the WHO guidelines (2000), the lifetime is assumed to be 50 years, and in a lifetime of 70 years, the first 20 years without smoking probably do not make a large contribution.

It may be remarked that there is a significant synergy phenomenon between the exposure to asbestos and tobacco smoke. In fact, the reason of this synergy has been intensively studied for almost 50 years. Although, as it was mentioned previously, this synergy is probably due to carcinogenic compounds absorbed on the fibers, some fundamental questions still remain unclear. Besides, as it has been found, the cancer risk is slightly lower for women than for men but this fact cannot be satisfactorily explained on the basis of the existing knowledge.

Using the above relations, the cancer risk for occupants of the buildings covered by the asbestos-cement sheets can be estimated. This risk ranges from 10^{-5} to 10^{-4} for nonsmokers and smokers, respectively.

Summarizing the exposure of the general population to asbestos, it can be concluded that in buildings constructed without asbestos materials, concentration of airborne asbestos fibers is generally below 1,000 F m^{-3} (respirable fibers, longer than 5 μm), while in the buildings with friable asbestos, including corroded asbestos-cement materials, concentrations vary irregularly: usually levels lower than 1,000 F m^{-3} are found, but in some cases the concentration level reaches 10,000–15,000 F m^{-3}. Such exposure creates the life cancer risk of mostly about 10^{-6}, however in some buildings it is increased up to 10^{-4} (for non-smoking people). This risk for smokers is increased roughly ten times. Maintenance staff and cleaners may be exposed to higher level of asbestos fibers—around 10^{6} F m^{-3} (Brown 1987; HEI 1991; Pastuszka 1997). Similar concentration levels were found in Israel (Ganor et al. 1992) and recently in India (Ansari et al. 2007). Such level of airborne asbestos in these areas/in buildings can create a dramatically high risk level of about 10^{-2} for people living there, what means that every hundredth nonsmoking person during his/her 70 years-life will suffer from lung cancer.

Although the cancer risk is mostly relatively easy to calculate, it should be noted that the prognosis of adverse health effects of the exposure to some kinds of asbestos-containing materials seems to be much more complicated than it could be expected. For example, OSHA and the USEPA categorize products containing encapsulated asbestos differently from other asbestos-containing materials. These products, which include some mastics, coatings, and adhesives, are excluding from the American federal regulations. This is due to the encapsulation of asbestos fibers in a solid matrix, which limits their potential for airborne release (Paustenbach et al. 2004). This limitation was confirmed by a number of studies which have characterized exposures associated with the replacement of asbestos-containing gaskets and packing materials, including different work tasks of vetting durations and involving the use of different tools (Boelter et al. 2011; Madl et al. 2014). However, there is additional intrigue; hypothesis indicates that if under some condition there is inhalation of some fibers, the presence of the encapsulating medium both inside and outside the fiber may significantly reduce, or even eliminate, its adverse effects (OSHA 1994; EPA 1999; Paustenbach et al. 2004).

Sometimes, during environmental catastrophes or due to other reasons, when the emission of asbestos fibers can be extremely high, new approaches to the prediction of the health effects are needed, especially to calculate the cancer risk for people exposed to the emitted fibers. There are two main problems. First is related with receiving a huge dose of asbestos fibers during a short time, and it is not clear if such specific exposure does not modify the classical dose-risk relationship obtained for a different, long-time exposure. The second problem is caused by the great amount of short fibers inhaled. Despite some papers suggesting that short asbestos fibers could also be carcinogenic, the common opinion about the carcinogenicity of long fibers is based on the fact that due to the activity of the macrophages only parts of long fibers can remain in the lung for a period of time long enough to generate the cascade of processes leading to cancer (what has already been mentioned in this chapter). However, if the amount of short fibers deposited in the lung quickly increases, the efficiency of biochemical destruction of fibers by macrophages (phagocitosis process) certainly rapidly decreases. Therefore, it may be concluded that some of the short asbestos fibers deposited in the lung will not be totally destroyed and will remain there for a long time. If this picture is true, the cancer risk calculated by using equation (5-2 or 5-4) should be elevated because of the additional contribution of the inhaled short fibers.

The above analysis is not detached from reality. For example, such a situation appeared during the catastrophic collapse of the high-rise office buildings (towers) associated with the WTC, in New York (Fig. 5-13), when the huge amount of asbestos-containing materials found inside the buildings, was rapidly broken and the cloud of fibers was emitted into the air. The group of scientists from HP Environmental, Inc., USA, and some other companies (Jim Millete, MVA, Inc.; Edward Dantsker, Anabell Environmental, Inc.; Tom McKee, Scientific Laboratories, Inc.; George Pineda, ET Environmental, LLC; Hugh Granger, HP Environmental, Inc.; Piotr Chmielinski, HP Environmental, Inc.; Brent Sharrer, HP Environmental, Inc.) studied the nature of chemical and particle characteristics of settled dust residues (Fig. 5-14) produced

Fig. 5-13. Satellite overview of the sampling area.

Fig. 5-14. Surface residues, Building A, One World Financial Center, New York, NY, USA (Chmielinski 2001. Private communication. With permission).

by the fire and catastrophic collapse of these towers, as well as, the concentration of asbestos in the indoor and outdoor air (Fig. 5-15).

Table 5-3 contains the results obtained by Chmielinski, Granger and co-workers. It is an appropriate illustration of the discussed problem of exposure to short fibers.

It can be seen that the concentration of respirable fibers longer than 5 μm, obtained by using the phase contrast microscopy, was mainly about 10^4 F m^{-3} but in two sites it was more than 10^5 F m^{-3}. However, the results of the concentration of the asbestos

Fig. 5-15. Sampling equipments used in the study of the air pollution in the ruins of WTC, New York, NY, USA (Chmielinski 2001. Private communication. With permission).

structures ≤ 5 μm reaching the level of 10^6 per cubic meter seem to be especially interesting. If the previous analysis of the importance of short fibers deposited in the lung is correct, the cancer risk calculated only on the basis of inhaled dose of long fibers was certainly significantly underestimated for persons exposed to asbestos inside and around the ruins of the WTC towers.

Other fibrous particles

As has been shown previously, fibers emitted from the asbestos-cement facades of the buildings migrate into indoor air, significantly elevating the concentration of fibrous aerosol in flats and creating a health risk, including a lung cancer risk. Because the carcinogenic properties of asbestos are most probably due to its fiber geometry, it can also be concluded that other fibers with the same characteristics may be carcinogenic (WHO 2000). Besides, different, non-asbestos fibers can also be treated as vehicles for fine, and ultrafine particles containing various toxic and carcinogenic compounds.

The micrograph showing airborne glass fibers (Fig. 5-16) seems to support this hypothesis. Precise demonstration of the carcinogenic properties of non-asbestos fibers could have very important implications in the future because during the last five years in a number of buildings in many countries, including Poland, the asbestos-cement sheets have been replaced by different isolating materials, mainly glass wool and other synthetic fibers. In fact, synthetic vitreous fibers (SVS), also called man-made mineral fibers (NEHC 1997), have been used extensively in residential and industrial settings for more than one century. They are used primarily for thermal and acoustical isolation, liquid and gas filtration, industrial textiles, and for reinforcing other materials. SVS include a very broad variety of inorganic fibrous substances with an amorphous (vitreous, i.e., non-crystalline) molecular structure. Traditionally, they have been arbitrarily divided

Table 5-3. Concentration of airborne asbestos in the damaged buildings A and B of WTC, New York, NY, USA (Chmielinski and Granger 2001. With permission).

	Transmission Electron Microscopy (EPA Method)		Phase Contrast Microscopy (NIOSH 74400 Method)
Location	Asbestos (> 5 μm) Concentration (structures/cc)	Asbestos (≤ 5 μm) Concentration (structures/cc)	Asbestos Concentration (F/cc)
Building A, Loading dock	< 0.007	0.255	0.049
Building A, Loading dock east freight elevator	0.155	0.464	0.053
Building A, 3rd floor, east side, office space	< 0.007	0.117	0.046
Building A, 11th floor, east side, office space	< 0.007	0.078	0.064
Building A, 36th floor, office 3619, NE corner	< 0.007	0.008	< 0.005
Building B, 3rd floor, office 3S-630	< 0.158	< 0.158	0.062
Building B, 3rd floor, north staging area, section 1	0.167	5.19	NA (1)
Building B, 6th floor, NE corner	< 171	3.43	0.247 (2)
Building B, 6th floor, SW corner	0.436	5.01	0.536 (2)
Building B, 9th floor, NE corner, 9E, 348	< 0.008	0.031	0.016
Building B, 9th floor, SW corner, 9W, 024	< 0.008	0.047	0.034

Note on laboratory report (SCILAB)

1) Analyst was unable to quantify fiber concentration because of overloading of particulate material on the filter.
2) Moderate to heavy particulate matter is present on the filter which may obscure some fibers causing results with a low bias.

into three general categories based on their composition and application (Hesteberg and Hart 2001): fiberglass (including glass wool and the thicker glass filament), mineral wool (rock, stone, and slag wool), and refractory ceramic fibers. However, in the last few years, these three categories have become antiquated; the categories are useless for a number of new "hybrid" SVS formulations and are irrelevant for hazard classification (Hesteberg and Hart 2001). The commercial production of these materials became especially important when the adverse health effects associated with asbestos prompted the search for a substitute material.

It should be noted that fiber-containing materials become dangerous only when the microscopic fibers are emitted from these materials into the air. Such phenomenon

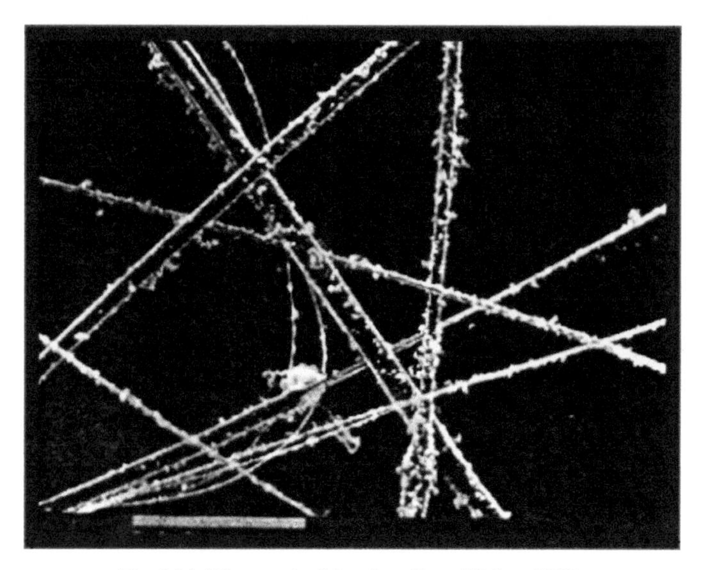

Fig. 5-16. Micrograph of the glass fibers (Walton 1995).

appears if fiber-containing materials become friable (crumbling), for example due to the atmospheric weather, or if building maintenance, repair, renovation or other activities (vibration and vandalism) disturb these materials. Therefore, it can be concluded that the important factor that generates, or considerably increases the emission of fibers from these fiber-containing materials used in buildings is the mechanical impact. Also in the industry (for example, in the factories producing isolating materials such as mineral wools or glass fibers), the health risk of workers exposed to fiber-containing materials is directly related with the condition of these materials and their property to emit airborne fibers due to mechanical contact (impact or/and vibration). Although the phenomenon of the mechanically generated emission of fibers from fiber-containing non-asbestos materials is still not satisfactorily recognized, recently Pastuszka (2010) determined the emission rate of fibers released from the new/fresh and used/worn ceramic fiber material, glass wool, and man-made mineral fiber material due to the mechanical impaction.

A simple experimental set-up, previously successfully used for the study of fibers emission from mechanically impacted asbestos-cement sheets has been applied to measure the emission of fibers from the selected materials. The samples of the studied materials having a surface area between 0.02 and 0.03 m^2 have been placed inside the cabinet adapted for the fiber emission experiments. The falling weight generated the emission of fibers from the samples of fiber-containing materials. The impaction energy was 0.4 J and 1.0 J. Such low impaction energy values were selected to simulate both: vibrations of the fiber-containing building facades caused by the turbulence of wind, and weak mechanical impacts made by workers during the renovation of the thermal isolation of buildings and industrial devices. An additional reason for this kind of preparation of the experiment was to keep the same conditions as in the previous study on the emission of fibers from the asbestos materials. The increasing concentration of fibers in the cabinet volume was measured by the Laser Fiber Monitor (MIE, Inc., Billerica, MA).

The emission rate was defined in this study as the number of fibers emitted from the unit mass of the investigated material due to the impaction of the unit impaction energy. This factor was calculated using the following equations (Pastuszka 2010):

$$\epsilon_m = \Delta C_{max} V/(m E) \tag{5-3}$$

where: ϵ_m [$g^{-1}J^{-1}$]—is the mass emission factor (using the traditional units [fibers/(g J)])

ΔC_{max} [fibers/m^3]—is the highest increase in the measured concentration of fibers inside the cabinet after impaction of 10 balls or one iron weight

V [m^3]—is the volume of the cabinet

m [g]—is the mass of the sample

E [J]—is the impaction energy

The emission rate was determined for the following, new/fresh and used man-made fiber-containing materials: glass wool, ceramic fibers, and man-made mineral fibers. The length distributions of the emitted fibers were also investigated.

The new/fresh samples were the production residuals while all samples of the used/worn fiber-containing materials were prepared from the materials previously used as a thermal isolating medium. The glass wool has been used for about 20 years as a thermal wrapper of the hot water pipe-line crossing over the ground. Other materials have been used in the industrial plants.

Table 5-4 shows the averaged values of the mass emission factor ϵ_m for three types of fiber-containing materials, fresh and used/worn: glass wool, ceramic fibers, and man-made mineral fibers.

Table 5-4. The mass-oriented emission rate of long (L > 5 μm) and short (L ≤ 5 μm) airborne fibers released from three kinds of fiber-containing materials due to the mechanical impact (Pastuszka 2010).

Numbers of samples studied	Material	ϵ_m [$g^{-1}J^{-1}$]*	
		L > 5 μm	L ≤ 5 μm
	Glass wool		
4	a) new (fresh)	6.9 x 10	2.0 x 10
4	b) old (used)	4.5 x 10	2.2 x 10
	Ceramic fiber		
5	a) new (fresh)	1.0×10^2	1.3 x 10
7	b) old (used)	3.0×10^3	2.4×10^2
	Man-made mineral fibers		
3	a) new (fresh)	2.2	2.2
4	b) old (used)	7.0×10^4	5.7×10^4

*In traditional units [fibers/g ·J].

It can be seen that the mechanically induced emission of long (L > 5 μm) and short (L ≤ 5 μm) fibers from the new/fresh materials is the lowest for the man-made mineral fibers. It should be noted that the emission factor of short fibers for all fresh materials is between 1 and 10 fibers/(g J) while the emission rate of long fibers shows more significant differences and ranged from 2.2 fibers/(g J) for material containing man-made mineral fibers through 69 fibers/(g J) for glass wool, up to 100 fibers/(g J) for ceramic fiber material.

It is interesting that the increase of the emission factor of used fiber-containing material highly depends on the type of material. The analysis of Table 5-4 indicates that the mass-oriented emission factor of the used mineral fiber material is ten thousand times higher in comparison with a fresh material while the emission rate of glass wool does not change with long term exploitation. The used ceramic fiber material has the emission factor ten times higher than the emission rate of the fresh material.

Figures from 5-17 to 5-19 show the length distributions of airborne fibers emitted during the mechanical impaction.

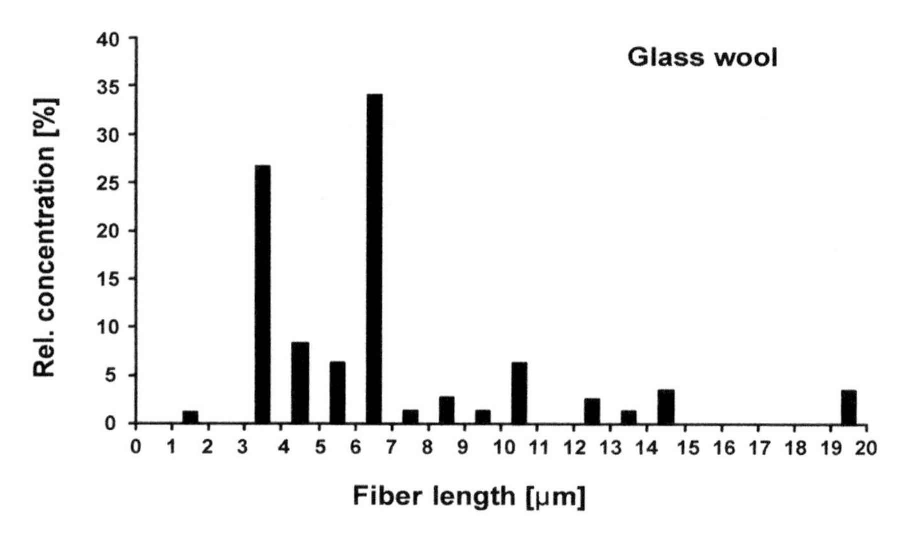

Fig. 5-17. Length distribution of the fibrous aerosol generated from the glass wool by the mechanical impaction (Pastuszka 2010).

These length distributions have been averaged for the data obtained from both new and old (used) materials. Although these distributions differ from each other it can be found that every distribution seems to be two-modal with a first peak appearing for short fibers (L ≤ 5 μm) and the second one observed in the 5–8 μm length range. In this context it is important to compare these results with the length distribution of airborne fibers emitted from asbestos-cement sheets (Fig. 5-12) which was one-modal, having a peak concentration level in the 4–5 μm length range.

It would be interesting to precisely describe how the exploitation conditions and exploitation time influence the emission factor. To find these relationships further studies are needed.

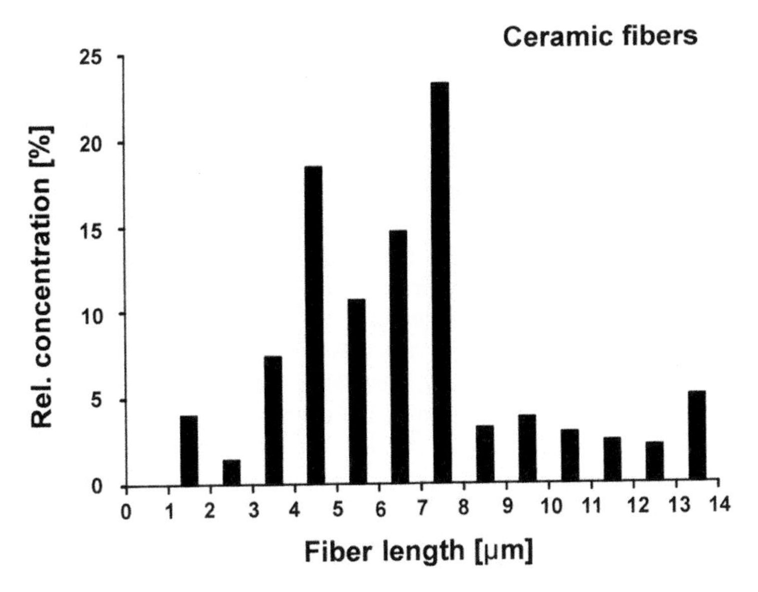

Fig. 5-18. Length distribution of the fibrous aerosol generated from the ceramic fiber-containing material by the mechanical impaction (Pastuszka 2010).

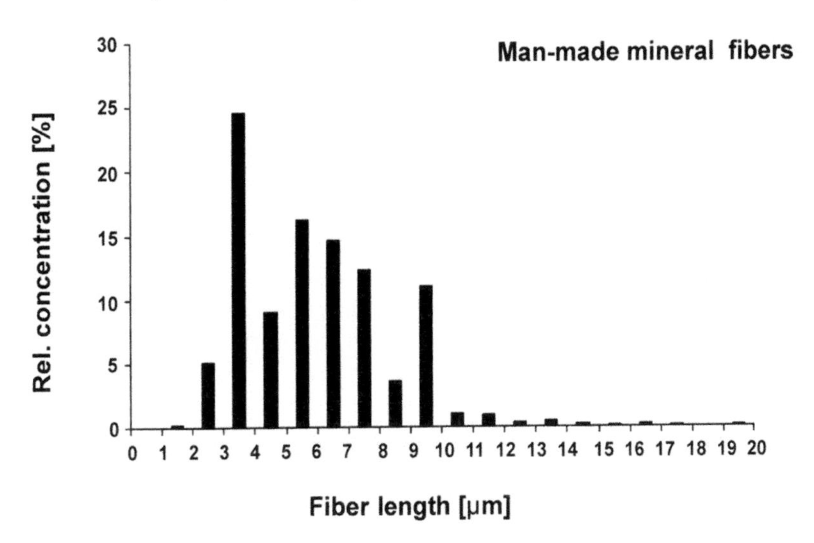

Fig. 5-19. Length distribution of the fibrous aerosol generated from a man-made mineral fiber-containing material by the mechanical impaction (Pastuszka 2010).

Summary of the existing knowledge

1. Inhaled asbestos fibers have been precisely shown to induce fibrosis, lung cancer, mesothelioma, and probably other kinds of intestinal cancer.
2. The health effects, their intensity, and carcinogenic risk depend on the dose, i.e., mainly on the concentration of airborne asbestos and on the time of exposure.

3. The most hazardous fibers (carcinogenic) are those longer than 5 μm and of diameter up to 3–4 μm.
4. The indoor concentration of fibrous aerosol between less than 100 up to 300 fibers/m^3 could probably be assumed as the background value of respirable fibers, longer than 5 μm in the urbanized and industrialized areas.
5. Because of the presence of asbestos fibers in the outdoor air, even in buildings constructed without asbestos materials, airborne asbestos fibers are detected but their concentration is generally below 1,000 F/m^3 (respirable fibers, longer than 5 μm) which creates the life-period cancer risk mostly at about 10^{-6} (acceptable level) for the buildings' occupants.
6. In the buildings with friable asbestos, including corroded asbestos-cement materials, concentrations vary irregularly: usually levels lower than 1,000 F m^{-3} are found, but in some cases the concentration level reaches more than 10,000 F m^{-3} which elevates the life-period cancer risk up to 10^{-4} (for non-smoking people).
7. Maintenance staff and cleaners may be exposed to a higher level of asbestos fibers—around 10^6 F m^{-3}. Similar concentration levels were found in some specific areas in the world. Such level of airborne asbestos in these areas, especially in flats, can create a dramatically high cancer risk level at about 10^{-2} for people living there.
8. The synergic effect between exposure to asbestos and tobacco smoke has been well documented. The cancer risk for smokers exposed to airborne asbestos is roughly ten times higher compared to non-smoking people exposed to asbestos of the same level.
9. Asbestos-cement slabs are susceptible to emitting fibrous particles generated by impact. Factors such as vibrations of the slabs caused by the turbulence/gust of wind can cause emission of fibers from the elevation of buildings made from asbestos-cement sheets.
10. The emission factor for long fibers increases with deterioration of the surface quality of the asbestos-cement facades, changing from 2.7 x 10^3 (Fibers) · $m^{-2}J^{-1}$ for the slabs with very good surface to 6.9 x 10^3 (Fibers) · $m^{-2}J^{-1}$ for the slabs with worn surface.
11. The length-distribution of fibers emitted from asbestos-cement slabs seems to be one-modal with the maximum ranging from 4–5 μm of fiber length and is similar to the length-distribution of airborne fibers obtained in the buildings with asbestos-cement facades.
12. Inside the buildings with asbestos-cement facades, the mean concentration of respirable fibers longer than 5 μm, is higher than in reference dwellings by approximately three times.
13. The carcinogenetic mechanism of asbestos fibers is still poorly recognized. There are three basic hypothesis concerning:

    a) chemical properties of asbestos (for example, presence of iron or other transition metals on fibers, ability of asbestos fibers to generate free radicals)
    b) carcinogenic compounds absorbed on the fibers
    c) physical properties of the asbestos fibers (mainly the fiber geometry)

14. It is possible that not only one but two or three carcinogenic mechanisms are responsible for the occurrence and development of cancer. If the mechanism (b) or/and (c) are true and sufficient to initiate the cascade of the processes leading to cancer, other, non-asbestos, respirable and long fibers should also be considered as potential carcinogens.
15. Various insulating materials made from natural or artificial fibers are susceptible to emitting fibrous particles generated by impact.
16. The susceptibility of worn/used fiber-containing materials to emitting fibrous particles due to a mechanical impaction is significantly diverse.
17. The length-distribution of fibers emitted from materials containing man-made fibers is two-modal, with the first maximum appearing for short fibers ($L \leq 5$ µm) and with the second maximum in the range of fibers of 5–8 µm long.
18. The method of the estimation of the emission rate, described in this chapter, could be useful in the assessment of the health risk related to human contact with the fiber-containing materials.

Aims of the future studies

1. The influence of the physical properties of the inhaled fibers on the support of the migration of chemical compounds contained in the fibers, especially on their surface, into the lung cells, seems to be the fundamental problem for understanding the fiber toxicity and carcinogenicity.
2. Explanation of the synergy effects between airborne fibers and other pollutants (spherical particles and gases) is probably crucial for understanding the total health effects resulting from the exposure to a mixture of various airborne pollutants, otherwise being a typical situation in the environment.

Acknowledgements

The author is grateful to Dr. Hugh Granger and Piotr Chmielinski, M.Sc., CIH, from the HP Environmental, Inc. Virginia, USA, for the results of their study of the air pollution in the ruins of WCT, New York, NY, USA. The author also thanks Dr. Ing. Marta Różkowicz, from the Central Institute of Mining, Katowice, Poland, for her micrographs of asbestos fibers.

References

Ansari, F.A., I. Ahmad, M. Ashqin, M. Yunus and Q. Rahman. 2007. Monitoring and identification of airborne asbestos in unorganized sectors, India. Chemosphere 68: 716–723.

Ben-Shlomo, R. and U. Shanas. 2011. Genetic ecotoxicology of asbestos pollution in the house mouse Mus musculus domesticus. Environ. Sci. Pollut. Res. 18: 1264–1269.

Bignon, J., J. Peto and R. Saracci [eds.]. 1989. Non-Occupational Exposure to Mineral Fibers (IARC Scientific Publications No. 90). International Agency for Research on Cancer, Lyon, France.

Boelter, F., C. Simmonds and P. Hewett. 2011. Exposure data from multi-application, multi-industry maintenance of surfaces and joins sealed with asbestos-containing gaskets and packing. J. Occup. Environ. Hyg. 8: 194–209.

Bornemann, P. and U. Hildebrandt. 1986. On the problem of environmental pollution by weathering products of asbestos cement. Staub Reinhalt. Luft 46: 487–489.

Brown, S.K. 1987. Asbestos exposure during renovation and demolition of asbestos-cement clad buildings. Am. Ind. Hyg. Assoc. J. 48: 478–486.

Brown, S.K. and J.A. Hoskins. 1992. Contamination of indoor air with mineral fibres. Indoor Environment. 1: 61–68.

Burmeister, B., T. Schwerdtle, I. Poser, E. Hoffmann, A. Hartwig, W.U. Müller, A.W. Rettenmeier, N.H. Seemayer and E. Dopp. 2004. Effects of asbestos on initiation of DNA damage, induction of DNA-strand breaks, P53-expression and apoptosis in primary, SV40-transformed and malignant human mesothelial cells. Mutation Research 558: 81–92.

Chmielinski, P. and H. Granger. 2001. Personal communication.

Corn, M. 1994. Airborne concentrations of asbestos in non-occupational environments. Ann. Occup. Hyg. 38: 495–502.

Crossman, Jr., R.N., G. Williams, Jr., J. Lauderdale, K. Schosek and R.F. Dodson. 1996. Quantification of fiber releases for various floor tile removal methods. Appl. Occup. Environ. Hyg. 11: 1113–1124.

Crump, K.S. and D.B. Farrar. 1989. Statistical analysis of data on airborne levels collected in an EPA survey of public buildings. Regul. Toxicol. Pharmacol. 10: 51–62.

[EPA] US Environmental Protection Agency. 1988. Integrated Risk Information System: Asbestos. http://www.epa.gov/iris/subst/0371.htm (accessed 28.09.2007).

[EPA] US Environmental Protection Agency. 1999. Asbestos-containing material in school. 40CFR 763.

Ganor, E., A. Fischbein, S. Brenner and P. Froom. 1992. Extreme airborne asbestos concentrations in a public building. Br. J. Ind. Med. 49: 486–488.

Gasparrini, A., A.M. Pizzo, G. Gorini, A.S. Constantini, S. Silvestri, C. Ciapini, A. Innocenti and G. Berry. 2008. Prediction of mesothelioma and lung cancer in a cohort of asbestos exposed workers. Eur. J. Epidemiol. 23: 541–546.

Gerde, P. 1987. A Model for the Uptake of Polycyclic Aromatic Hydrocarbons by the Human Lung—Influence of Inhaled Mineral Fibers. Department of Chemical Engineering, Royal Institute of Technology, Stockholm. Sweden.

Gilbertson, K.E., W.H. Finlay, C.F. Lange, M.J. Brett, D. Vick and Y.S. Cheng. 2005. Generation of fibrous aerosols from thin films. J. Aerosol Sci. 36: 933–937.

Gold, J., H. Amandusson, A. Krozer, B. Kasemo, T. Ericsson, G. Zanetti and B. Fubini. 1997. Chemical characterization and reactivity of iron chelator-treated amphibole asbestos. Environ. Health Perspect. 105(Suppl. 5): 1021–1030.

[HEI] Health Effects Institute. 1991. Asbestos in Public and Commercial Buildings: A literature Review and Synthesis of Current Knowledge. Cambridge, MA, USA.

Hesteberg, T.W. and G.A. Hart. 2001. Synthetic vitreous fibers: a review of toxicology research and its impact on hazard classification. Crit. Rev. Toxicol. 31: 1–53.

Janeczek, J., T. Noszczyk and A. Obminski. 2001. Contribution of asbestos in the total count of tespirable fibers in ambient air in Sosnowiec (in Polish). Ochrona Powietrza i Problemy Odpadów 35: 226–228.

Kadiiska, M.B., A.J. Ghio and R.P. Mason. 2004. ESR investigation of the oxidative damage in lungs caused by asbestos and air pollution particles, Spectrochimica Acta Part A 60: 1371–1377.

Kane, A.B. 2003. Asbestos bodies: clues to the mechanism of asbestos toxicity? Human Pathology 34: 735–736.

Kazan-Allen, L. 2005. Asbestos and mesothelioma: worldwide trends. Lung Cancer 49 S1: S3–S8.

Kim, Y.-Ch,, W.-H. Hong and Y.-L. Zhang. 2015. Development of a model to calculate asbestos fibers from damaged asbestos slates depending on the degree of damage. J. Cleaner Production. 86: 88–97.

LaDou, J. 2004. The asbestos cancer epidemic. Environ. Health. Persp. 112: 285–290.

Lee, R.J. and D.R. Van Orden. 2008. Airborne asbestos in buildings. Regul. Toxicol. Pharmacol. 50: 218–225.

Madl, A.K., D.M. Hollins, K.D. Devlin, E.P. Donovan, P.J. Dorpat, P.K. Scott and A.L. Perez. 2014. Airborne asbestos exposures associated with gasket and packing replacement: a simulation study and meta-analysis. Regul. Toxicol. Pharmacol. 69: 304–319.

Maxim, L.D. and E.E. McConnell. 2001. Interspecies comparisons of the toxicity of asbestos and synthetic vitreous fibers: a weight-of-the-evidence approach. Regul. Toxicol. Pharmacol. 33: 319–342.

Moolgavkar, S.H., J. Turim, R.C. Brown and E.G. Luebeck. 2001. Long man-made fibers and lung cancer risk. Regul. Toxicol. Pharmacol. 33: 138–146.

Mossman, B.T. and J.E. Craighead. 1981. Mechanisms of asbestos carcinogenesis. Environ. Res. 25: 269–280.

Mossman, B.T., P.J. Borm, V. Castranova, D.L. Costa, K. Donaldson and S.R. Kleeberger. 2007. Mechanisms of action of inhaled fibers, particles and nanoparticles in lung and cardiovascular diseases. Particle and Fibre Toxicology 4: 4 doi:10.1186/1743-8977-4-4.

[NEHC] Navy Environmental Health Center. 1997. Man-Made Vitreous Fibers. Navy Environmental Health Center, Norfolk, Virginia, USA.

[NIOSH] National Institute of Occupational Safety and Health. 1994. Asbestos and other fibers by PCM. *In*: P.M. Eller [ed.]. NIOSH Manual of Analytical Methods, Method 7400, 4th, NIOS, Washington, D.C.

[OSHA] Occupational Safety and Health Administration. 1994. Occupational exposure to asbestos; final rule. Federal Register 59: 40964–41158.

Pastuszka, J.S. 1997. Asbestos fibers in the indoor environment. pp. 261–280. *In*: G.A. Peters and B.J. Peters [eds.]. Sourcebook on Asbestos Diseases. LEXIS Publishing, Charlottesville, Virginia, USA.

Pastuszka, J.S. 2009. Emission of airborne fibers from mechanically impacted asbestos-cement sheets and concentration of fibrous aerosol in the home environment in Upper Silesia, Poland. Journal of Hazardous Materials 162: 1171–1177.

Pastuszka, J.S. 2010. Emission of airborne fibers from mechanically impacted non-asbestos fiber-containing materials: preliminary results. Archives of Environmental Protection 36: 3–12.

Pastuszka, J.S., J.A. Sokal and R.L. Górny. 1993. Fibrous and particulate pollution of indoor air in Upper Silesia, a highly industrialized Polish region. Preliminary results. pp. 17–22. *In*: P. Kalliokoski, M. Jantunen and O. Seppänen [eds.]. Proceedings of Indoor Air`93, Vol. 4. Helsinki, Finland.

Pastuszka, J.S., A. Kabała-Dzik and K.T. Paw U. 1999. A study of fibrous aerosols in the home environment in Sosnowiec, Poland. Sci. Total Environ. 229: 131–136.

Pastuszka, J.S., K.T. Paw U., A. Kabała-Dzik, N. Kohyama and J.A. Sokal. 2000. Respirable airborne fibers in the home environment in Upper Silesia, Poland, compared with Davis, California. J. Aerosol Sci. 31(Suppl. 1): 484–485.

Paustenbach, D.J., A. Sage, M. Bono and F. Mowat. 2004. Occupational exposure to airborne asbestos from catings, mastics, and adhesives. J. Expo. Anal. Environ. Epidemiol. 14: 234–244.

Peters, G.A. and B.J. Peters [eds.]. 1997a. The Treatment and Prevention of Asbestos Diseases, Volume 15 of the Sourcebook on Asbestos Diseases: Medical, Preventive and Socio-Economic Aspects. LEXIS Publishing, Charlottesville, Virginia, USA.

Peters, G.A. and B.J. Peters [eds.]. 1997b. Asbestos and Cancer, Volume 16 of the Sourcebook on Asbestos Diseases: Medical, Preventive and Socio-Economic Aspects. LEXIS Publishing, Charlottesville, Virginia, USA.

Podgórski, A. and L. Gradoń. 1998. Mechanics of a deformable fibrous aerosol particle: general theory and application to the modeling of air filtration. pp. 193–218. *In*: K. Spurny [ed.]. Advances in Aerosol Filtration. Lewis Publishers, Boca Raton, Florida, USA.

Różkowicz, M. 2007. Preparation of the complex method of monitoring of airborne asbestos. Ph.D. Thesis, Central Mining Institute, Katowice, Poland.

Spencer, J.W., M.J. Plisko and J.L. Balzer. 1999. Asbestos fiber release from the brake pads of overhead industrial cranes. Appl. Occup. Environ. Hyg. 14: 397–402.

Spurny, K.R. 1989. On the release of asbestos fibers from weathered and corroded asbestos cement products. Environ. Res. 48: 100–116.

Sullivan, P.A. 2007. Vermiculite, respiratory disease, and asbestos exposure in Libby, Montana: update of a cohort mortality study. Environ. Health Perspect. 115: 579–585.

Suzuki, Y., S.R. Yuen and R. Ashley. 2005. Short, thin asbestos fibers contribute to the development of human malignant mesothelioma: pathological evidence. Int. J. Hyg. Environ.-Health 208: 201–210.

Topinka, J., P. Loli, P. Georgiadis, M. Dušinská, M. Hurbánková, Z. Kovácikoví, K. Volkovová, A. Kažimirová, M. Baranckovà, E. Tetrai, D. Oesterle, T. Wolff and S.A. Kyrtopoulos. 2004. Mutagenesis by asbestos in the lung of ë-lacI transgenic rats. Mutat. Res. 553: 67–78.

[USEPA] United States Environmental Protection Agency, Office of Solid Waste and Emergency Response. Memorandum of June 9, 2002 on the subject: Testing Carpet, the Asbestos Reservoir, Washington, D.C., USA.

Walton, W.H. 1991. Airborne dust. pp. 55–77. *In*: D. Liddell and K. Miller [eds.]. Mineral Fibers and Health. CRC Press, Boca Raton, Florida, USA.

[WHO] World Health Organization. 1987. Asbestos and other natural mineral fibers. *In*: Environmental Health Criteria, No. 53. Geneva, Switzerland.

[WHO] World Health Organization. 2000. Air Quality Guidelines for Europe, Second Edition. Regional Office, Copenhagen, Denmark.

Wozniak, H., E. Więcek, W. Pelc, D. Dobrucka, M. Król and B. Opalska. 1994. Respirable mineral fibers in the atmospheric air of city of Wroclaw (in Polish). Medycyna Pracy 3: 239–247.

Xu, A., X. Huang, Y.C. Lien, L. Bao, Z. Yu and T.K. Hei. 2007. Genotoxic mechanisms of asbestos fibers: role of extranuclear targets. Chem. Res. Toxicol. 20: 724–733.

Yada, K. 1971. Study of microstructure of chrysotile asbestos by high resolution electron microscopy. Acta Cryst. A27: 659–664.

Zurer, P.S. 1985. Asbestos, the fiber that's panicking America. C&EN Special Report. March 4: 28–41. American Chemical Society, Washington D.C., USA.

6

Fungi, Bacteria, and Other Biopollutants

Aino Nevalainen, Martin Täubel* and *Anne Hyvärinen*

Definitions

Biological particles are a diverse group of particulate matter present almost everywhere, in the atmospheric environment, soils and waters, and in indoor environments. Common to all of these particles is the biological origin: microbial, plant, or animal sources. Examples of microbial particles are moulds and yeasts, bacteria, and viruses; particles of plant origin are pollen and algae, and various particles also originate from animals (including humans). Biological particles may be viable or non-living, dead material. The health effects and biological characteristics of these particles are also diverse; pollen and microbial spores being essential parts of the organism's multiplication system, some viruses and bacteria having dramatic potential as pathogenic organisms, and some components representing other types of health-related biological activity, such as plant and fungal allergens and bacterial endotoxin.

Biological particles are present in the atmosphere, and assessments of their concentrations have been made for a long time. Using culture-based sampling and analysis methods for fungi and bacteria, and direct counting methods for pollen, their occurrence has been monitored and their transport across regions and even continents documented. However, only the recent development of DNA based methods has really made it possible to comprehensively characterize the atmospheric microbiomes and their behavior. Thus, the importance of biological particles in the atmosphere has only been possible to recognize during the last decades. Research has suggested that they

National Institute for Health and Welfare, Department of Health Protection, Neulaniementie 4, FI-70701 Kuopio, Finland.
* Corresponding author: aino.nevalainen@thl.fi

can have a substantial influence on atmospheric chemistry, cloud development, and precipitation (Andreae and Rosenfeld 2008; Després et al. 2012).

The size range of airborne biological particles extends from nanometers, the size of cell fragments and viruses, to around 100 μm and more (Després et al. 2012; Reponen et al. 2011), the size of some pollen, plant debris, and aggregates of biological material. Part of the biological particles occurs as individual particles in the air. Examples of such particles are pollen and fungal spores. Traces of plant and their allergen material may also be carried by other particles. Fungal and bacterial spores may occur as individual particles but also as aggregates or they may be carriers of microbial products, such as mycotoxins or bacterial equivalents. Fungal and bacterial material may also be present in air as smaller particles than the original cells (Górny et al. 2002; Green et al. 2006).

Fungi and bacteria may occur as viable particles, which means that they are metabolically active and able to reproduce once the environmental conditions for nutrients, moisture, and temperature are good enough, or they may occur as non-viable (dead) particles, or in a dormant state which is something in between these statuses. Even dead or dormant cells may carry biologically active constituents, such as endotoxin, glucan, allergens, or toxins; among the characteristics that have importance to health, only the ability to cause infections is known to be bound to the viability of the causal organism.

Sources, description, and occurrence of biological particles

Fungi—Introduction

Fungi are eukaryotic organisms that form their own biological Kingdom, having features both from micro-organisms (bacteria, viruses, protozoa) and those of plants. Fungi are classified in groups based on their method of growth and multiplication. Yeasts are a group of single cell fungi that differ from other fungi; they multiply by dividing cells only. Filamentous fungi form filaments (hyphae) that aggregate to form fungal mycelium. This can often be seen by naked eye, usually known as mold or mildew. Fungal mycelium can transport nutrients and water and thus allow the growth in sites where these supplies are scarce. Mycelium also develops reproductional structures and releases spores that may travel long distances to find new sites for growth. In general, fungal spores are present everywhere, and their germination and growth are regulated by the environment, i.e., availability of nutrients and water, adequate temperature, and a proper surface.

Fungi contribute about one quarter of the total biomass in the planet. Their basic ecological role is to aerobically decay organic material in nature, which is a crucial basis for the nutrient cycle, and to provide plants the necessary nutrients for their growth. Fungi consist of different groups: yeasts, molds and mildews, commonly known as microfungi, and mushrooms and other fungi. Microfungi are common both in outdoor and indoor environments. Fungal spores and also other fungal material, such as hyphal fragments, are easily released into the air. Their importance to mankind is both potentially harmful and beneficial. They may act as human pathogens or allergens, plant pathogens destroying crops and forests, and damage buildings as do many molds and rot fungi. However, fungi are also used in biotechnological industries as producers of enzymes, antibiotics and other pharmaceuticals, and in production of cheese, wine, beer, milk products, and baking products.

Some structural components of fungal cells have been identified as biologically active agents and are therefore interesting from the health point of view. Beta-(1->3)-D-glucans (Douwes 2005) are regular constituents of fungal cells and potential agents contributing to the health effects linked with fungal exposures. They have inflammatory potential (Douwes 2005) but may also provide protective effects against respiratory allergies similarly to bacterial endotoxin (Yossifova et al. 2007; Gehring et al. 2007). Fungi that contain known allergens are, e.g., *Cladosporium*, *Aspergillus*, *Alternaria*, and *Fusarium* species. Another structural component of fungal cells, ergosterol, which is the common sterol for all fungi, is used as a chemical marker in assessment of total fungal biomass, but its possible role in health risks of fungal exposures is poorly known.

On top of intracellular components of fungal cells, fungi also have great potential for producing extracellular secondary metabolites. As their primary metabolism aims to provide proteins, carbohydrates, nucleic acids, and lipids for their own growth, secondary metabolism produces volatile compounds, also known as MVOC (microbial volatile organic compounds) and non-volatile compounds often designated as mycotoxins. MVOC compounds are typically organic acids, alcohols, esters, and aldehydes and their mixtures are often recognized as odor of mold, earth, or cellar. These volatile compounds are usually not acutely toxic, but they may have irritative effects (Korpi et al. 2009). Instead, the group of non-volatile secondary metabolites, mycotoxins, together with bacterial exotoxins, represent a more potential group of agents with possible health effects, and they are introduced as a separate section below.

Traditional methods of analyzing fungi are largely based on either direct microscopy or on their cultivation on laboratory media and subsequent morphological identification by microscopy. The identification relies upon the characteristic hyphal and reproductive structures of each fungus. With these methods, those fungi that can be cultivated on the given media and growth conditions can be found and quantified. Examples of common fungal genera in outdoor and indoor air are listed in Table 6-1.

Sources and occurrence of fungi

Essential sources of environmental fungi are all natural environments that provide surface and substrate for their growth, that is organic material and enough moisture. Therefore, plants and soils of the planet are the major sources of all fungal material. The more vegetation, the more organic cycle of nutrients and the more fungi that do most of the decaying work of organic material. However, even the driest deserts and coldest areas of the world have their natural mycobiota although the scarce biological action in such extreme environments means that they are no major sources of reproducing fungi.

All waters contain fungi as well, but they are not nearly as important growth sites for fungi as are all kinds of plants and soils. From these sources, fungal spores, fragments, and other particles end up into the atmosphere where they can travel long

Table 6-1. Examples of fungal genera commonly occurring in outdoor and indoor environments as detected with culture methods. Many fungal genera, particularly *Aspergillus* and *Penicillium*, are represented by a number of different species.

Acremonium
Alternaria
Aspergillus
Aureobasidium
Botrytis
Chaetomium
Cladosporium
Epicoccum
Eurotium
Exophiala
Fusarium
Geomyces
Geotrichum
Mucor
Oidiodendron
Paecilomyces
Penicillium
Phialophora
Phoma
Rhizopus
Rhodotorula
Scopulariopsis
Sporobolomyces
Stachybotrys
Trichoderma
Tritirachium
Ulocladium
Wallemia

distances to the other side of the globe (Smith et al. 2012). Fungal spores are found in atmospheric layers (Polymenakou 2012; Després et al. 2012). Wind, rain, and other weather factors influence their traffic in the same way as that of other atmospheric particles. Concentrations and composition of the mycobiota at a given monitoring site is not stable but have relatively large variation due to geographical, weather, and light conditions (Bowers et al. 2013). Examples of concentrations of cultivable fungi in outdoor air are shown in Table 6-2.

The fungal biomass in the environment that cannot be cultivated under laboratory conditions is greater both quantitatively and qualitatively than the "viable" or "cultivable" fungal material. The ratio of viable and non-viable fungal biomass depends on the environment, i.e., whether the sample has been taken from soil, air, water, house dust, or another source. The fact that only a part of all microbes can be cultivated has been known for long by analyzing fungal content of samples, e.g., by direct microscopy without cultivation. For example, in aerobiological monitoring of pollen and fungal allergens, certain types of spores and fungal groups can be quantified by this way. However, the development of DNA-based methods during the past decades has made a methodological revolution in making it possible to characterize the fungal biomass independently from its viability. The methods for microbial analyses are further discussed later in this chapter.

Table 6-2. Examples of concentrations of viable fungi in various environments.

Environment	Concentration CFU m^{-3}	Reference
Outdoor air	10^2–10^4	1, 2, 3
Homes in different countries	10^1–10^3	4, 5, 6, 7
School buildings	< 10^1–10^3	8, 9, 10
Waste collection work	< 10^2–10^5	11
Composting facility	10^3–10^7	12
Wood chip handling	10^3–10^7	13
Swine confinement buildings	10^2–10^5	14
Demolition work of moldy structures	10^3–10^5	15

1. Macher et al. 1991, 2. Kuo and Li 1994, 3. Ren et al. 1999, 4. Haas et al. 2007, 5. Frankel et al. 2012, 6. Reponen et al. 1992, 7. Pastuszka et al. 2000, 8. Meklin et al. 2002, 9. Gravesen et al. 1983, 10. Levetin et al. 1995, 11. Moller-Nielsen et al. 1997, 12. Durand et al. 2002, 13. Alwis et al. 1999, 14. Rautiala et al. 2003, 15. Rautiala et al. 1996.

The estimates of the portion biological particles represent in atmospheric particle mass vary. For example, biological particles > 0.2 μm have been assessed to comprise 5–50% of outdoor particles (Jaenicke 2005). Fungal spores have been assessed to occur in number and mass concentrations ~ 10^4 m^{-3} and ~1 μg m^{-3}, respectively, and are estimated to account for up to ~ 10% of organic carbon and ~ 5% of PM10 at urban or suburban locations (Després et al. 2012).

Fungi in indoor environments

Indoor environments are where humans today mostly spend their time, and therefore most contacts with airborne biological particles take place probably in these environments. This is critical from the point of view of health effects assessment. In general, outdoor air is the main source of fungi commonly occurring in indoor environments (Gravesen 1979; Burge 1990). Depending to some extent on the climate, geography, and the ventilation system of the building, normal mycobiota of an indoor environment resembles mainly that of the local outdoor air. Outdoor fungi enter the indoor environments not only through open windows and doors but also through cracks in the structures and via ventilation system as do other outdoor particles. Fungi are also carried indoors on clothing and shoes of occupants, and on fur of pets (Pasanen et al. 1989).

On top of the background concentrations and mycobiota originating from outdoors, there are also intramural sources of fungi that make their contribution to the concentrations and species diversity of indoor air and to the exposures of humans. Some sources of indoor fungi are presented in the following.

Normal household activities have an effect on indoor fungal concentrations and species. Measured as viable fungi, increase of fungal concentrations of orders of magnitude was observed as a result of handling of firewood or opening a cellar door. Increased concentrations were also shown as a result of cleaning and handling fruit and unwashed root vegetables. The major viable fungal genera observed in such concentration peaks were *Penicillium*, *Cladosporium*, and *Aspergillus*, in addition to yeasts (Lehtonen et al. 1993). Temporary wetting of spots on various surfaces where fungal spores have accumulated may also allow the germination and hyphal growth of rapidly growing fungi, creating microcolonies that can produce spores into the indoor air. This phenomenon may only take hours or days (Pasanen et al. 1991). Common sources of indoor fungi are also sites or areas in the building structures where moisture can be accumulated for one reason or another, e.g., by leaks, condensation, flooding, or capillary movement of moisture. Such moisture causes growth of fungi and bacteria, which is recognized as mold with naked eye. Growth of mold on indoor surfaces or in building structures is associated with many health effects (WHO 2009) and considered therefore an undesired phenomenon which acts as a source of fungal particles, volatile metabolites, and toxic substances (Adan and Samson 2011).

It has been discussed whether a wall-to-wall carpet acts as a source or as a sink of indoor fungi. Evidently, both are possible and the phenomenon depends both on the characteristics of the carpet and on the fungal species in question. In an experimental study, Shorter (2012) showed that both PM2.5 and fungal spores were aerosolized from a hard floor more easily than from a carpet surface. Almost no reaerosolization took place from a carpet loaded with low levels of dust, but when higher loading was used, differences in resuspension between fungal spore types were noticed. Small and spherical *Penicillium* spores were more easily aerosolized than larger *Alternaria* spores (Shorter 2012).

Microbial diversity in house dust has been characterized in a number of studies using cultivation methods, and in a few studies using non-culturable methods. Especially studies using culture-independent methods have shown that in indoor environments, there are various sources and reservoirs for microbial material. According to the review on microbes in house dust by Rintala et al. (2012), the studies using culturable methods have identified the fungal genera *Penicillium*, *Aspergillus*, *Cladosporium* and approximately 20 other genera as the most commonly isolated fungal genera in house dust. As expected, culture-independent studies have shown that both fungal and bacterial diversity is in fact far more extensive, with up to 500–1000 different species present in house dust (Rintala et al. 2004; Pitkäranta et al. 2008; Amend et al. 2010; Adams et al. 2013).

While indoor air concentrations of viable fungi usually remain inside the range 10^1–10^3 CFU/m^3, there are occupational environments where fungal exposures may exceed these with several orders of magnitude (Tsapko et al. 2011; Gora et al. 2009). Very large concentrations have been reported in agricultural environments, food industry, wood processing, waste sorting, and other occupations where organic material is being handled or processed. Examples of fungal concentrations in such environments are shown in Table 6-2.

These concentrations are also larger than those in normal outdoor air, and they are linked with many occupational diseases, such as allergic alveolitis.

Bacteria—Introduction

Bacteria are ubiquitous microorganisms present in soils, waters, plants, and other living organisms including humans (Starr et al. 1981) and they are also easily spread into the air (Després et al. 2012). Bacteria are prokaryotic organisms which replicate asexually by cell division. Some bacteria produce endospores that are resistant to environmental stresses and that may be activated for replication under suitable environmental conditions. Bacteria may also stay in a dormant state in order to ensure survival (Colwell 2000). Bacteria can be classified into different groups, such as gram negative and gram positive bacteria, depending on their ability to retain Gram stain, and into rod-shaped bacteria or spherical cocci based on morphological criteria. Actinomycetes are separated from other gram positive bacteria by their ability to form a mycelium and sporulate similarly to fungi.

Bacterial cells are typically of the size range from 1 to a few micrometers. Bacteria may occur in air as single cells and as aggregates or attached on other particles. Bacterial spores are usually around 1 μm in their size, being smaller than those of fungi.

Bacteria are found in the atmospheric environment and also in the troposphere layers (Maki et al. 2013). Their concentrations have large temporal variability, the concentrations being lower in winter than in other seasons (Bertolini et al. 2013). Mean concentrations of bacteria may be more than $10^4\,m^{-3}$ over land areas, corresponding to 0.03% of the organic carbon content of the aerosol (Bauer et al. 2002), while the concentrations over sea are much lower (Prospero et al. 2005). Analyzed with cultural methods, concentrations of airborne bacteria may vary between 10^2–10^5 CFU m^{-3}. The bacteria types most often reported are *Firmicutes*, *Proteobacteria*, and *Actinobacteria*. However, conclusions on the relative abundance of various bacteria in the atmosphere are difficult due to high short-term variability and biases from the different detection methods (Després et al. 2012).

Bacteria, similarly to fungi, can be analyzed and characterized by culture based methods, which allow the studying of various biochemical and physiological characteristics of individual organisms. Culture based methods are also used for quantification and phenotypical description of bacterial communities. However, only a part of all bacteria can be cultivated in laboratory conditions (Rappe and Giovannoni 2003), and therefore cultural methods may seriously underestimate the real magnitude of bacterial concentration in a given sample or environment. Non-culture-based methods which may be based on DNA analysis, use of chemical markers of bacterial biomass, or on immunological bioassays cover both viable and non-viable material and thus give a more accurate estimate of bacterial biomass or diversity.

Bacteria in indoor environments

For bacteria of indoor environments, outdoor air is an important source. However, major sources of indoor bacteria are humans and animals, such as pets. The important role of humans as sources of indoor bacteria is based on the fact that human skin is practically covered with beneficial, normal bacteria that protect the skin from harmful agents from outside. The outer layers of skin are continuously renewed, approximately every 4th day. As the skin scales that are then emitted into the individual's intimate environment carry bacterial colonies, the presence of humans is always seen as elevated levels of bacteria in any indoor environment. This human effect on indoor bacteria has been shown both using cultural methods, and by using non-cultural methods such as chemical markers of bacteria (Fox et al. 2005). The cultivable bacterial flora of the indoor environment is dominated by Gram-positive genera, such as *Staphylococcus*,

Corynebacterium, and *Lactococcus* (Rintala et al. 2012), which are also common bacteria on human skin. Studies on indoor sources of bacteria have indeed shown that bed dust is typically dominated by species originating from the user of the mattress, whereas floor dust reflects more also outdoor sources (Täubel et al. 2009). Recent studies of indoor air using DNA based methods have come to similar conclusions as concerns the role of human occupants on indoor bacterial communities (Hospodsky et al. 2012; Meadow et al. 2013; Dunn et al. 2013).

High concentrations of bacteria are present in occupational environments where organic material is being handled, similarly to fungi and other biological particles (Gora et al. 2009). As fungi, bacteria, and other organisms occur together in various organic dusts, the exact role of each type of microorganisms in development of associated occupational diseases is not yet clear. Spore-forming actinomycetes, especially thermophilic species such as *Thermoactinomyces vulgaris*, are strongly associated with allergic alveolitis caused by moldy hay (doPico 1986).

Bacterial endotoxins

Bacterial endotoxin is a well-known component of the outer membrane of gram negative bacteria. It consists chemically of proteins, lipids, and lipopolysaccharide (LPS) but the term "endotoxin" is often used to emphasize its immunotoxic properties, which are also utilized in the quantification of endotoxin by the immunochemical *Limulus* assay (Heinrich et al. 2003). Water soluble LPS contains a hydrophobic lipid A and a long covalently linked hydrophilic heteropolysaccharide. Each genus, species, and strain of gram negative bacteria has a unique LPS with varying biological effects, although bacteria belonging to the same family typically have structurally similar LPS (Morrison and Ulevich 1978). Endotoxins are stable and heat-resistant molecules. As they are normal constituents of gram negative bacteria, which are present in practically all environments, also endotoxins are present everywhere. Endotoxins can occur in the air both as pure endotoxin, as parts of bacterial cells and carried by other particles. These compounds have significant inflammatory potential which may lead to various respiratory health effects among exposed individuals.

Microbial toxins in indoor environments

As presented above, bacterial endotoxins and glucans of fungi are normal constituents of bacterial and fungal cells and they have immunotoxic potential that is probably contributing to the health effects associated with exposures to biological particles and dusts. Another type of toxic compound linked with bacteria and fungi are microbial toxins that are produced as secondary metabolic products of these organisms. Toxins produced by fungi are called mycotoxins, but many species of both fungi and bacteria have the potential of producing highly bioactive and toxic secondary metabolites.

Mycotoxins are fungal secondary metabolites that pose a potential health risk to humans and/or animals due to the unusual toxicity of many such metabolites. The occurrence of toxigenic fungi in indoor environments is well documented as well as their ability to proliferate on various building materials, provided they are moist

(Nielsen 2003; Jarvis and Morey 2001). Indoor environments with moisture damage or dampness generally provide good growth conditions for microbes, which is also seen as higher microbial levels in such environments. However, toxigenic molds are present also in 'normal' buildings without moisture problems and thus the occurrence of mycotoxins is not only limited to damp indoor environments. Among the building associated toxigenic fungi are various *Aspergillus* and *Penicillium* species, *Chaetomium globosum*, *Wallemia sebi*, *Eurotium* spp., *Trichoderma* spp., and *Stachybotrys chartarum*, to give a few examples. As for bacteria, the detection of elevated levels of culturable *Streptomyces* spp. in buildings has been linked to moisture damage. Many species of *Streptomyces* are potent producers of bioactive metabolites with antimicrobial and immune-suppressing qualities and these bacteria are widely used in the pharmaceutical industry.

Secondary metabolites of microbes are different from primary metabolites—such as amino acids, sugars, carbohydrates—in that these compounds do not act directly in the process of microbial growth. The ecological function of secondary metabolites is either in initiation of growth and differentiation, or they act in competitive interactions between microorganisms to ensure survival, and facilitate the entry of the infectious fungus into the plant cell. There is a wide range of chemical structures and different biological activities within fungal and bacterial secondary metabolites, such as sporogenic factors, pigments, antibiotics, insecticides, or herbicides, and these compounds are produced via organized sets of genes following controlled mechanisms.

To date, several hundreds of different mycotoxins have been identified and characterized, the majority of which refers to food and feed contaminants in agricultural settings; however, up to 20,000 mycotoxins are estimated to be present in the environment. Examples of common and well characterized mycotoxins are sterigmatocystin and aflatoxin, ochratoxin, and macrocyclic trichothecenes. Mycotoxins are non-volatile, low molecular weight natural products that are typically very stable and get airborne attached to spores, fragments, and particulate matter. One mycotoxin may be produced by different fungal strains, and one fungal strain may produce different mycotoxins. Almost all of our knowledge on the modes of action of mycotoxins relates to ingestion exposure, where for example inflammatory, immune-suppressive, cytotoxic, and carcinogenic effects in various organs have been described.

There is an extensive literature documenting the potential of mycotoxin and bacterial toxin production under laboratory conditions but only few reports so far document the occurrence of toxins on naturally infested building materials (e.g., Nielsen et al. 1999; Gravesen et al. 1999; Bloom et al. 2007, 2009; Täubel et al. 2011). The dispersal of mycotoxins into indoor air may occur by intact spores that are released from the mycelium, or by fungal fragments (Brasel et al. 2005) that are released in much higher numbers than spores, as shown by Kildesø et al. (2003) and Gorny et al. (2002). Thus, methods to assess human exposure to airborne mycotoxins must apply relevant techniques to cover also the smallest particles.

Many studies from the last decade have indeed shown the presence of mycotoxins and bacterial toxins in indoor air and settled house dust, suggesting potential exposure to humans in these indoor environments (e.g., Engelhart et al. 2002; Brasel et al. 2005; Bloom et al. 2007, 2009; Täubel et al. 2011; Peitzsch et al. 2012; Polizzi et al. 2009; Kirjavainen et al. 2015).

Amoebae

Amoebae are single-cell eukaryotic protozoans that constitute a diverse taxonomic group. The term "amoeba" is a practical term that refers to cells that are able to move and engulf food particles by producing projections of the cytoplasm. Amoebae are present in all aquatic environments all over the world (Rodriguez-Zaragoza 1994), whether natural or man-made. They are found, e.g., in drinking water, swimming pools, and eyewash stations from where amoebae can also be aerosolized and inhaled by humans.

Most amoebae are also able to exist in durable resting forms, cysts, in which the organism endures adverse conditions. Amoebae have a role in the ecosystem in the mineralizing of nitrogen, carbon, and phosphorus due to their importance as bacterial predators (Yli-Pirilä 2009). Not only do bacteria act as a source of nutrition for amoebae, but bacteria can also utilize amoebae as a vehicle for survival, replication, or as a means of transmission from one host to another. For example, legionellae are able to avoid digestion in the amoebae, but they can replicate inside amoebae until the amoebae burst (Newsome et al. 1998). This is one potential although indirect health effect of amoebae; they facilitate the spread of *Legionella* bacteria. Other possible health effects of amoebae include infections, e.g., in the eye.

Amoebae have been found in samples of mold growth associated with moisture damage of buildings (Yli-Pirilä 2009). This is an example of potential indoor exposure to amoebae. Experiments with associated bacteria and fungi showed that amoebae not only are members of the microbial network present in moisture-damaged, moldy building materials but they also have many interactions with other microbes present. Amoebae may increase the growth of other organisms, and render the microbes more cytotoxic, thus possibly modifying the health effect potential of microbes (Yli-Pirilä et al. 2006, 2007).

Viruses

Viruses are microorganisms that only can replicate within a host cell. Therefore, they cannot grow on nonliving substrates although they can survive and stay virulent on surfaces for days, in some cases even weeks (Casanova et al. 2010). Airborne survival time of viruses is much shorter (McDevitt et al. 2008). Viruses are usually specific to a certain species or group and they can infect bacteria, humans, animals, and plants. Viruses that replicate within bacterial cells are called bacteriophages.

Viruses are the smallest of microorganisms; their size range is 20–300 nm but they are often attached to other suspended particles (Yang and Marr 2011). The airborne behavior of viruses and the importance of their spread by the airborne route have not been well known until in recent years when molecular methods have been applied to the detection of airborne viruses. Viruses can be collected with bioaerosol collection devices and identified with molecular methods, but their airborne behavior has also been studied by modeling due to the technical difficulties of actual sampling and identification (Myatt et al. 2010).

Pollen

Pollen grains are part of the reproduction system of plants that are stable in their structure and transmitted long distances as airborne; in optimal weather conditions even to other parts of the world. Their sizes vary between 10 and 100 μm and therefore, a lot of pollen grains end up on the ground in nearby areas of the source. Pollen allergens and other fragments of the pollen grain may also occur in smaller particle sizes in the range 30 nm–5 μm (Miguel et al. 2006; Matikainen and Rantio-Lehtimäki 1998). Pollen are also distributed by insects. Pollen may carry important allergens, major allergenic plants being, e.g., many grasses, ragweed, and trees such as birch and olive tree. Due to the great influence of pollen allergens on people's health, occurrence of pollen is being monitored worldwide through monitoring networks, programs, and databases.

House dust mites

House dust mites are small arachnids that occur in mattresses and other furniture, provided the RH of their microenvironment is > 55%. They feed on, e.g., human skin scales and fungal spores and are typical contaminants of indoor environments. They produce allergenic proteins in their feces which can be aerosolized and are considered important causal agents of indoor environmental allergies. The most common species are *Dermatophagoides pteronyssinus* and *D. farinae*. Their major allergens are called *Der p* I and *Der f* I, respectively (Chapman et al. 1987). Other allergenic mite species are storage mites such as *Acarus siro*, *Glycyphagus domesticus*, and *Tyrophagus putrescentiae* that may proliferate in food storage facilities in institutions, grocery stores, and even homes. In fact, while house dust mites are common in mattresses and soft furniture, the mites that are found within a mold growth on a wall, may more often be storage mites (Charpin et al. 2010).

Exposure, a prerequisite to health effects

Introduction

Studying the health effects of environmental agents is always bound to the assessment of exposure, which is defined as the contact between the human and the environmental agent in question. For documentation of a causal connection, usually a dose-response relationship between the agent and the effect is needed. The concept of dose-response relationships is essential in the toxicological paradigm but cannot be applied as such to immunological and allergic reactions that are common in relation to health effects of biological exposures. In allergic conditions, the dose is often not critical for the reaction which is strongly bound to the individual characteristics of the exposed person. In any case, assessment of exposure and the studies on its determinants are essential in the pursuit of understanding the health effect of environmental agents.

Exposure assessment of environmental biological agents includes both qualitative and quantitative elements. There is a wide selection of methods available that can be used for collecting the samples from the environment and for detecting and analyzing the content. Depending on the aims of the investigation, the focus may be on characterization of the microbial communities present in the sample, on quantitation of fungal or bacterial biomass, on specific quantitation of a species, or a group of microbes or some other approach.

Sampling and characterization of the microbial community with culturing methods covers both qualitative and quantitative aspects but the method is selective, favoring those species for which the sampling technique and culture medium used are optimal. Culturing methods do not detect non-viable microbial material. Sequencing methods give an overall picture of the microbial DNA present, but for more precise quantification of a species or genus, additional analysis with qPCR is needed. In cases where quantification of microbial biomass is of interest, analyses of chemical markers such as ergosterol for assessment of fungal biomass is a good choice.

Most biological particles, fungi, bacteria, and other such material are present everywhere, in soil, water, air, and in built environments. They are part of the natural background of atmospheric particles and they are generally transported long distances around the world. Their occurrence in the atmosphere is regulated by geographical and climatic conditions as well as local events and circumstances. Humans get exposed to these particles both via outdoor air and via indoor air. People spend most of their time indoors, therefore, it is evident that the exposures that take place indoors make the major contribution of total exposures to biological particles, given that indoor concentrations are not considerably lower than outdoors. Whether the importance of indoor exposures for health is greater than the exposures outdoors is not quite clear. For example, pollen are important allergens and exposure to them takes place mainly outdoors, while the health effects of bio-particles that are linked with moisture and mold damage of buildings have their origin in the indoor environments.

Evaluation of the health effects of environmental stressors comprises of assessment of exposure and of evaluation of the health effects of the measured concentrations. Exposure assessment is based on specific sampling techniques and procedures optimized for each agent and on subsequent analysis that allows quantification of the measured agent. Sampling and analysis techniques of biological particles have been specifically developed for various biological particles. It should be emphasized that no single methodology allows the quantitative and qualitative determination of all biological particles in parallel, but each method is selective and focuses on certain characteristics of the measured material, be it enumeration of a group of microbes, or quantification of material with certain immunological activity, for example, endotoxin.

Sampling of biological particles

Sampling of biological particles from air is based on similar principles as any aerosol sampling. Air samplers can be designed for the detection of culturable or non-culturable material. Sampling devices may be passive samplers using natural aerosol convection, diffusion or gravity, or active samplers using stationary or personal pumps (Nevalainen et al. 1992; An et al. 2004). Various impactors, impingers, centrifugal samplers, and filters are in common use. Biological particles can be collected on culture medium, i.e., agar plates in impactors, or into a liquid medium in an impinger. These collection methods are followed by culturing of the collected microorganisms into visible colonies that can be counted and identified. Centrifugal sampling into a viscous liquid has been developed to avoid bouncing and stress to the collected cells, typical phenomena in impactors (Lin et al. 1999). Sampling on filters is a natural choice in those cases where the collected material is not cultured but suspended for further analysis with chemical, immunochemical, or DNA based methods. Also, electrostatic sampling has been developed (e.g., Yao and Mainelis 2006). Bioaerosol samplers are extensively presented and evaluated in, e.g., Reponen et al. (2011).

Active sampling of biological particles aims at an accurate assessment of air concentrations. Especially with indoor environmental sampling, this usually means a labor-intensive approach as each sampling site must be visited by the sampling staff. In population-scale studies, this is often beyond the economic resources of the study. Recent progress in epidemiological research that aims to link health effects with environmental exposures in home environments has led to the development of low cost sampling methods that allow studying of, e.g., hundreds or thousands of homes in the same study. For such studies, standardized methods to collect house dust have been developed, as house dust is considered an important reservoir of microbial material of the indoor environment. House dust sampling has also been used for exposure assessment of indoor toxicants (Lioy et al. 2002). Collection of house dust samples can be done, e.g., by vacuuming house dust from floors, mattresses or other surfaces, or by passive collection of settling dust (Kaarakainen et al. 2009). Although the passive collection methods or different ways of sampling house dust are not quite representative in the sense of aerosol sampling, these sampling methods have been shown useful and as having cost-benefit advantages especially in large population studies. A major part of the health data that is related to indoor microbial exposures, is linked with house dust sampling as the method of exposure assessment (e.g., Ege et al. 2011).

Analysis of biological particles

The analysis of a collected sample may be based on cultivation of viable fungi or bacteria, or on non-cultivation methods.

Fungi and bacteria are cultivated using different culture media. They have different properties for their content of nutrients, water activity, and growth suppressants. With the composition of the culture medium and by regulating growth temperature, humidity, light, time, and availability of oxygen, the conditions can be set to an optimal state for certain groups or types of microorganisms which facilitates their observation. This approach is typically used when certain pathogenic species are looked for. However, in environmental sampling, the whole range of the microorganisms present in the sample is usually of interest. Therefore, non-selective media for fungi and bacteria are used, such as malt extract agar. Non-selective growth media for environmental microbes are usually low in their nutrient content which imitates the typically poor nutritional conditions of their natural environment.

Even with the non-selective growth media, only a part of the species present can be grown into visible colonies. Those cells that are somehow damaged or for which the given medium and growth conditions are not optimal will probably not grow, and those with too long germination or division time are not seen as the space on the growth medium is usually limited and occupied fast by the most easily growing species. Furthermore, there are plenty of microbes that cannot be cultured in laboratory conditions. In fact, it has been assessed that only 1–10% of all the microbial cells present in environmental samples can be quantified and identified with cultural methods.

While cultivation based methods have historically been widely used in indoor microbial assessments and reference data are extensive, these approaches suffer from well-known limitations, which has triggered the rise of alternative, molecular methods. These are DNA based methods, analyses of chemical markers of microbial biomass, and immunochemical methods. Some agents can also be directly counted from a sample with a microscope, for example, pollen or certain types of fungal spores. A main factor driving this development is that a vast majority of microbial mass around us is not cultivable and/or not viable. From a health perspective—with the health of people being the main reason for why we aim to assess exposure—both not cultivable and dead microbes and their constituents are relevant in addition to the viable portion of the exposure. Cultivation based approaches in indoor assessments have typically been applied in combination with impaction-based, active air sampling, with short-term air sampling as such being problematic already. Active air sampling in the field and the need for cultivation/identification of microbes is an equipment intensive, laborious process (during sample collection) that requires time and expertise (in the laboratory analyses) and thus is only poorly applicable to larger-scale epidemiological studies. In the following, we provide a short overview of some of the methods that are applied in epidemiological study settings that aim to clarify the impacts of indoor microbial exposures on human health.

DNA-based methods

It is long known that the majority of micro-organisms is uncultivable under laboratory conditions, with estimates of 10% to less than 0.01% of the microbiome being captured with cultivation, depending on the habitat. For samples from indoor air, the ratio of total fungi and bacteria—assessed via microscopic counting—to viable fungi and bacteria, was estimated with 100:1 (Toivola et al. 2002). A way out of this dilemma of the 'uncultivable majority' (Pitkäranta 2012) was offered with the rise of polymerase chain reaction and other DNA based technologies that would allow to circumvent the need of cultivation prior to quantification and identification of microbes. Today, cultivation-independent methods, that is DNA targeting methods as well as immuno-/bioassays and chemical analytical approaches for quantifying microbial cell wall components, are routinely used in the characterization and monitoring of indoor microbial exposures.

The initial step in all DNA-based methods is the extraction of DNA directly from a given environmental sample. There is a multitude of different extraction methods, but typically microbial cells first get suspended from the sample matrix (in indoor studies, e.g., an air filter, indoor dust, or a building material); the cell envelopes are then mechanically or/and enzymatically disrupted to free the intracellular DNA, which then is further purified for subsequent analyses steps.

Common to most of the currently used approaches that aim at characterizing microbial communities or at quantifying microbial exposure is the amplification of a target gene fragment, that occurs ubiquitously in all microbes of interest, e.g., all bacteria or all fungi and contains phylogentically informative regions that allows differentiation of microbes on different taxonomic levels. These target sequences are usually amplified using a polymerase chain reaction and are then utilized in different DNA based methods, including quantitative PCR, fingerprinting, and sequencing approaches. Most commonly used for this purpose are the 16S ribosomal RNA gene in bacteria (16S rDNA) and the nuclear internal transcribed spacer region (ITS or nucITS) located between 18S, 5.8S, and 28S rRNA genes in fungi. Highly conserved regions within these 16S rDNA or ITS sequences allow the designing of oligonucleotide primers and probes that will target bacteria and fungi on a higher taxonomic level, for example, using 'universal' primers that will allow detection of the majority of organisms within the respective kingdom, or group specific primers. More specific stretches within the 16S rDNA and ITS regions—specific for a given group of closely related microbes—can be used to target a certain bacterial or fungal genus or species of interest. In essence, the differences in variability/conservation within these 'barcode' gene sequences, as they are often referred to, allow targeting of microbial groups at various taxonomic levels.

QPCR methods

The characteristic of the barcode gene sequences, i.e., targeting of microbial groups at different levels of resolution, is widely exploited in quantitative PCR (qPCR) methods that have been recently applied for the assessment of indoor exposure to microbes. Displaying the accumulation of a fluorescently labeled product during PCR amplification in 'real time' and comparing it to a standard curve based on amplifications of known concentrations of the target gene sequence allows the quantification of the original amount of the target gene or target organism (number of spores) in an environmental sample.

QPCR method is sensitive and fast; DNA extraction and qPCR analyses of an environmental sample can be performed within hours, where cultivation based methods required days to weeks. The specificity of the qPCR depends on the qPCR primers and probes that are applied. Since the initial applications of qPCR as a tool to assess exposure to the toxigenic fungus *Stachybotrys chartarum* in indoor environments more than a decade ago (Haugland et al. 1999), a large number of qPCR assays has been designed for various indoor fungal species, genera, or groups, and a number of bacterial genera and species thought to be relevant in indoor environments. A good part of these efforts on qPCR assay development (Haugland and Vesper 2002) has been facilitated by the US Environmental Protection Agency (US EPA) and has been complemented by individual research groups (e.g., Rintala and Nevalainen 2006; Torvinen et al. 2010; Kärkkäinen et al. 2010; Yamamoto et al. 2010).

QPCR analysis in the context of indoor environmental research and monitoring has been applied on building material samples, house dust, and indoor air collected via filtration of air. Generally, higher prevalence is observed for individual microbial species and overall microbial levels measured with qPCR are usually orders of magnitude higher than those found with cultivation based approaches. This confirms what has already been known, i.e., that cultivation-based approaches clearly underestimate both diversity and quantities of microbial exposures indoors. QPCR has been shown to be a useful tool in research investigating the effects of indoor microbial exposures on human health. This method is also used in practical situations in order to identify abnormal microbial conditions in buildings, usually in the context of moisture damage and indoor dampness. However, especially in order to be a useful diagnostic tool for building practitioners, profound knowledge is needed on what is a normal situation in a given country/climatic region. Unlike for cultivation-based approaches, extensive reference datasets thus far do not exist. Another technical limitation is that targeting, for example, a specific fungal species in a sample will provide quantitative information for only this one targeted species, but one stays blind for any other fungal taxa in a sample. This is different from cultivation based approaches, where with one cultivation a multitude of different fungal taxa can be quantified.

DNA fingerprinting methods

Different DNA fingerprinting methods have been applied on some occasions in indoor environmental studies, typically with the aim to follow changes or identify differences in bacterial or fungal communities, or to use the DNA fingerprints as a proxy for microbial diversity in a sample (Ege et al. 2011). Examples for these methods are terminal fragment length polymorphism (tRFLP), single-strand conformational polymorphism (SSCP), temperature- and denaturing gradient gel electrophoresis (TGGE and DGGE, respectively). What all these methods have in common is that they start from a pool of rather general PCR amplicons (e.g., universal bacterial or universal fungal amplicons) obtained from an environmental sample, with single amplicons then being separated based on differences in DNA sequences, producing a fingerprint of different bacterial or fungal species in a given sample. This method is useful in comparing microbial communities in tens or even hundreds of different samples; however, due to limitations in sensitivity and resolution of these methods, only major changes in the more abundant microbial groups can be reliably observed. Working with DNA fingerprinting methods is rather laborious, especially when one attempts to identify a specific microbial group that is associated with a change in the microbial community. This may be the reason why this method has not been widely applied in indoor exposure studies.

Sequencing methods

Similar to DNA fingerprinting, most sequencing methods currently applied utilize a pool of typically 'universal' bacterial or fungal PCR amplicons from an environmental sample. However, rather than profiling these amplicons into a fingerprint, sequencing approaches attempt obtaining the DNA sequences of each single PCR fragment in a given sample. Traditional sequencing methods using Sanger chemistry still were both costly and time-consuming. The so-called "next-generation" sequencing approaches, including different methods such as 454 pyrosequencing or Illumina sequencing platforms, have improved sample throughput and greatly reduced costs. Processing of hundreds of samples and obtaining thousands to millions of sequence reads per sample can be conveniently done within a matter of hours. The enormous amount of information that is produced with these novel sequencing techniques has created pressure on the development of bioinformatics tools that would be able to process mass of information in a (biologically) meaningful way. While great progress has been made in this field it still feels that ways to analyze and interpret next generation sequencing data lag behind the technological possibilities to produce more and more extensive and detailed datasets.

DNA-targeting approaches and especially sequencing methods have revolutionized our view on microbes and their role in environmental and human microbial ecosystems and their role in human health and disease. These methods have indeed proven to overcome limitations of cultivation-based approaches by being less selective and by allowing processing of large sample quantities in highly automated processes. However, DNA-approaches also have technical limitations and have introduced a new type of uncertainties that need to be considered in applications of such methods.

Chemical markers

Composition of the microbial communities can be studied by determining the amounts of chemical markers that present overall proportions of different microbial groups. For example, ergosterol is sterol specific and universally present in all fungal cells, and by determining the ergosterol content of the sample, fungal biomass can be quantified (Saraf et al. 1997). Similarly, muramic acid is present in bacterial cells, more so in Gram positive bacteria, and it is used as a marker of (Gram positive) bacterial biomass (Sebastian and Larsson 2003). Furthermore, 3-hydroxy-fatty acids that are found in lipid A component of lipopolysaccharide of Gram negative bacteria, can be specifically used as a marker of Gram negative bacteria (Saraf et al. 1997).

These compounds are covalently linked to various structures in the cell membranes, and therefore, prior to analysis, samples are hydrolyzed to cleave these linkages, purified by extractions, derivatised and finally analyzed, using gas chromatography-tandem mass spectrometry (GC-MS/MS).

Bioassays

Bioassays can be used for quantification of microbial biomass by determining their specific biological activity. A commonly used bioassay is the *Limulus* amebocyte lysate assay for endotoxin, which is the method used to describe the biological activity of bacterial LPS. The assay can also be used to determine the fungal beta (1->3)-D-glucan. The assay is based on enzymatic cascade of amebocytes isolated from horseshoe crab *Limulus* which has the specific capacity to react with bacterial LPS or fungal β-(1->3)-D-glucan in a dose-response pattern.

Endotoxin concentrations are expressed as either endotoxin units, based on the immunological activity in the *Limulus* assay, or as SI units ng m^{-3}, where the amount of LPS has been calculated either from the bioassay or from gas chromatography-mass spectrometry analysis of the bacterial cell fragments that are proportional to amounts of the corresponding macromolecular structures (Larsson 1994).

Allergens, fungal glucans and extracellular polysaccharides are detected with specific immunochemical methods where a known antibody is bound with the antigen that is being analyzed. Biological particles are often analyzed with enzyme-linked bioassay (ELISA).

Metrics and units to describe the quantity of microbial material in a sample

As there are many different ways to analyze the biological content of an environmental sample, there also are a number of different ways of expressing the concentrations of biological particles or biological material in a sample. Only a few of such units belong to the metric system. Therefore, comparison of concentrations or exposure levels and summarizing the results from different studies may be difficult due to lack of common metrics of the determinations.

When culturable methods are used to quantify fungi or bacteria, the unit is usually CFU m^{-3}, 'colony forming units per cubic meter of air' or CFU g^{-1} of dust or building material. The term 'colony forming unit' is based on the fact that the microbial colonies that are counted from the cultivation media surface may originally have been either single cells or aggregates of several cells. In some direct-calculation methods where bioparticles can be counted directly from the collection surface using a microscope, the used unit may be simply # m^{-3}.

In DNA based sequencing techniques, the number of taxonomic units that have been identified in a given sample, are many times expressed as 'OTU's', or operational taxonomic units per sample. Such a operational taxonomic unit is defined based on sequence similarity; a similarity of 97% in the DNA sequence of the bacterial 16S rRNA gene, for example, is commonly used to group sequences into OTUs. Such a level of sequence similarity is believed to compare close to the common species concept that is however also based on morphology, physiology, and other characteristics of organisms, and therefore, 'species' and 'OTUs' are not fully comparable. QPCR results can be calculated to give the amount of the target organism as specific DNA amount or cell equivalents per sample mass.

When chemical methods are used to assess the microbial biomass, the concentration of the marker is usually expressed as mass units per sampled volume of air or per sample mass, for example, for ergosterol in house dust, x ng g^{-1}. An expression of x mmol g^{-1} may also be used. In case of samples that have been vacuumed from a surface, the concentration may be expressed as surface loading per sampled area. When immunological assays such as *Limulus* assay for endotoxin are used, typically the unit is 'international endotoxin units'.

References

Adams, R.I., M. Miletto, J.W. Taylor and T.D. Bruns. 2013. Dispersal of microbes: Fungi in indoor air are dominated by outdoor air and show dispersal limitation short distances. ISME J. 7: 1262–1273.

Adan, O.C.G. and R.A. Samson. 2011. Introduction. pp. 15–38. *In*: O.C.G. Adan and R.A. Samson [eds.]. Fundamentals of Mold Growth in Indoor Environments and Strategies for Healthy Living. Wageningen Academic Publishers, Wageningen, The Netherlands.

Alwis, K.U., J. Mandryk and A.D. Hocking. 1999. Exposure to biohazards in wood dust: bacteria, fungi, endotoxins, and beta(1->3)-D-glucans. Appl. Occup. Environ. Hyg. 14: 598–608.

Amend, A.S., K.A. Seifert, R. Samson and T.D. Bruns. 2010. Indoor fungal composition is geographically patterned and more diverse in temperate zones than in the tropics. Proc. Natl. Acad. Sci. 12107: 13748–13753.

An, H.R., G. Mainelis and M. Yao. 2004. Evaluation of a high-volume portable bioaerosol sampler in laboratory and field environments. Indoor Air 14: 385–393.

Andreae, M.O. and D. Rosenfeld. 2008. Aerosol-cloud-precipitation interactions. Part 1. The nature and sources of cloud-active aerosols. Earth Sci. Rev. 89: 13–41.

Bauer, H., A. Kasper-Giebl, M. Löflund, H. Giebl, R. Hitzenberger, F. Zibuschka and H. Puxbaum. 2002. The contribution of bacteria and fungal spores to the organic carbon content of cloud water, precipitation and aerosols. Atmos. Res. 64: 109–119.

Bertolini, V., I. Gandolfi and R. Ambrosini. 2013. Temporal variability and effect of environmental variables on airborne bacterial communities in an urban area of Northern Italy. Appl. Microbiol. Biotechnol. 97: 6561–6570.

Bloom, E., K. Bal, E. Nyman, A. Must and L. Larsson. 2007. Mass-spectrometry-based strategy for direct detection and quantification of some mycotoxins produced by *Stachybotrys* and *Aspergillus* spp. in indoor environments. Appl. Environ. Microbiol. 73: 4211–4217.

Bloom, E., E. Nyman and A. Must. 2009. Molds and mycotoxins in indoor environments—a survey in water-damaged buildings. J. Occup. Environ. Hyg. 6: 671–678.

Bowers, R.M., N. Clements and J.B. Emerson. 2013. Seasonal variability in bacterial and fungal diversity of the near-surface atmosphere. Environ. Sci. Technol. 47: 12097–12106.

Brasel, T.L., J.M. Martin, C.G. Carriker, S.C. Wilson and D.C. Straus. 2005. Detection of airborne *Stachybotrys chartarum* macrocyclic trichotecenes in the indoor environment. Appl. Environ. Microbiol. 72: 7376–7388.

Burge, H. 1990. Bioaerosols: Prevalence and health effects in the indoor environment. J. Allergy Clin. Immunol. 86: 687–701.

Casanova, L.M., S. Jeon, W.A. Rutala, D.J. Weber and M.D. Sobsey. 2010. Effects of air temperature and relative humidity on coronavirus survival on surfaces. Appl. Environ. Microbiol. 76: 2712–2717.

Chapman, M.D., P.W. Heymann, S.R. Wilkins, M.J. Braun and T.A.E. Platts-Mills. 1987. Moniclonal immunoassays for major dust mite (*Dermatophagoides*) allergens, *Der p* I and *Der f*I, and quantitative analysis of the allergen content of mite and house dust extracts. J. Allergy Clin. Immunol. 80: 184–194.

Charpin, D., P. Parola, I. Arezki, C. Charpin-Kadouch, A. Palot and H. Dumon. 2010. House-dust mites on wall surfaces of damp dwellings belong to storage mite genus. Allergy 65: 274–275.

Colwell, R. 2000. Viable but nonculturable bacteria: a survival strategy. J. Infect. Chemother. 6: 121–125.

Després, V.R., J.A. Huffman, S.M. Burrows, C. Hoose, A.S. Safatov, G. Buryak, J. Fröhlich-Nowoisky, W. Elbert, M.O. Andreae, U. Pöschl and R. Jaenicke. 2012. Primary biological aerosol particles in the atmosphere: a review. Tellus B 64: 15598.

doPico, G.A. 1986. Health effects of organic dusts in the farming environment. Report of diseases. Am. J. Ind. Med. 10: 261–265.

Douwes, J. 2005. (1->3)-beta-D-glucans and respiratory health: a review of scientific evidence. Indoor Air 15: 160–169.

Dunn, R.R., N. Fierer, J.B. Henley, J.W. Leff and H.L. Menninger. 2013. Home life: factors structuring the bacterial diversity found within and between homes. PLoS ONE 8:3 e64133.

Durand, K.T.H., M.L. Muilenberg, H.A. Burgeand and N.S. Seixas. 2002. Effect of sampling time on the culturability of airborne fungi and bacteria sampled by filtration. Ann. Occup. Hyg. 46: 113–118.

Ege, M., M. Mayer, A.-C. Normand, J. Genuneit, W.O. Cookson, C. Braun-Fahrländer, D. Heederik, R. Piarroux, E. von Mutius and the GABRIELA/TR22 study group. 2011. Exposure to environmental microorganisms and childhood asthma. N. Engl. J. Med. 364: 701–709.

Engelhart, S., A. Loock, D. Skutlarek, H. Sagunski, A. Lommel, H. Färber and M. Exner. 2002. Occurrence of toxigenic Aspergillus versicolor isolates and sterigmatocystin in carpet dust from damp indoor environments. Appl. Environ. Microbiol. 68: 3886–3890.

Fox, A., W. Harley, C. Feigley, D. Salzberg, C. Toole, A. Sebastian and L. Larsson. 2005. Large particles are responsible for elevated bacterial marker levels in school air upon occupation. J. Environ. Monit. 7: 450–456.

Frankel, M., G. Beko, M. Timm, S. Gustavsen, E.W. Hansen and A.M. Madsen. 2012. Seasonal variations in indoor microbial exposures and their relation to temperature, relative humidity, and air exchange rate. Appl. Environ. Microbiol. 78: 8289–8297.

Gehring, U., J. Heinrich, G. Hoek, M. Giovanangelo, E. Nordling, T. Bellander, J. Gerritsen, J.C. de Jongste, H.A. Smit, H.E. Wichmann, M. Wickman and B. Brunekreef. 2007. Bacteria and mould components in house dust and children's allergic sensation. Eur. Respir. J. 29: 1144–1153.

Gora, A., B. Mackiewicz, P. Krawczyk, M. Golec, S. Skórska, G. Cholewa, L. Larsson, M. Jarosz, A. Wójcik-Fatla and J. Dutkiewicz. 2009. Occupational exposure to organic dust, microorganisms, endotoxin and peptidoglycan among plant processing workers. Ann. Agric. Environ. Med. 16: 143–150.

Górny, R.L., T. Reponen, K. Willeke, D. Schmechel, E. Robine, M. Boissier and S.A. Grinshpun. 2002. Fungal fragments as indoor air biocontaminants. Appl. Environ. Microbiol. 68: 3522–3531.

Gravesen, S. 1979. Fungi as a cause of allergic disease. Allergy 34: 135–154.

Gravesen, S., L. Larsen and P. Skov. 1983. Aerobiology of schools and public institutions—part of a study. Ecol. Dis. 2: 411–413.

Gravesen, S., P.A. Nielsen, R. Iversen and K.F. Nielsen. 1999. Microfungal contamination of damp buildings—examples of risk constructions and risk materials. Environ. Health Perspect. 107 Suppl. 3: 505–508.

Green, B.J., E.R. Tovey, J.K. Sercombe, F.M. Blachere, D.H. Beezhold and D. Schmechel. 2006. Airborne fungal fragments and allergenicity. Med. Mycol. 44: S245–S255.

Haas, D., J. Habib, H. Galler, W. Buzina, R. Schlacher, E. Marth and F.F. Reinthaler. 2007. Assessment of indoor air in Austrian apartments with and without visible mold growth. Atmos. Environ. 41: 5192–5201.

Haugland, R. and S. Vesper. 2002. Method of identifying and quantifying specific fungi and bacteria. US Patent 6387652.

Haugland, R.A., S.J. Vesper and L.J. Wymer. 1999. Quantitative measurement of *Stachybotrys chartarum* conidia using real tim e detection of PCR products with the TaqMan™ fluorogenic probe system. Mol. Cell. Probes. 13: 329–340.

Heinrich, J., B. Hölscher, J. Douwes, K. Richter, A. Koch, W. Bischof, B. Fahlbusch, R.W. Kinne, H.E. Wichmann and INGA Study Group. 2003. Reproducibility of allergen, endotoxin and fungi measurements in the indoor environment. J. Expo. Anal. Environ. Epidemiol. 13: 152–160.

Hospodsky, D., J. Qian, W.W. Nazaroff, N. Yamamoto, K. Bibby, H. Rismani-Yazdi and J. Peccia. 2012. Human occupancy as a source of indoor air bacteria. PLoS ONE 7: e34867.

Jaenicke, R. 2005. Abundance of cellular material and proteins in the atmosphere. Science 308: 73.

Jarvis, J.Q. and P.R. Morey. 2001. Allergic respiratory disease and fungal remediation in a building in a subtropical climate. Appl. Occup. Environ. Hyg. 16: 380–388.

Kaarakainen, P., H. Rintala, A. Vepsäläinen, A. Hyvärinen, A. Nevalainen and T. Meklin. 2009. Microbial content of house dust samples determined with qPCR. Sci. Total Environ. 407: 4673–4680.

Kärkkänen, P., M. Valkonen, A. Hyvärinen, A. Nevalainen and H. Rintala. 2010. Determination of bacterial load in house dust using qPCR methods, chemical markers and culture. J. Environ. Monit. 12: 759–768.

Kildesø, J., H. Würtz, K.F. Nielsen, P. Kruse, K. Wilkins, U. Thrane, S. Gravesen, P.A. Nielsen and T. Schneider. 2003. Determination of fungal spore release from wet building materials. Indoor Air 13: 148–155.

Kirjavainen, P.V., M. Täubel, A.M. Karvonen, M. Sulyok, P. Tiittanen, R. Krska, A. Hyvärinen and J. Pekkanen. 2015. Microbial secondary metabolites in homes in association with moisture damage and asthma. Indoor Air doi: 10.1111/ina.12213.

Korpi, A., J. Jarnberg and A.-L. Pasanen. 2009. Microbial volatile organic compounds. Crit. Rev. Toxicol. 39: 139–193.

Kuo, Y.M. and C.S. Li. 1994. Seasonal fungus prevalence inside and outside of domestic environments in the subtropical climate. Atmos. Environ. 28: 3125–3130.

Larsson, L. 1994. Determination of microbial chemical markers by gas chromatography-mass spectrometry—potential for diagnosis and studies on metabolism *in situ*. Acta Pathol. Microbiol. Immunol. Scand. 102: 161–169.

Lehtonen, M., T. Reponen and A. Nevalainen. 1993. Everyday activities and variation of fungal spore concentration in indoor air. Int. Biodeter. Biodegrad. 31: 25–39.

Levetin, E., R. Shaughnessy, E. Fisher, B. Ligman, J. Harrison and T. Brennan. 1995. Indoor air quality in schools: exposure to fungal allergens. Aerobiologia 11: 27–34.

Lin, X.J., T.A. Reponen, K. Willeke, S.A. Grinshpun, K.K. Foarde and D.S. Ensor. 1999. Long-term sampling of airborne bacteria and fungi into a non-evaporating liquid. Atmos. Environ. 33: 4291–4298.

Lioy, P.J., N.C.G. Freeman and J.R. Millette. 2002. Dust: a metric for use in residential and building exposure assessment and source characterization. Environ. Health Perspect. 110: 969–983.

Macher, J.M., F.Y. Huang and M. Flores. 1991. A two-year study of microbiological indoor air quality in a new apartment. Arch. Environ. Health 46: 25–29.

Maki, T., M. Kakikawa, F. Kobayashi, M. Yamada, A. Matsuki, H. Hasegawa and Y. Iwasaka. 2013. Assessment of composition and origin of airborne bacteria in the troposphere over Japan. Atmos. Environ. 74: 73–82.

Matikainen, E. and A. Rantio-Lehtimäki. 1998. Semiquantitative and qualitative analysis of pre-seasonal airborne birch pollen allergens in different particle sazes—Background information for allergen reports. Grana 37: 293–297.

McDevitt, J.J., D.K. Milton, S.N. Rudnik and M.W. First. 2008. Inactivation of poxviruses by upper-room UVC light in a simulated hospital room environment. PLoS ONE 3(9): 1–9.

Meadow, J.F., A.E. Altrichter, S.W. Kembel, J. Kline, G. Mhuireach et al. 2013. Indoor airborne bacterial communities are influenced by ventilation, occupancy and outdoor air source. Indoor Air: DOI: 10.1111/ina.12047.

Meklin, T., T. Husman, A. Vepsäläinen, M. Vahteristo, J. Koivisto, J. Halla-aho, A. Hyvärinen, D. Moschandreas and A. Nevalainen. 2002. Indoor air microbes and respiratory symptoms of children in moisture damaged and reference schools. Indoor Air 12: 175–183.

Miguel, A.G., P.E. Taylor, J. House, M.M. Glovsky and R.C. Flagan. 2006. Meteorological influences on respirable fragment release from Chinese elm pollen. Aerosol Sci. Technol. 40: 690–696.

Moller-Nielsen, E., N.O. Breum, B. Herbert Nielsen, H. Wurtz, O. Melchior Poulsen and U. Midgaard. 1997. Bioaerosol exposure in waste collection: a comparative study on the significance of collection equipment, type of waste and seasonal variation. Ann. Occup. Hyg. 41: 325–344.

Morrison, D.C. and R.J. Ulevich. 1978. The effects of bacterial endotoxins on host mediating systems. Am. J. Pathol. 93: 527–617.

Myatt, T.A., M.H. Kaufman, J.G. Allen, D.L. Macintosh, M.P. Fabian and J.J. McDevitt. 2010. Modeling the airborne survival of influenza virus in a residential setting: the impacts of home humidification. Environ. Health 9: 55.

Nevalainen, A., J. Pastuszka, F. Liebhaber and K. Willeke. 1992. Performance of bioaerosol samplers: collection characteristics and sampler design consideration. Atmos. Environ. 26A: 531–540.

Newsome, A.L., T.M. Scott, R.F. Benson and B.S. Fields. 1998. Isolation of an amoeba naturally harboring a distinctive *Legionella* species. Appl. Environ. Microbiol. 64: 1688–1693.

Nielsen, K.F. 2003. Mycotoxin production by indoor molds. Fungal Genet. Biol. 39: 103–117.

Nielsen, K.F., S. Gravesen, P.A. Nielsen, B. Andersen, U. Thrane and J.C. Frisvad. 1999. Production of mycotoxins on artificially and naturally infested building materials. Mycopathologia 145: 43–56.

Pasanen, A.L., P. Kalliokoski, P. Pasanen, T. Salmi and A. Tossavainen. 1989. Fungi carried from farmers' work into farm homes. Am. Int. Hyg. Assoc. J. 50: 631–633.

Pasanen, A.L., H. Heinonen-Tanski, P. Kalliokoski and M.J. Jantunen. 1991. Fungal microcolonies on indoor surfaces—an explanation for the base level fungal counts in indoor air. Atmos. Environ. 26B: 117–120.

Pastuszka, J.S., K.T. Paw U, D.O. Lis, A. Wlazło and K. Ulfig. 2000. Bacterial and fungal aerosol in indoor environments in Upper Silesia, Poland. Atmos. Environ. 34: 3833–3842.

Peitzsch, M., M. Sulyok, M. Täubel, V. Vishwanath, E. Krop, A. Borrás-Santos, A. Hyvärinen, A. Nevalainen, R. Krska and L. Larsson. 2012. Microbial secondary metabolites in school buildings inspected for moisture damage in Finland, The Netherlands and Spain. J. Environ. Monit. 14: 2044–2053.

Pitkäranta, M. 2012. Molecular profiling of indoor microbial communities in moisture damaged and non-damaged buildings. PhD Thesis, University of Helsinki, Finland, 2012.

Pitkäranta, M., T. Meklin, A. Hyvärinen, L. Paulin, P. Auvinen and A. Nevalainen. 2008. Analysis of fungal flora in indoor dust by ribosomal DNA sequence analysis, quantitative PCR and culture. Appl. Environ. Microbiol. 74: 233–244.

Polizzi, V., B. Delmulle, A. Adams, A. Moretti, A. Susca, A.M. Picco, Y. Rosseel, R. Kindt, J. Van Bocxlaer, N. De Kimpe, C. Van Peteghem and S. De Saeger. 2009. JEM Spotlight: Fungi, mycotoxins and microbial volatile organic compounds in mouldy interiors from water-damaged buildings. J. Environ. Monit. 11: 1849–1858.

Polymenakou, P.N. 2012. Atmosphere: a source of pathogenic or beneficial microbes? Atmosphere 3: 87–102.

Prospero, J.M., E. Blades and G. Mathison. 2005. Interhemispheric transport of viable fungi and bacteria from Africa to the Caribbean with soil dust. Aerobiologia 21: 1–19.

Rappe, M.S. and S.J. Giovannoni. 2003. The uncultured microbial majority. Ann. Rev. Microbiol. 57: 369–394.

Rautiala, S., T. Reponen, A. Hyvärinen, A. Nevalainen, T. Husman, A. Vehviläinen and P. Kalliokoski. 1996. Exposure to airborne microbes during the repair of moldy buildings. Am. Ind. Hyg. Assoc. J. 57: 279–284.

Rautiala, S., J. Kangas, K. Louhelainen and M. Reiman. 2003. Farmer's exposure to airborne microorganisms in composting swine confinement buildings. Am. Ind. Hyg. Assoc. J. 64: 673–677.

Ren, P., T.M. Jankun and B.P. Leaderer. 1999. Comparisons of seasonal fungal prevalence in indoor and outdoor air and in house dusts of dwellings in one Northeast American county. J. Exposure Anal. Environ. Epidemiol. 9: 560–568.

Reponen, T., A. Nevalainen, M. Jantunen, M. Pellikka and P. Kalliokoski. 1992. Normal range criteria for indoor bacteria and fungal spores in a subarctic climate. Indoor Air 2: 26–31.

Reponen, T., K. Willeke, S. Grinshpun and A. Nevalainen. 2011. Biological particle sampling. pp. 549–570. *In*: P. Kulkarni, P.A. Baron and K. Willeke [eds.]. Aerosol Measurement, Principles, Techniques and Applications, 3rd Ed. John Wiley & Sons, Inc., Holbroken, New Jersey, USA.

Rintala, H. and A. Nevalainen. 2006. Quantitative measurement of streptomycetes using real-time PCR. J. Environ. Monit. 8: 745–749.

Rintala, H., A. Hyvärinen, L. Paulin and A. Nevalainen. 2004. Detection of streptomycetes in house dust—comparison of culture and PCR methods. Indoor Air 14: 112–119.

Rintala, H., M. Pitkäranta and M. Täubel. 2012. Microbial communities associated with house dust. *In*: Allen I. Laskin, S. Sariaslani and G.M. Gadd [eds]. Adv. Appl. Microbiol. 78: 75–120.

Rodriquez-Zaragoza, S. 1994. Ecology of free-living amoebae. Crit. Rev. Microbiol. 20: 225–241.

Saraf, A., L. Larsson, H. Burge and D. Milton. 1997. Quantification of ergosterol and 3-hydroxy fatty acids in settled house dust by gas chromatography-mass spectrometry: comparison with fungal culture and determination of endotoxin by a *Limulus* amebocyte lysate assay. Appl. Environ. Microbiol. 63(7): 2554–9.

Sebastian, A. and L. Larsson. 2003. Characterization of the microbial community in indoor environments: a chemical-analytical approach. Appl. Environ. Microbiol. 69(6): 3103–3109.

Shorter, C. 2012. Fungi in New Zealand homes: Measurement, aerosolization & association with children's health. PhD thesis, University of Otago, New Zealand.

Smith, D.J., D.A. Jaffe, M.N. Birmele, D.W. Griffin, A.C. Schuerger, J. Hee and M.S. Roberts. 2012. Free tropospheric transport of microorganisms from Asia to North America. Micr. Ecol. 64: 973–985.

Starr, M.P., H. Stolp, H.G. Trüper, A. Balows and H.G. Schlegel [eds.]. 1981. The Prokaryotes, A Handbook on Habitats, Isolation and Identification of Bacteria. Spinger-Verlag, Berlin, Heidelberg, New York.

Täubel, M., H. Rintala, M. Pitkäranta, L. Paulin, S. Laitinen, J. Pekkanen, A. Hyvärinen and A. Nevalainen. 2009. The occupant as a source of house dust bacteria. J. Allergy Clin. Immunol. 124(4): 834–840.

Täubel, M., M. Sulyok, V. Vishwanath, E. Bloom, M. Turunen, K. Järvi, E. Kauhanen, R. Krska, A. Hyvärinen, L. Larsson and A. Nevalainen. 2011. Co-occurrence of toxic bacterial and fungal secondary metabolites in moisture-damaged indoor environments. Indoor Air 21: 368–375.

Toivola, M., S. Alm, T. Reponen, S. Kolari and A. Nevalainen. 2002. Personal exposures and microenvironmental concentrations of particles and bioaerosols. J. Environ. Monit. 4: 166–174.

Torvinen, E., P. Torkko and H. Rintala. 2010. Real-time PCR detection of environmental mycobacteria in house dust. J. Microbiol. Methods 82: 78–84.

Tsapko, V.G., A.J. Chudnovets, M.J. Sterenbogen, V.V. Papach, J. Dutkiewicz, C. Skórska, E. Krysinska-Traczyk and M. Golec. 2011. Exposure to bioaerosols in the selected agricultural facilities of the Ukraine and Poland—a review. Ann. Agric. Environ. Med. 18: 19–27.

WHO. 2009. WHO Guidelines for Indoor Air Quality: Dampness and Mould. World Health Organization, Copenhagen, Denmark, 228 pp.

Yamamoto, N., M. Kimura, H. Matsuki and Y. Yanagisawa. 2010. Optimization of a real-time PCR assay to quantitate airborne fungi collected on a gelatin filter. J. Biosci. Bioeng. 109: 83–88.

Yang, W. and L.C. Marr. 2011. Dynamics of airborne influenza A viruses indoors and dependence on humidity. PLoS ONE 6(&): e21481.

Yao, M.S. and G. Mainelis. 2006. Utilization of natural electrical charges on airborne microorganisms for their collection by electrostatic means. J. Aerosol Sci. 37: 513–527.

Yli-Pirilä, T. 2009. Amoebae in moisture-damaged buildings. PhD dissertation. National Institute for Health and Welfare, Kuopio, Finland.

Yli-Pirilä, T., J. Kusnetsov, M.-R. Hirvonen, M. Seuri and A. Nevalainen. 2006. Effects of amoebae on the growth of microbes isolated from moisture-damaged buildings. Can. J. Microbiol. 52: 383–390.

Yli-Pirilä, T., K. Huttunen, A. Nevalainen, M. Seuri and M.-R. Hirvonen. 2007. Effects of co-culture of amoebae with indoor microbes on their cytotoxic and proinflammatory potential. Environ. Toxicol. 22: 357–367.

Yossifova, Y.Y., T. Reponen, D.I. Bernstein, L. Levin, H. Kalra, P. Campo, M. Villareal, J. Lockey, G.K. Hershey and G. LeMasters. 2007. House dust (1-3)-beta-D-glucan and wheezing in infants. Allergy 62: 504–513.

7

Health Effects of Fungi, Bacteria, and Other Bioparticles

Aino Nevalainen, Martin Täubel* and *Anne Hyvärinen*

Introduction

Most health effects associated with biological particles are various respiratory conditions and skin reactions. Their mechanisms vary from infections, irritation symptoms, and allergic diseases to toxic or immunotoxic reactions and other conditions with less evident pathophysiology. It is often the case that epidemiological and clinical evidence of the association between biological agents and the adverse health effect is strong, but the causal link between the exposing agents and the disease are poorly known. Furthermore, the mechanistic pathways leading to a certain condition are often not well known.

The health outcomes where the causal connection to certain microorganisms is evident and better understood are infections and allergies. For example, *Legionella* bacteria cause an infection known as legionellosis, and the allergen *Der p*I is the main causal agent for house dust mite allergy. For other health effects of biological particles, the links between the agent and the health condition have been shown as time- and space-specific associations but usually the causal connection and the pathophysiological mechanisms by which the health effects develop are not well understood.

Furthermore, in the cases of non-infectious and non-allergic reactions, the exposures usually are complex mixtures of various micro-organisms, their components and metabolic products, and the roles of individual molecules, microbial species, or other agents in development of adverse health effects are yet to be identified.

National Institute for Health and Welfare, Department of Health Protection, Neulaniementie 4, FI-70701 Kuopio, Finland.
* Corresponding author: aino.nevalainen@thl.fi

Infections and allergies

Pathogenic bacteria and viruses are the main infectious agents that cause disease in humans. Less common are the infections that are caused by certain fungi or amoebae. The fight against infectious diseases caused by bacteria and viruses is based on vaccinations, general hygiene, and use of antibiotics.

Examples of common infectious agents are rhinoviruses that cause 'common colds', noroviruses that cause gastric infections, and pathogenic bacteria, such as certain streptococci, staphylococci, pneumococci, and clostridia. In case of infectious diseases, the causal agent is usually detectable and the route of infection is most often human to human contact, either directly or indirectly through surfaces or via an airborne route. In hospitals and institutions, specific bacterial strains that are resistant to the known antibiotics, form an increasing challenge to the health care systems worldwide.

While infections are usually spread via human to human route, there is one well known infectious agent, i.e., the bacterium *Legionella*, the source of which is the indoor environment and water system within it (Tobin et al. 1980). Legionellae proliferate typically in man-made water systems, i.e., cooling systems and humidifiers (Arnow et al. 1982; Kusnetsov 1997). Humans may get infected by inhaling aerosolized legionella bacteria, and the infection may be lethal if proper antibiotic treatment is not available.

Allergies are conditions where the contact to the allergen causes an immunological reaction via IgE-mediated pathway in the exposed individual, leading to symptoms of varying severity. A prerequisite to allergy is atopy, a genetic tendency towards such reaction to allergens. Allergies are usually very specific and they can be treated with medication and avoidance policies. While allergies to food items are common, the most important environmental allergens are pollen and fungi in outdoor air; and indoors, cat, dog, house dust mites, cockroaches, and some plants. In occupational settings, some bacterial proteins and glycoproteins, such as proteolytic enzymes produced by, e.g., *Bacillus subtilis* may act as allergens and have importance as causative agents of an occupational disease (Göthe et al. 1972).

Indoor dampness and mold

Environmental exposure to microbes and biological material is often characterized by the complexity of the exposure, that may consist of a myriad of various microbial species, particles, cell components, microbial products, and other biological material, be it hay, straw, wood dust, dirt, or other matrix depending on the exposure situation. One situation that has gained a lot of attention during the last decades has been the problem of moisture, dampness, and mold in buildings and associated health effects.

Population studies in the late 1980s and later revealed that residential moisture, dampness, and mold growth are a risk factor for many respiratory diseases and other symptoms (IOM 2004; WHO 2009). The issue has been thoroughly studied since then, and is today recognized as one of the most common building-related health risks.

Dampness and moisture problems of the building may be caused by several factors. Inadequate ventilation, leaks, condensation, and capillary raise of moisture from the ground are among the major reasons of moisture accumulation in the building structures. Then, whenever a surface or material of the building is moist for prolonged time, microbial growth will start, which includes bacteria, fungi, and other organisms. When this growth becomes visible, it is recognized as 'mold' or 'mildew' that often has a characteristic odor of 'cellar' or 'earth'. It is not exactly known why such mold growth is harmful for health, but its emissions into indoor air include spores, fragments, and other particles from the mycelium, volatile compounds known as MVOC (microbial volatile organic compounds), and non-volatile microbial products, i.e., mycotoxins and bacterial toxins. In addition, chemical substances with non-biological origin may be emitted into indoor air as a result of moisture-induced disruption of materials.

Exposure in a moisture- or mold-affected indoor environment is verified by studying the building. Any signs of dampness or moisture, i.e., damp spots, discolored wood, loosening tiles, or peeling paint may be visible to the naked eye and tell about the risk of mold growth behind the surface if the growth is not visible. While visual observation of the telltale signs of moisture is the first step of the building investigation, it usually takes many kinds of professional skills to detect, locate, and assess such problems. Microbial sampling of indoor air is one of those methods. Sampling strategy and choice of methods need to be planned carefully as the fungal levels are affected by not only mold growth in the building but by many other factors as well.

Human health effects associated with dampness and mold

The health outcomes associated with moisture, dampness, and mold of buildings have been evaluated by working groups of WHO (WHO 2009) and Institute of Medicine (IOM 2004) and later on, reviewed by Mendell et al. (2011) and Kanchongkittiphon et al. (2015). The health effects that have been best documented include irritant effects of the respiratory system and mucous membranes, increased and prolonged respiratory infections, and exacerbation of previous asthma and onset of new asthma. Other, less strongly documented effects include ODTS (organic dust toxic syndrome), allergic alveolitis, sarcoidosis, and autoimmune diseases including rheumatic conditions. A systematic review by Tischer et al. (2011) concluded that visible mold in the home increases the risk of wheezing (adjusted odds ratio (aOR) 1.68, 95% confidence interval (95%CI) 1.48–1.90) and asthma (aOR 1.49, 95%CI 1.28–1.72).

Health effects are reported among both children and adults, among atopic and non-atopic individuals, and in all kinds of buildings, i.e., residences, schools, offices, hospitals, and other public buildings. Not all individuals react in the same way, and recognition and identification of a situation where dampness and mold are behind a health condition may be challenging. There are no good diagnostic tools to verify if someone's symptoms or disease is related to exposure to indoor mold but a timely occurrence of exposure and recovery of symptoms during absence from the building are important diagnostic criteria. Individual factors also affect the recovery of the symptoms which may take place slowly, even for years (Rudblad et al. 2002).

The health outcomes that WHO (2009) recognized as relatively well documented are listed in Table 7-1.

Table 7-1. Health effects observed in damp buildings. The respiratory health effects for which WHO (2009) found (a) sufficient evidence or (b) limited or suggestive evidence of an association with indoor dampness-related factors.

(a) Asthma exacerbation, upper respiratory tract symptoms, cough, wheeze, asthma development, dyspnea, current asthma, respiratory infections
(b) bronchitis, allergic rhinitis

Agents that may potentially have causal role in dampness-related health effects

The pathophysiological pathways that lead to the symptoms and diseases in moldy buildings are still poorly known. Although many symptoms mimic allergic reactions, IgE-mediated allergies are only seldom observed and therefore this mechanism does not explain all the symptom findings (Taskinen et al. 2001). Fungi do contain and produce allergens and antigens, but most well-characterized human allergens come from fungi that occur mainly outdoors (Miller 2011). Other possible mechanistic aspects are, e.g., irritation, inflammation, and toxic mechanisms.

Inflammation is a local response of tissues to injury or irritation, serving as a mechanism initiating the elimination of noxious agents and of damaged tissue (Huttunen et al. 2003). There are indications that inflammation would play a role in the development of symptoms in damp and mold-problem buildings. Many microbial agents are capable of inducing a number of inflammatory responses both *in vitro* and *in vivo* (Miller et al. 2010); probably the best known example is bacterial endotoxin (Heine et al. 2001). Fungi also contain a cellular component in their cell walls with observed inflammatory potential, i.e., glucans. Glucans of various types of fungi have been shown to induce an inflammatory response as summarized by Miller (2011).

Microbial strains, especially bacteria, isolated from mold-damaged indoor environments have shown remarkable proinflammatory activity. For example, Gram positive bacteria *Bacillus cereus* and *Streptomyces californicus* and the Gram negative bacterium *Pseudomonas fluorescens* were compared in an experimental set-up to the fungi *Stachybotrys chartarum*, *Aspergillus versicolor*, and *Penicillium spinulosum*, all isolated from indoor environments (Huttunen et al. 2003). The inflammatory potency was observed especially in the various bacteria, and the potency decreased in the order *Ps. fluorescens* > *Str. californicus* > *B. cereus* > *S. chartarum* > *A. versicolor* > *P. spinulosum*. It is noteworthy that while the Gram negative *Pseudomonas fluorescens* showed the highest potency for inflammation—probably due to its endotoxin—the Gram positive *Streptomyces* also had remarkable potency. Cytotoxicity is measured in *in vitro*-studies for assessment of a toxic mechanism. When potency-to-cause cytotoxicity was measured in the experimental set-up described, the rank order was *Ps. fluorescens* > *S. chartarum* > *Str. californicus* > *A. versicolor* > *B. cereus* > *P. spinulosum* (Huttunen et al. 2003). Inflammatory potential of airborne particles was shown to be associated with the amount of microbial material in air samples in a study focusing on personal exposures (Roponen et al. 2003).

Increase of proinflammatory markers in nasal lavage fluid of individuals exposed to a moisture-problem school environment have been reported (Hirvonen et al. 1999). This increase was observed during the exposure period, a decrease to control levels of the same markers during a break in the exposure (summer vacation), and an increase again at the end of the following exposure period. Interestingly, this phenomenon could not, however, be shown in a similar time-series study in sawmill workers although their microbial exposure levels are much higher (Roponen et al. 2002). Inflammation was also the main pathway observed in a murine model studying effects of mycotoxins such as atranone C, brevianamide, cladosporin, mycophenolic acid, neoechinulin A and B, sterigmatocystin, and TMC-120A. The doses used in the study were within the estimated range of possible indoor-related human exposure (Miller et al. 2010).

In addition to inflammation, toxic mechanisms also appear to be involved in the development of symptoms in mold-infested indoor environments. This is not, however, a well known aspect of indoor-related exposures where the main route of exposure is assumed to be inhalation, as most documentation concerning acutely toxic secondary metabolites comes from studies focusing on dietary dosing.

Protective effects of microbial exposures

Consistent observations in population studies in different countries have revealed that children of farmers have less allergies than urban children (Adler et al. 2005; Braun-Fahrländer et al. 1999; von Mutius and Vercelli 2010; Ege et al. 2011). It has been postulated that the more intense and diverse microbial exposure in the farming environment could serve as a protective quality by challenging the developing immune system of infants with a natural microbial material. So far, no individual microbial species or factor has been shown to have a critical role in this phenomenon.

The agricultural environment does not, however, protect from the occupational health risks that have been identified with many work situations. As mentioned before, in these cases, exposure levels are usually very high, i.e., several orders of magnitude higher than in the indoor air of the home, be it a farming home or an urban home.

As the adverse health effects of damp and moldy indoor environments and the beneficial protective effect in a farming environment are compared in parallel, an interesting paradox can be suggested. In both these domestic environments, microbial levels are elevated compared to "normal" urban homes with no moisture or mold problem (Hyvärinen et al. 2001; Green et al. 2003; van Strien et al. 2004). In homes that have observed dampness, moisture, or mold, there is a risk for adverse health effects such as irritative, allergic, or other respiratory health effects including an increased risk for asthma (e.g., Norbäck et al. 2013). In farming homes, a protective effect from allergies is seen (Genuneit 2012) while they have higher microbial concentrations due to many natural sources in their immediate surroundings (Ege et al. 2011). However, the farming environment does not necessarily protect from adverse effects of moisture damage exposure (Karvonen et al. 2009).

It is evident that microbial concentrations as such do not explain either the beneficial or the adverse health effects observed in domestic environments. However, when this phenomenon that appears as a paradox at a first glance is analyzed in more detail, it is clear that it is a question of different environments with different sources of microbial exposures. Microbial exposures are not a uniform concept but a highly variable range of microbial species, their cellular components, and metabolic products, with a wide array of potential biological effects. Not only do microorganisms have health-relevant potential but also other agents, such as pollen, protozoa, and animal allergens may contribute to the health effects of biological particles. The knowledge on the complex interactions with these agents and other air pollutants will increase in the future, facilitated with the significant methodological progress in the assessment of biological exposures.

Occupational diseases linked with biological particles

In occupational environments where organic material is being handled and processed, exposure levels to organic dusts and biological particles may be very high. Such occupational environments are found in, e.g., agriculture, food processing, animal care, tobacco processing, and waste handling and processing. Measurements have shown that the concentrations of viable fungi and bacteria in such environment may be 10^5–10^7 CFU m^{-3}, orders of magnitude higher than the concentrations in any other indoor environment (Table 7-1). The organic dusts in such occupational environments are usually mixtures of fungi, bacteria, and other microorganisms, combined with plant, animal, and waste material depending on the industry in question. Several types of occupational diseases are known to occur in such environments.

Hypersensitivity pneumonitis or allergic alveolitis is an occupational disease that may lead to irreversible impairment of lung function. It has been linked with exposures to very large concentrations of microbes and organic dust in, e.g., agricultural work, sawmills, and composting facilities (Kotimaa et al. 1984; Lacey and Crook 1988; Malmberg et al. 1993). The disease is named according to the occupation, e.g., farmer's lung, wood trimmer's disease, cheese washer's disease, or bird breeder's lung. The disease has even been observed in situations with much lower levels of exposure, such as office environments with dampness, moisture, and mold problems (Jarvis and Morey 2001).

The onset of hypersensitivity pneumonitis has traditionally been connected with fungi of the genera *Acremonium*, *Aspergillus*, *Aureobasidium*, *Cladosporium*, *Mucor*, *Penicillium*, *Rhizopus*, *Sporobolomyces*, and *Trichoderma* (doPico 1986; Lacey and Crook 1988; Richerson 1994; Levetin et al. 1995). Among bacteria, actinomycetes, especially species of *Streptomyces* and thermophilic species growing at temperatures 40–60°C, such as *Saccharopolyspora rectivirgula* and species of *Thermoactinomyces* have been associated with the disease in dairy farms and other occupational settings with high levels of organic dust (Kotimaa et al. 1984).

In agricultural environments, a disease called organic dust toxic syndrome (ODTS) has been described (doPico 1986). It is a non-infectious febrile illness resembling influenza with major symptoms being fever and malaise (Emanuel et al. 1975). It typically occurs after a heavy inhalation exposure to organic dust.

The risk of asthma has been shown in many occupational settings with exposure to biological dusts. Asthma has also been connected with exposure to moisture- and mold-affected indoor environments. Visible mold or mold odor was shown to be a risk factor for asthma in work places (Jaakkola et al. 2002). Clusters of asthma cases have also been reported in moisture damaged schools (Patovirta et al. 2004), hospitals (Seuri et al. 2000), offices (Yossifova et al. 2011), and other buildings (Bornehag et al. 2001).

References

Adler, A., I. Tager and D.R. Quintero. 2005. Decreased prevalence of asthma among farm-reared children compared with those who are rural but not farm-reared. J. Allergy Clin. Immunol. 115: 67–73.

Arnow, P.M., T. Chou, D. Weil, E.N. Shapiro and C. Kretzschmar. 1982. Nosocomial legionnaire's disease caused by aerosolized tap water from respiratory devices. J. Infect. Dis. 146: 460–467.

Bornehag, C.G., G. Blomquist, F. Gyntelberg, B. Järvholm, P. Malmberg, L. Nordvall, A. Nielsen, G. Pershagen and J. Sundell. 2001. Dampness in buildings and health. Nordic interdisciplinary review of the scientific evidence on associations between exposure to "dampness" in buildings and health effects (NORDDAMP). Indoor Air 11: 72–86.

Braun-Fahrländer, Ch., M. Gassner, L. Grize, U. Neu, H. Sennhauser, H.S. Varonier, J.C. Vuille, B. Wutrich and the Scarpol team. 1999. Prevalence of hay fever and allergic sensitization in farmers' children and their peers living in the same rural community. Clin. Exp. Allergy 29: 28–34.

doPico, G.A. 1986. Health effects of organic dusts in the farming environment. Report of diseases. Am. J. Ind. Med. 10: 261–265.

Ege, M., M. Mayer, A.-C. Normand, J. Genuneit, W.O. Cookson, C. Braun-Fahrländer, D. Heederik, R. Piarroux, E. von Mutius and the GABRIELA/TR22 study group. 2011. Exposure to environmental microorganisms and childhood asthma. N. Engl. J. Med. 364: 701–709.

Emanuel, D.A., F.J. Wenzel and B.R. Lawton. 1975. Pulmonary mycotoxicosis. Chest 67: 293–297.

Genuneit, J. 2012. Exposure to farming environments in childhood and asthma and wheeze in rural populations: a systematic review with meta-analysis. Pediatr. Allergy Immunol. 23: 509–518.

Göthe, C.J., A. Westlin and S. Sundquist. 1972. Air-borne *B. subtilis* enzymes in the detergent industry. Int. Arch. Arbeitsmed. 29: 201–208.

Green, C.F., P.V. Scarpino and S.G. Gibbs. 2003. Assessment and modeling of indoor fungal and bacterial concentrations. Aerobiologia 19: 159–169.

Heine, H., E.T. Rietschel and A.J. Ulmer. 2001. The biology of endotoxin. Mol. Biotechnol. 19: 279–296.

Hirvonen, M.R., M. Ruotsalainen, M. Roponen, A. Hyvärinen, T. Husman, V.M. Kosma, H. Komulainen, K. Savolainen and A. Nevalainen. 1999. Nitric oxide and proinflammatory cytokines in nasal lavage fluid associated with symptoms and exposure to moldy building microbes. Am. J. Respir. Crit. Care Med. 160: 1943–1946.

Huttunen, K., A. Hyvärinen, A. Nevalainen, H. Komulainen and M.-R. Hirvonen. 2003. Production of proinflammatory mediators by indoor air bacteria and fungal spores in mouse and human cell lines. Environ. Health Perspect. 111: 85–92.

Hyvärinen, A.M., M. Vahteristo, T. Meklin, M. Jantunen, A. Nevalainen and D. Moschandreas. 2001. Temporal and spatial variation of fungal concentrations in indoor air. Aerosol Sci. Technol. 35: 688–695.

IOM (Institute of Medicine). 2004. Damp Indoor Spaces and Health. National Academies Press, Washington DC, USA.

Jaakkola, M.S., H. Nordman, R. Piipari, J. Uitti, J. Laitinen, A. Karjalainen, P. Hahtola and J.K. Jaakkola. 2002. Indoor dampness and molds and development of adult-onset asthma: a population-based incident case-control study. Environ. Health Perspect. 110: 543–547.

Jarvis, J.Q. and P.R. Morey. 2001. Allergic respiratory disease and fungal remediation in a building in a subtropical climate. Appl. Occup. Environ. Hyg. 16: 380–388.

Kanchongkittiphon, W., M.J. Mendell, J.M. Gaffin, G. Wang and W. Phipatanakul. 2015. Indoor environmental exposures and exacerbation of asthma: an update to the 2000 review by the Institute of Medicine. Environ. Health Perspect. 123(1): 6–20.

Karvonen, A.M., A. Hyvärinen, M. Roponen, M. Hoffmann, M. Korppi, S. Remes, E. von Mutius, A. Nevalainen and J. Pekkanen. 2009. Confirmed moisture damage at home, respiratory symptoms and atopy in early life: a birth-cohort study. Pediatrics 124: e329–e338.

Kotimaa, M., K. Husman, E.O. Terho and M. Mustonen. 1984. Airborne molds and actinomycetes in the work environment of farmer's lung patients in Finland. Scand, J. Work Environ. Health 10: 115–119.

Kusnetsov, J. 1997. Isolation, occurrence and prevention of Legionella in Finnish cooling water systems. PhD dissertation. Publications of the National Public Health Institute A5/1997, Kuopio, Finland.

Lacey, J. and B. Crook. 1988. Fungal and actinomycete spores as pollutants of the workplace and occupational allergens. Ann. Occup. Hyg. 32: 515–533.

Levetin, E., R. Shaughnessy, E. Fisher, B. Ligman, J. Harrison and T. Brennan. 1995. Indoor air quality in schools: exposure to fungal allergens. Aerobiologia 11: 27–34.

Malmberg, P., A. Rask-Andersen and L. Rosenhall. 1993. Exposure to microorganisms associated with allergic alveolitis and febrile reactions to mold dust in farmers. Chest 103: 287–293.

Mendell, M.J., A.G. Mirer, K. Cheung, M. Tong and J. Douwes. 2011. Respiratory and allergic health effects of dampness, mold and dampness-related agents: a review of the epidemiologic evidence. Environ. Health Perspect. 119: 748–756.

Miller, J.D. 2011. Health effects from mold and dampness in housing in western societies: early epidemiology studies and barriers to further progress. pp. 183–210. *In*: O.C.G. Adan and R.A. Samson [eds.]. Fundamentals of Mold Growth in Indoor Environments. Wageningen Academic Publishers, Wageningen, The Netherlands.

Miller, J.D., M. Sun, A. Gilyan, J. Roy and T.G. Rand. 2010. Inflammation-associated gene transcription and expression in mouse lungs induced by low-molecular weight compounds from fungi from the built environment. Chem. Biol. Interact. 183: 113–124.

Norbäck, D., J.P. Zock, E. Plana, J. Heinrich, C. Svanes, J. Sunyer, N. Künzli, S. Villani, M. Olivieri, A. Soon and D. Jarvis. 2013. Mould and dampness in dwelling places and onset of asthma: the population-based cohort ECRHS. Occup. Environ. Med. 70: 325–331.

Patovirta, R.L., T. Husman, U. Haverinen, M. Vahteristo, J.A. Uitti, H. Tukiainen and A. Nevalainen. 2004. The remediation of mold damaged school—a three-year follow-up study on teachers' health. Cent. Eur. J. Public Health 12: 36–42.

Richerson, H.B. 1994. Hypersensitivity pneumonitis. pp. 139–160. *In*: R. Rylander and R.R. Jacobs [eds.]. Organic Dusts: Exposure, Effects and Prevention. CRC Press, Inc., Boca Raton, FL, USA.

Roponen, M., M. Seuri, A. Nevalainen and M.K.R. Hirvonen. 2002. Fungal spores as such do not cause nasal inflammation in mold exposure. Inhal. Toxicol. 14: 541–549.

Roponen, M., M. Toivola, S. Alm, A. Nevalainen, J. Jussila and M.R. Hirvonen. 2003. Inflammatory and cytotoxic potential of the airborne particle material assessed by nasal lavage and cell exposure methods. Inhal. Toxicol. 15: 23–28.

Rudblad, S., K. Andersson, G. Stridh, L. Bodin and JE. Juto. 2002. Slowly decreasing mucosal hyperreactivity years after working in a school with moisture problems. Indoor Air 12: 138–144.

Seuri, M., K. Husman, H. Kinnunen, M. Reiman, R. Kreus, P. Kuronen, K. Lehtomäki and M. Paananen. 2000. An outbreak of respiratory diseases among workers at a water-damaged building—a case report. Indoor Air 10: 138–145.

Taskinen, T., S. Laitinen, A. Hyvärinen, T. Meklin, T. Husman, A. Nevalainen and M. Korppi. 2001. Mold-specific IgG antibodies in relation to exposure and skin test data in school children. Allergol. Internat. 50: 239–245.f.

Tischer, C., C.M. Chen and J. Heinrich. 2011. Association between domestic mould and mould components, and asthma and allergy in children: a systematic review. Eur. Resp. J. 38: 812–24.

Tobin, J.O'H., J. Beare, M.S. Dunnill, S. Fisher-Hoch, M. French, R.G. Mitchell, P.J. Morris and M.F. Muers. 1980. Legionnaire's disease in a transplant unit: isolation of the causative agent from shower baths. Lancet 2(8186): 118–121.

van Strien, R.T., R. Engel, O. Holst, A. Bufe, W. Eder, M. Waser, C. Braun-Fahrländer, J. Riedler, D. Nowak, E. von Mutius and the ALEX study team. 2004. Microbial exposure of rural school children,

as assessed by levels of N-acetyl-muramic acid in mattress dust, and its association with respiratory health. J. Allergy Clin. Immunol. 113: 860–867.
von Mutius, E. and D. Vercelli. 2010. Farm living: effects of childhood asthma and allergy. Nat. Rev. Immunol. 10: 861–868.
WHO. 2009. WHO Guidelines for Indoor Air Quality: Dampness and Mould. World Health Organization, Copenhagen, Denmark, 228 pgs.
Yossifova, Y.Y., J.M. Cox-Ganser, J.H. Park, S.K. White and K. Kreiss. 2011. Lack of respiratory improvement following remediation of a water-damaged office building. Am. J. Ind. Med. 54: 269–272.

PART 2

Study Design to Receive the Information about the Influence of Air Pollutants on Human Health

8

Short Introduction to Inhalation Exposure Methodology

Jozef S. Pastuszka

Background

Epidemiological studies may show a relation between air pollutants (e.g., airborne particulate matter) and some adverse health effects (e.g., cardiopulmonary morbidity and mortality). However, an inhalation exposure study is necessary in order to identify the mechanisms triggered by air pollution and causing these health effects. For example, Saunders et al. (2002) found the process of modulation of neurotic behavior in F-344 rats by temporal disposition of benzo(a)pyrene. The existing knowledge on the nature of these mechanisms likely relates to the insidious and multifactorial nature of the underlying pathophysiologic processes (Bartoli et al. 2010). The experimental nature of the inhalation exposure studies conducted on animals and sometimes on people offers a chance for controlled investigations. A well designed laboratory simulating real or predicted exposure is commonly believed to be a reliable source of information on the adverse health effects of inhaling various air pollutants. The results of such experiments, apart from determining the dose-effect relationships, can contribute to the extension of current knowledge. For example, Kleinman et al. (2000) studied toxicity of chemical components of ambient fine particulate matter inhaled by aged rats. Another pulmonary toxicity study in rats with PM10 and PM2.5 carried out by Zhang et al. (2011) revealed differential responses related to the scale and composition of these airborne particles.

Silesian University of Technology, Division of Energy and Environmental Engineering, Department of Air Protection, 22B Konarskiego St., 44-100 Gliwice, Poland.
E-mail: Jozef.Pastuszka@polsl.pl

Delivering a well-controlled and characterized dose of air pollutants for a specified period of time is the fundamental function of an inhalation exposure system. Exposure duration may last between minutes (single acute exposures) and several years (chronic exposures). However preferable in most inhalation studies, the elimination or limitation of exposure through skin, eyes or other non-respiratory pathways, such as contaminated food, is not always possible.

Design and operating principles

A number of authors have previously reviewed the design and operating principles of inhalation exposure systems (for example, Silver 1946; Lippmann 1970; Dorato and Wolff 1991; Karg et al. 1992; Phalen et al. 1994). The above mentioned, along with some of the newest articles, are a recommended source for finding more information on this subject. The amount of human or animal subject in contact with the studied polluted air is the basis for general classification of exposure systems (Table 8-1).

The possibility of simultaneous exposures of many subjects is the main advantage of whole-body exposures. A well designed and well maintained large chamber is quite humane for long term exposures as it allows the subjects to move around freely. The whole-body exposure chambers also have some disadvantages, such as the occurrence of substantial eye, skin, and gastrointestinal exposures. Apart from being

Table 8-1. Summary of major advantages and disadvantages associated with various modes of exposure (from Phalen et al. 1994).

Mode of exposure	Advantages	Disadvantages
Chambers	• large number of subjects	• dermal, eye, and oral exposure in addition to inhalation
	• suitable for chronic studies	• large amounts of test material required
	• minimum restraint	• large air cleaning requirement which is expensive
	• labor efficient	• excreta can interact with pollutants
Head-only	• good for repeated exposure	• can be stressful
	• minimal skin contamination	• losses can be large
	• more efficient dose delivery	• neck seal problems
	• better control of dose	• labor intensive
Nose/mouth-only	• no skin contamination	• can be stressful
	• can be used for repeated exposures	• needs good face seal
	• uses much less material exposures can be pulsed personnel and facility	• labor intensive
	• contamination minimized	• technically difficult
Lung-only	• precision of dose	• anesthesia or tracheotomy bypasses nose (could be an advantage) artifacts in deposition and response
	• uses less material (efficient)	• technically most difficult

expensive, the chambers may require large amounts of study material and throughput air. Contamination of the air inside the chamber with animal fur, dander, excretions, or food is frequent. There are, however, several successful designs of the chambers. They include chambers designed for long-term housing and chambers which allow controlled exercise during short inhalation exposures.

Head-only exposure systems have an advantage over whole-body exposure chambers due to the fact that contamination of the air with animal fur, dander, excretions, or food, so frequent in whole-body chambers, is diminished. They also facilitate monitoring and control of the inhaled study material. In terms of disadvantages, head-only exposure systems limit head movement and may require maintaining seals around the neck and may therefore be stressful for the subject.

There are several advantages of nose-only or mouth-only exposures. These systems eliminate skin and eye exposure and therefore they provide exact monitoring and identification of the inhaled and exhaled air contaminant concentrations. The so called 'nasal provocation tests' are popular among these types of studies with the use of human subjects as well (see, for example, Braga et al. 2004).

Some systems are designed to provide continuous measurement of breathing patterns of small animals during exposure. Nevertheless, whole-body or head-only exposure systems are less stressful for the subjects than nose-only or mouth-only exposure systems. Moreover, establishing and maintaining tight seals against the nose, mouth, or face is difficult.

Very precise delivery of exposure study materials even in very large doses over a short period of time is possible through direct lung-only (or some parts of the lung) exposure. In case of partial-lung exposure it is possible to use the unexposed parts of the lung as control tissue to evaluate the effects of the tested substance. However, for such exposures, the study material is delivered in an artificial way and there is a probability that sedation, anesthesia, or surgery will be required to carry out the study.

The necessity to house the animals when they are subject to long-term exposures causes the exposure systems to become more complex. Monitoring of the study material ideally involves sampling from the 'breathing zone' of the subjects. The purity of air in the exposure system is very important and, apart from the study material, any unintentional pollution with exhaled water vapor, ammonia, or carbon dioxide should be avoided. The temperature and humidity should be maintained at an appropriate level, comfortable for the subject. Stressors such as noise, vibration, light, and confinement should also be taken into account when studying conscious subjects so as to make the exposure environment humane.

Phalen et al. (1994) have described various types of exposure systems in their excellent paper. Below are some of the most important issues brought up in the paper, related with selecting and applying of these systems.

In a static system, an agent is introduced into a chamber as a bolus, and the air is subsequently mixed. Static systems have many limitations, the limited volume of the chamber being one of the main issues and reasons for which these systems are generally not used to determine inhalation toxicity (Phalen et al. 1994).

Another type of exposure is whole body exposure, for which stainless steel and glass chambers are usually used.

There are many head-only exposure systems' designs depending on the size and kind of studied animal species. A body tube is used for small animals, with the head of the animal protruding from the tube. Scheimberg et al. (1973) described exposures of monkeys to aerosols in which a small individual helmet exposure chamber was used. Bowes et al. (1990) described a "head dome" which allows inhalation exposures of people during physical exercises. Head-only exposure systems allow the performance of repeated brief exposures. They also limit the number of possible ways in which the study material may enter the subject. However, avoiding inhalation exposure in such systems is nearly impossible.

Masks, catheters in the nose or mouth, or individual containers with one end to the exposure atmosphere, limit the inhalation exposure to nasal or oral cavity and are usually used for nose- or mouth-only exposure systems. Their general design considerations are similar to those of the head-only systems. However, mask design presents specific problems. The mask ought to be tight but comfortable for the subject and either collect or drain saliva. Therefore, successful masks and nasal tubes are most often handcrafted.

Different techniques are used for lung-only or partial lung-only exposures. An oral tracheal tube is usually applied to anaesthetized animals to introduce air pollutants for intra-tracheal exposure.

Modern inhalation studies

The development of modern exposure study designs brought more complexity, realism, and better definition into the field of study. We can observe a growing frequency of studies on mixtures of agents as well as application of better methodology for characterizing the study materials. A novel exposure system for large animals was developed by Bartoli et al. (2010). Permanent tracheotomies and a cuffed endotracheal tube are used for large animals to breathe through from a closed inhalation exposure system. These devices are applied for each exposure. With the use of this modern equipment and new research technique, Bartoli recently documented evidence of acute cardiovascular alterations in chronically instrumented or anesthetized large animals exposed to concentrated ambient particles.

References

Bartoli, C.R., J.J. Godleski and R.L. Verrier. 2010. Mechanism mediating adverse effects of air pollution on cardiovascular hemodynamic function and vulnerability to cardiac arrhythmias. Air Qual. Atmos. Health, DOI 10.1007/s11869-010-0091-6.

Bowes, S.M., R. Frank and D.L. Swift. 1990. The head dome: a simplified method for human exposures to inhaled air pollutants. Am. Ind. Hyg. Assoc. J. 51: 257–260.

Braga, C.R., M.C. Rizzo, C.K. Naspitz and D. Sole. 2004. Nasal provocation test (NPT) with isolated and associated dermatophagoides pteronyssinus (Dp) and endotoxin lipopolysaccharide (LPS) in children with allergic rhinitis (AR) and nonallergic controls. J. Investig. Allergol. and Clin. Immunol. 14(2): 142–148.

Dorato, M.A. and R.K. Wolff. 1991. Inhalation exposure technology, dosimetry, and regulatory issues. Toxicol. Pathol. 19: 373–383.

Karg, E., Th. Tuch, G.A. Ferron, B. Haider, V.G. Kreyling, J. Peter, L. Ruprecht and J. Heyder. 1992. Design, operation and performance of whole body chambers for long-term aerosol exposure of large experimental animals. J. Aerosol Sci. 23: 279–290.

Kleinman, M.T., C. Bufalin, R. Rasmussen, D. Hyde, D.K. Bhalla and W.J. Mautz. 2000. Toxicity of chemical components of ambient fine particulate matter (PM2.5) inhaled by aged rats. J. Appl. Toxicol. 20: 357–364.

Lippmann, M. 1970. Experimental inhalation studies equipment and procedures. pp. 55–76. *In*: M.G. Hanna, Jr., P. Nettesheim and J.R. Gilbert [eds.]. Inhalation carcinogenesis. US Atomic Energy Commission, Oak Ridge, Tennesse, USA.

Phalen, R.F., M.T. Kleinman, W.J. Mautz and R.T. Drew. 1994. Inhalation exposure methodology. pp. 59–82. *In*: P.G. Jenkins, D. Kayser, H. Muhle, G. Rosner and E.M. Smith [eds.]. The Proceedings of the International Symposium of the Respiratory Toxicology and Risk Assessment. Wissenschaftliche Verlagsgesellschaft mbH, Stuttgart, Germany.

Saunders, C.R., A. Ramesh and D.C. Shokley. 2002. Modulation of neurotic behavior in F-344 rats by temporal disposition of benzo(a)pyrene. Toxicological Letters 129: 33–45.

Scheimberg, J., O.P. McShane, S. Carson, H.E. Swann and A.M. Blair. 1973. Inhalation of a powdered aerosol medication by non-human primates in individual space-types exposure helmets (abstract). Toxicol. Appl. Pharmacol. 25: 478.

Silver, S.D. 1946. Constant flow gassing chambers: principles influencing design and operation. J. Lab. Clin. Med. 31: 1153–1161.

Zhang, W., T. Lei, Z.-Q. Lin, H.-S. Zhang, D.-F. Yang, Z.-G. Xi, J.-H. Chen and W. Wang. 2011. Pulmonary toxicity study in rats with PM10 and PM2.5: differential responses related to scale and composition. Atmos. Environ. 45: 1034–1041.

9

Introduction to Environmental Epidemiology

*Kinga Polańska** and *Wojciech Hanke*[a]

Types of epidemiological studies

Two basic kinds of observational epidemiological studies have been conducted to determine risks associated with air pollution: descriptive (such as ecological studies) and analytical (such as: cross-sectional, case-control, and cohort studies) (Table 9-1). These two study approaches differ primarily in the supportive evidence they provide about a possible causal association. Unlike an analytical study, an *Ecological Study* does not link individual outcome events to the individual exposure or confounding characteristics, and it does not link individual exposure and confounding characteristics with one another. In an ecological study, information about exposure and disease is available only for groups of people, and critical information can be lost in the process of aggregating these data. Results from ecological studies are difficult to interpret, and serious errors may occur when it is assumed that inferences from an ecological analysis pertain either to the individuals within the group or to individuals across the groups. In the time series analysis, associations between air pollution metrics and morbidity or mortality are analysed. Log-linear regression models express the expected total number of events on each day as a function of the exposure level and potential confounding variables.

Analytical Studies can provide necessary information to help evaluate the causality of an association and estimate magnitude of the risk. For each person included in the study, information concerning their exposure and disease status as well as confounding characteristics is obtained.

Nofer Institute of Occupational Medicine, Department of Environmental Epidemiology, 8 Saint Teresa St., 90-950 Łódź, Poland.

[a] E-mail: wojt@imp.lodz.pl

* Corresponding author: kinga@imp.lodz.pl

Table 9-1. Main study designs used in the environmental epidemiology.

Type of study	Study population	Exposure data	Outcome data	Confounding factors	Temporal exposure—disease relationship	Advantages	Disadvantages
Ecological study	Total number of exposed and unexposed individuals and the total number of the diseased and nondiseased individuals.	Exposure data (for populations) from existing records (at a country or regional level).	Outcome data (for populations) from existing records (at a country or regional level).	Data about confounding factors are unavailable.	Hard to establish.	Can be done quickly and inexpensively, often using available information.	Lack of ability to control the effects of potential confounding factors. Inability to link exposure with a disease in the individuals. Can be difficult to distinguish cause and effect.
Time series analysis	Associations between air pollution metrics and morbidity or mortality are analysed. Log-linear regression models express the expected total number of events on each day as a function of the exposure level and potential confounding variables.	Exposure data from existing records (at a country or regional level).	Health effects registers/ hospital admission data.	Smooth functions of time and weather are the main confounders.	Sometimes hard to establish.	No need for control population.	Good quality data for air pollution. The possibility of health outcome misclassification. Strict case definitions required.
Cross-sectional survey	Individuals in whom both exposure status and disease status are measured in the study subjects at one point of time or over a short period of time.	Actual assessment based on questionnaires or measurements.	Actual assessment based on questionnaires or measurements.	Can be assessed using a questionnaire.	Hard to establish.	Often based on a sample of the general population. Tend to be carried out over a relatively short period of time. Can be utilized as cohorts in the future.	A series of prevalent cases will have a higher proportion of cases with a disease of long duration than a series of incidental cases.

Table 9-1. contd....

Table 9-1. contd.

Type of study	Study population	Exposure data	Outcome data	Confounding factors	Temporal exposure—disease relationship	Advantages	Disadvantages
Prospective cohort study	Exposed and non exposed persons.	Assessed at baseline and updated at a follow-up.	Assessed at a follow-up.	Easy to obtain.	Easy to establish.	Efficient for study of rare exposures. Can yield information on multiple exposures and multiple outcomes. Minimizes bias in the ascertainment of exposure.	Time-consuming and expensive. Often requires a large sample size. Inefficient for evaluation of rare diseases. Validity of the results can be seriously affected by losses in follow-up. Changes over time in a diagnostic method.
Retrospective cohort study	Exposed and non exposed persons.	Retrospective data based on existing records.	Mostly for the mortality analysis. Data based on existing records.	Usually difficult to measure because of the lack of existing data.	Sometimes hard to establish.	Less expensive than a prospective study.	Susceptible to bias both in assessment of exposure and outcome. Changes over time in a diagnostic method.
Case-control study	Persons with the disease (cases) and persons without the disease (controls).	Retrospective data, obtained mostly based on a questionnaire.	Established at the beginning of the study.	Can be obtained based on a questionnaire.	Sometimes hard to establish.	Relatively quick and inexpensive. Requires a smaller number of subjects. Well-suited to the evaluation of diseases with long latent periods. Optimal for evaluation of rare diseases. Can examine multiple etiologic factors for a single disease.	Inefficient for evaluation of rare exposures, unless the attributable risk is high. Particularly prone to bias (selection bias and recall bias).

Case-crossover	Subject's characteristics and exposures at the time of a health event (case period) are compared with another time period when that subject was a noncase (control period).	Retrospective data, obtained mostly based on a questionnaire.	Established at the beginning of the study.	Control for stable subject-specific covariates is not necessary.	Sometimes hard to establish.	No need for recruitment of the controls.	Weather factors such as temperature and precipitation should be controlled as possible confounders. Recall bias may possibly occur when the information is not "recorded" and is recalled by the individuals.

In a *Cross-Sectional Study*, both exposure status and disease status are measured in the study subjects at one point of time or over a short period of time. Although a cross-sectional study can be suggestive of possible risk factors for a disease, when an association is found in such a study, given the limitations in establishing a temporal relationship between exposure and outcome, to establish etiologic relationships we rely on case-control and cohort studies.

In a *Case-Control Study*, persons with a given disease (cases) and persons without the disease (controls) are selected, and thereafter, the proportions of cases and controls who have been exposed to potential risk factors are determined and compared. The cases are usually selected from among persons seeking medical care for the disease(s) under the study. Most frequently, persons seeking medical care for conditions believed to be unrelated to the cases' diagnoses are used as the group of controls. Another option is to select controls as probability samples from the population which the cases come from. In a case-control study, multiple exposures can be evaluated, and a relatively small number of study participants is needed to obtain reasonably precise estimates of risk associated with environmental exposures. It is important to remember that this type of study is particularly prone to bias compared with other analytic designs, especially selection bias and recall bias regarding the exposure and confounding variables. In addition, in some situations, temporal relationship between exposure and a disease may be difficult to establish.

In a *Cohort Study*, the investigator selects a group of exposed and non exposed individuals and follows both groups to compare the incidence of the disease (or rate of death due to the disease) in these two groups. The cohort study can be either retrospective or prospective. A multiple exposure and disease end-points can be evaluated, but the fact that large numbers of people must be studied, especially for environmental exposures, is an important disadvantage. Because of a lengthy latent period for cardiovascular diseases and cancer, a long follow-up period is required for a prospective cohort, and this is usually not feasible. In a retrospective cohort study, the investigator identifies a cohort of individuals based on their characteristics in the past and then, reconstructs their subsequent disease experience up to some defined point in the more recent past, or occasionally in the future.

A study design that has been used more and more frequently in recent years is a *Nested Case-Control Study*—a design in which a case-control study is nested in the cohort one. In this type of a study, a population is identified and followed over time. At the time the population is identified, baseline data are obtained from questionnaires or measurements. The population is then followed for a period of years and then a case-control study is carried out using persons in whom the disease developed (cases) and a sample of those in whom the disease did not develop (controls). As interviews (or measurements) are performed at baseline, the advantage of this study design is a clear temporal sequence and elimination of recall bias. Such studies allow researchers to select more detailed information for the subpopulation.

The other type of a study used for the assessment of impact of air pollution on health outcomes is a *Case Crossover Analysis*, in which instead of obtaining information from 2 groups (cases and controls), the exposure information is obtained from the same case group but during two different periods of time. In the first period, exposure is measured immediately before the disease onset. In the second period, exposure is measured at an earlier time (supposed to represent background exposure in the same person). Exposure among cases just prior to the disease onset is then compared to the exposure among the same cases at an earlier time. Each case and its matched control (himself), therefore, automatically match with regard to many characteristics (age, sex, socio-economic status, etc.).

Application to air pollution studies: conclusions from the literature review

1) Sources of air pollution emissions relevant to health include: traffic, household use of biomass fuels for heating, power generation (oil and coal), and biomass combustion—most notably residential wood combustion. However, because of considerable intercorrelations between particle constituents in the ambient air, the detection of associations in epidemiological studies is difficult and usually not sufficient to judge about causality. Greater understanding of the most 'responsible' pollutants for causing the greatest diseases burden should be done in relation to the sources (traffic vs. low emission), chemicals, and susceptible groups. As a result the most hazardous pollutants can be identified and effective control measures introduced (WHO 2013).
2) Effects of 24-hour exposures to air pollution have been observed after 0 to 5 days also in regions with relatively low annual averages of pollution and often at low concentrations (PM10 below 10 $\mu g\ m^{-3}$). A larger annual declines in lung function were linked with higher levels of PM10. The ability to separate long-term health effects of PM10 from other pollutants, such as NO_2 or PM2.5, has been limited so far.
3) The relationship between PM2.5 and lung cancer has been examined in several studies and all of them have provided positive findings. Possible misclassification of exposure and lack of possibility to control for important confounders (smoking, SES) did not allow for definite conclusions regarding living in the area close to heavy traffic and the risk of lung cancer.
4) A large number of epidemiological studies have presented the association indicating an increased risk of cardiovascular diseases in populations exposed to air pollution. Air pollution cardiovascular health outcomes include mortality and morbidity due to: sudden death, arrhythmias, myocardial infarction, strokes, heart failure, and conventional cardiovascular well known risk factors such as diabetes and hypertension. No evidence of safe PM threshold has been observed.
5) The adverse effects of exposure to daily ozone concentrations, after adjustment for the effects of particles (PM10), have been reported in the case of both respiratory and cardiovascular hospital admissions. However, since high temperatures are associated with increased mortality, separation of the health effects of ozone from those of temperature is still problematic. The influence of long-term exposure to ozone on the increased risk of respiratory and cardiorespiratory mortality is well documented and strongly postulated for asthma incidence, asthma severity, hospital care for asthma and lung function growth.

6) As far as biological mechanism is concerned, a major uncertainty is the pathway(s) whereby air pollution in the lungs affects the circulatory system. Identification of the main mediators will provide an opportunity to start a process of developing measures to mitigate cardiovascular response effectively. In the meantime, continuous efforts should be made to keep the concentrations of air pollution as low as possible (WHO 2013).
7) Several "natural experiments" have demonstrated the benefits of reduction of air pollution exposure. They included an increased life expectancy resulting mostly from the decreases in cardiovascular and respiratory mortality.
8) Shorter exposure window, compared to time spans from the studies of chronic exposure to air pollution in children and adults, is an important advantage of the studies focusing on the impact of air pollution on birth outcomes. This can result in better evaluation of the exposures of interest and in the decrease of the possibility of other risk factors influencing the outcomes.
9) The results of the studies evaluating the impact of air pollution (with the measurement of different pollutants, using different methods of exposure assessment and using different study design) and preterm birth are mixed and inconclusive, although some of them indicate that such an association exists.
10) Most of the existing studies indicate the association between exposure to air pollution and some restrictions in fetal growth.

Reference

WHO Regional Office for Europe. 2013. Review of evidence on health aspects of air pollution—REVIHAAP Project Technical Report.

10

Epidemiological Methods Used to Receive Information on the Influence of Air Pollution on Human Health

*Kinga Polańska** and *Wojciech Hanke*[a]

Background

Environmental epidemiology focuses on identification of environmental exposures that contribute to or protect against injuries, illnesses, developmental conditions, disabilities and deaths, and on identification of public health and health care actions to avoid, prepare for, and effectively manage the risks associated with harmful exposures. Air pollution represents the most common type of such exposure and is defined as the introduction of chemicals, particulates, or biological materials that cause discomfort, disease, or death to humans in the atmosphere. Compared with other harmful factors that may affect human development and health, such as exposure to tobacco smoke, exposure to air pollution cannot be fully avoided and affects large numbers of individuals. In addition, although an individual hazard may be smaller than in the case of other harmful factors, the ubiquitous distribution makes it of interest from epidemiology and public health perspective.

The major strength of environmental epidemiological studies is their ability to assess the relationship between environmental exposures and wide-range of health conditions in humans. This, however, is a difficult task. The relationship between exposure to air pollution and health outcomes recognized as causative has to fulfill several conditions, i.e., (a) relationship in time, (b) strength of association, (c) dose-response relationship, (d) replication of the findings, (e) biologic plausibility, (f)

Nofer Institute of Occupational Medicine, Department of Environmental Epidemiology, 8 Saint Teresa St., 90-950 Łódź, Poland.

[a] E-mail: wojt@imp.lodz.pl

* Corresponding author: kinga@imp.lodz.pl

cessation of exposure, (g) specificity of the association, and (h) consistency with other knowledge and validity of the measures of exposure and biological outcomes. Identification of possible confounders and co-exposures and correction of risk estimates for their effects is of major importance.

This chapter reviews the major directions of research on health outcomes of air pollution in both adult and newborn populations simultaneously emphasizing important, methodological issues typical for studies of both groups of populations. Many, however not all, methodological difficulties have been overcome in epidemiological studies on adult populations. On the other hand, in the case of newborn studies, substantial progress is still expected. The role of problems related to exposure assessment, and confounders control will be presented. Finally, a short outline of the identified future research for each reviewed topic will be provided.

Health effects of air pollution on adults

Introduction

Severe pollution events, like the fog episode in London 1952, in which thousands of people died, brought health effects of pollution to public attention (Logan 1953). Since that time, a number of air pollution sources have been associated with different types of health effects. Most of the evidence accumulated so far is for an adverse effect on health of carbonaceous materials from coal combustion, which results in sulfate-contaminated particles. Sources of air pollution emissions relevant to health include: traffic, household use of biomass fuels for heating, power generation (oil and coal), and biomass combustion—most notably residential wood combustion. There are, typically, considerable intercorrelations between particle constituents in the ambient air, especially between constituents from the same source. This is one of the reasons why detection of associations in epidemiological studies is difficult and it is usually not sufficient to judge about causality.

Study designs—from a cross-sectional to time-series and cross-over design

The type of epidemiological studies include short-term, ultra acute, and long-term (prospective) models. Short-term studies link health events like deaths or hospitalization with daily changes of exposure to air pollution in a given area. Case cross-over design is related to a few hours observation and is convenient for detecting the ultra-acute effects. In the long-term studies, survival of individuals in relation to the changing levels of air pollution concentrations is measured in the area of their residence. In the prospective cohort studies, a sample of individuals is selected and followed over time. The Harvard Six Cities Study—a 15-year prospective study based on approximately 8000 individuals from six cities in the eastern United States (Dockery et al. 1993)—is an excellent example. Another example of a prospective study is the mortality experience of about 550,000 individuals from 151 cities in the United States, sponsored by the American Cancer Society (Pope et al. 2002). Those studies used individual-level data, so that other factors that affect mortality could be taken into account. In those and similar long-term studies, several causes of mortality were examined, including cardiopulmonary diseases and lung cancer.

Health outcomes of PM in adults

Respiratory outcomes: short- and long-term effects of air pollution

Single- and multicity studies, mostly from the United States, described associations between 24-hour average exposure to PM2.5 and hospital admissions as well as mortality due to cardiorespiratory health problems. Based on them, in 2005 a WHO 24-hour guideline for PM2.5 of 25 μg m^{-3} was proposed. Since that time, several new multicity studies have indicated increases of 0.4–1% per 10 μg m^{-3} in daily mortality associated with PM2.5 and PM10 (Ostro et al. 2006; Katsouyanni et al. 2009; Zanobetti and Schwartz 2009).

Effects of 24-hour exposure to air pollution have been observed after 0 to 5 days also in regions with relatively low annual averages of pollution (Burnett et al. 2004; Dominici et al. 2007). Short-term exposure to PM2.5 may lead to the early mortality of tens of thousands of individuals per year in the United States (Brook et al. 2010). For more information we refer to the previous reviews (Brook et al. 2010; EPA 2009; Rückerl et al. 2011; Anderson et al. 2007; WHO 2013).

In the case of long-term exposure to PM2.5, a significant number of prospective cohort studies have provided a well-documented evidence of the effect of such exposure on mortality. Effects have been observed at lower concentration (mean 14.10 ± 12.86 μg m^{-3}) levels than in the earlier studies (Pope et al. 2009).

In a large Canadian study, associations persisted at very low concentrations (mean 8.7 μg m^{-3}; interquartile range, 6.2 μg m^{-3}) (Crouse et al. 2012). In Chen's systematic review of six cohorts, the estimate was 6% increase in the non-acidental mortality for every 10 μg m^{-3} of PM2.5 (Chen et al. 2008). An identical estimate was obtained in the Dutch mortality cohort study for PM2.5 (Beelen et al. 2008a). Living close to busy roads was linked to the increased risk (Beelen et al. 2008; Gehring et al. 2006; Jerrett et al. 2005). A review of most of those and related studies can be found in the United States Environmental Protection Agency (EPA) integrated science assessment for PM (EPA 2009).

The ability to separate long-term health effects of PM10 from other pollutants, such as NO_2 or PM2.5, has been limited so far. However, in the Nurses' Health Study, independent associations of PM2.5 and coarse particles with total mortality have been observed (Puett et al. 2009). The findings were not confirmed in men neither for PM2.5 nor for coarse particles (Puett et al. 2011).

Lung cancer

The relationship between PM2.5 and lung cancer was examined in five studies and all of them provided positive findings (Chen et al. 2008). Based on this data, the pooled estimate was 21% increase in the risk of lung cancer per every 10 μg m^{-3} increase in PM2.5. Studies which controlled for smoking had lower estimates than those that did not do it—15% and 34% respectively. Possible misclassification of exposure and the lack of possibility to control for important confounders (smoking, SES) did not allow for definite conclusions regarding living in the area close to heavy traffic and the risk of lung cancer.

Respiratory function

In several studies associations of exposure to PM2.5 with asthma, chronic obstructive pulmonary disease and respiratory infections have been reported. In a Swiss Cohort Study on Air Pollution and Lung and Heart Diseases in Adults (SAPALDIA), based on a population of almost 5,000 adults, larger annual declines in lung function were linked with higher levels of PM10 (Downs et al. 2007).

A recent study of 481 adults with asthma has suggested that severity of this disease is associated with PM10 (Jacquemin et al. 2012). On the other hand, a meta-analysis of long-term exposure to PM10 and asthma prevalence has showed no such association (Anderson et al. 2013a,b).

Cardiovascular effects and biological mechanisms that relate short- and long-term PM2.5 exposure to cardiovascular outcomes

A large number of epidemiological studies have presented the association indicating an increased risk of cardiovascular diseases in populations exposed to air pollution. Air pollution cardiovascular health outcomes include mortality and morbidity due to: sudden death, arrhythmias, myocardial infarction, strokes, heart failure, and conventional cardiovascular well known risk factors such as diabetes and hypertension. The evidence has been presented in a Criteria Document published by the U.S. EPA (EPA 2009).

The dose response relationship between PM2.5 exposure and the risk of short-term and long-term cardiovascular effects has been identified. Pope and Dockery (Pope and Dockery 2006) estimated that cardiovascular deaths increase by 1% for every 10 $\mu g\ m^{-3}$ short term daily increase in PM2.5 concentrations. No evidence of safe PM threshold has been observed (Dominici et al. 2003; Schwartz et al. 2002).

The risk of several acute morbidity indicators such as myocardial infarctions, cardiac ischaemia, heart failure, arrhythmias, sudden cardiac death, strokes, peripheral arterial disease, and excess cardiovascular related hospitalization has been increasing with elevation of PM2.5 (Brook 2008). The risk of myocardial infarction rose by almost 50% in response to a 2h-long elevation of PM2.5 by 25 $\mu g\ m^{-3}$ (Peters et al. 2001).

Regarding chronic health effects, the dose-response mechanism, is well characterized. In a systematic review of four studies, PM2.5 was positively associated with total cardiovasculary mortality—a 14% increase of the risk per 10 $\mu g\ m^{-3}$ (Chen et al. 2008). The risk of coronary heart diseases was evaluated based on three cohorts. The pooled risk was 16% per 10 $\mu g\ m^{-3}$; however, considerable heterogeneity was observed (Chen et al. 2008).

There are three potential pathways which allow particulate matters to affect the cardiovascular system. The first relates to the autonomic nerve system mechanisms. Nano-sized soluble PM may be deposited in the respiratory system, enter pulmonary small vessels and be, in the next stage, transported through the cardiovascular system. They may stimulate the sympathetic nervous system and/or withdraw the parasympathetic nervous system, which then disrupts the cardiac autonomic balance.

The second postulated pathway is the release of pro-oxidative and/or pro-inflammatory agents from lungs to the cardiovascular system and their interaction with the vessel's structure. Short exposure to PM2.5 can mediate systematic oxidative stress and inflammation, while chronic effects may lead to building up of atherosclerosis or may be responsible for a chronic pro-inflammatory state. Chronic exposure may influence plaque stability and/or increase blood thrombogenecity resulting in myocardial infarctions or stroke. A series of studies have confirmed associations between various markers of atherosclerosis, including intima media thickness and coronary artery calcification, and the long-term average PM2.5 concentration and proximity to traffic in Europe (Bauer et al. 2010; Hoffmann et al. 2006, 2007).

The third pathway is possible only for nano-sized particles and assumes that such particles reach the cardiovascular system though pulmonary small vessels and act with vessels structure without producing oxidative stress or inflammation.

Diabetes

Air pollution may also promote insulin resistance (Brook et al. 2008). The evidence has been recently strengthened. This includes epidemiological studies in Germany (Krämer et al. 2010) and Denmark (Andersen et al. 2012a; Raaschou-Nielsen et al. 2013). Taking into account the large numbers of exposed people worldwide and the raising incidence of diabetes, these findings, if confirmed, are of great importance from the public health point of view.

Susceptibility issues—socio-economic disparities, genetic predisposition

Those with unstable cardiopulmonary diseases are most susceptible to the effects of short-term exposures to PM. However, apparently, even healthy people are susceptible to the effects of long-term exposure to PM, as exposure can potentially accelerate progression of a disease or it may even initiate it. Progression of a disease due to particle exposure may be associated with acceleration of inflammatory processes, whereas other mechanisms, such as changes in the autonomic nervous control of the heart, may play a role in triggering an acute exacerbation of the diseases (Brook et al. 2010). The fact that in epidemiological studies, effect estimates are higher for the long-term exposure than for the short-term exposure demonstrates that long-term effects are not merely the sum of short-term effects.

Health effects of ozone exposure

The 2005 global update of the WHO air quality guidelines finds support only for short-term effects of ozone on mortality and respiratory morbidity (WHO 2006). The adverse effects of exposure to daily ozone concentrations, after adjustment for the effects of particles (PM10), have been reported on both respiratory and cardiovascular hospital admissions; however, since high temperatures (Baccini et al. 2008) and heat waves, in particular (Kovats and Hajat 2008), are associated with increased mortality, separation of the health effects of ozone from those of temperature is still problematic. On the other hand, since 2005, several cohort analyses have been published on long-term ozone exposure and mortality. There is evidence from the American Cancer Society (Krewski et al. 2009) and other cohorts (Lipfert et al. 2006; Krewski et al. 2009; Jerrett et al. 2009; Smith et al. 2009; Zanobetti and Schwartz 2011) for an effect of long-term exposure to ozone on the increased risk of respiratory and cardiorespiratory mortality. Additionally, several new follow-up long-term exposure studies have reported adverse effects on the incidence of asthma, asthma severity, hospital care for asthma, and lung function growth.

Impact of other metrics on adults health

In addition to epidemiological and toxicological studies examining health effects of PM mass (PM2.5 and PM10), substantial research has been conducted in relation to other important air pollution components, i.e., black carbon, secondary organic aerosols, secondary inorganic aerosols as well as ultrafine (smaller than 0.1 μm) particles.

The WHO Regional Office for Europe has recently published a report that systematically evaluates the health significance of black carbon (Janssen et al. 2012). An association has been found between daily outdoor concentrations of black carbon and all-cause and cardiovascular mortality, and cardiopulmonary hospital admissions as well as between long-term black carbon concentration and all-cause and cardiopulmonary mortality.

Organic carbon has been included in epidemiological studies less often than black carbon. In most studies published after the 2009 EPA integrated science assessment, (total) organic carbon has been found to be associated with short-term changes in cardiovascular (Delfino et al. 2010a; Ito et al. 2011; Kim et al. 2012; Son et al. 2012; Zanobetti and Schwartz 2009) and respiratory health (Kim et al. 2008), or with changes in the levels of inflammatory markers (Hildebrandt et al. 2009). The WHO Regional Office for Europe report on black carbon concludes that black carbon per se may not be responsible for the observed health effects, but that black carbon could be interpreted as an indicator for a wide variety of combustion-derived chemical constituents (Janssen et al. 2012).

Secondary inorganic aerosols (sulfate and nitrate)

Sulfate is a major component of secondary inorganic particles that are formed from gaseous primary pollutants. It was noted in the 2009 EPA integrated science assessment that secondary sulfate had been associated with both cardiovascular and respiratory health effects in short-term epidemiological studies. Since that time, epidemiological evidence has continued to accumulate on the short-term effects of sulfate on both cardiovascular (Ito et al. 2011) and respiratory (Atkinson et al. 2010; Kim et al. 2012; Ostro et al. 2009) hospital admissions. Two studies have linked sulfate with cardiovascular mortality (Ito et al. 2011; Son et al. 2012). There is also some new evidence on the associations between daily increases in ambient sulfates and physiological changes related to cardiovascular diseases, such as ventricular arrhythmias and endothelial dysfunction (Anderson et al. 2010; Bind et al. 2012).

Nitrate is another indicator of emissions from combustion processes, including traffic exhausts that are rich in oxides of nitrogen. In a mortality study conducted in Seoul there was some evidence of cardiovascular, but not respiratory, effects of nitrate, and even more so of ammonium (Son et al. 2012). On the other hand, two studies on hospital admissions found evidence of respiratory, but not cardiovascular, effects of nitrate (Atkinson et al. 2010; Kim et al. 2012). However, sulfate was not associated with cardiovascular admissions. In a study conducted in California (Ostro et al. 2010), both nitrate and sulfate were associated with cardiopulmonary mortality. Sulfate and organic carbon have also shown consistent associations in multipollutant models.

Links between daily changes in ultrafine particles and markers of altered cardiac function, inflammation, and coagulation are suggested by some studies (Rückerl et al. 2011; Weichenthal 2012; Rich et al. 2012).

Proximity to roads as an indicator of exposure

The Health Effects Institute Panel identified an exposure zone within a range of up to 300–500 m from a highway or a major road as the area most highly affected by traffic emissions (WHO 2013). In general, PM2.5 does not exhibit the sharp distance-decay gradient evident for carbon monoxide, NO_2, or ultrafine particles.

Many studies have shown excess health risks of several cardiovascular (myocardial infarction, coronary artery calcification, cardiac function-left ventricular mass index) and respiratory outcomes (asthma, wheezing, asthma hospitalization, lung function reduction). In several of these studies, the adjustment has been made for socio-economic confounders (WHO 2013).

Roemer and van Wijnen (2001) published an epidemiological study that evaluated short-term effects of multiple air pollutants in proximity to roads and further away. These investigators obtained health data from a sample of more than 4000 Amsterdam residents who lived for 11 years along roads with more than 10,000 motorized vehicles per day. Higher levels of NO_2, nitric oxide, carbon monoxide, and black smoke at the traffic influenced sites were compared with the background sites. Black smoke and NO_2 were associated with the increased total mortality, however, definite conclusions regarding which of them have been responsible for the observed risk are missing.

The Committee on the Medical Effects of Air Pollutants (COMEAP) has recently concluded that the epidemiological evidence for associations between ambient levels of air pollutants and asthma prevalence at a whole community level was unconvincing; a meta-analysis confirmed the lack of such an association (Gowers et al. 2012).

In contrast, a meta-analysis of cohort studies has found an association between the incidence of asthma and within-community variations in air pollution. Similarly, a systematic review has suggested an association between asthma prevalence and exposure to traffic, although only in those living very close to heavily trafficked roads carrying many trucks, suggesting a possible role for diesel exhaust.

Some studies have examined the effects of air pollution and noise. After adjustment for noise, excess risks of cardiovascular mortality as well as hypertension due to air pollution in the proximity of roads generally remained.

Contributions of different sources

Coal combustion

A study in Washington, DC based on a source apportionment included a category for coal combustion, and evidence of an effect on total and cardiovascular mortality has been provided (Ito et al. 2006). In another study, selenium, an indicator element for emissions from coal combustion, was found to be associated with cardiovascular mortality and hospital admissions in New York City (Ito et al. 2011). Coal combustion emissions, measured by particulate and gaseous components concentrations, were able to exacerbate various allergic airway responses in the exposed adults (Barrett et al. 2011).

Oil combustion

The few epidemiological studies that have examined the effects of oil combustion source on respiratory and cardiovascular health provided conflicting results. However, an effect has been found for vanadium, an indicator element for emissions from oil combustion (Bell et al. 2010; de Hartog et al. 2009).

Biomass combustion

The source "biomass combustion" includes particles from residential wood combustion, wildfires, and burning agricultural residues. In low-income countries, as biomass is extensively used for heating and cooking, it is an important indoor air pollutant.

A systematic review of the health effects of particles from biomass combustion concluded that there was no reason to consider PM from biomass combustion less harmful than particles from other urban sources (Naeher et al. 2007). Experts admitted that so far research has been rather limited as no specific indicators of wood combustion were available.

The few studies, based on source apportionment, provided an opportunity to compare the short-term health effects of particles from biomass combustion with particles from traffic. No definite results have been obtained. In some locations, particles from biomass combustion were associated with a higher risk of cardiovascular and respiratory hospital admissions; in other locations, lower or no clear effect of particles from biomass combustion on cardiovascular health was observed.

A recent study conducted in a woodsmoke affected community provided evidence on the processes through which woodsmoke may affect cardiovascular health (Allen et al. 2011). Introduction of portable air filters resulted in an improved endothelial function and decreased inflammatory biomarkers.

Natural experiments

Improvements in the air quality in US

Based on the data from 51 cities examined in the American Cancer Society study (Pope et al. 2009), it has been concluded that reductions in PM2.5 concentration between 1980 and 2000 were strongly associated with an increased life expectancy. The authors made adjustments for changes in other risk factors (socio-economic, demographic, smoking) during the time of observation.

The Irish coal sale bans, 1990–1998

Clancy et al. (2002) reported a significant decrease in black smoke levels and SO_2 levels (71% and 34% respectively) as a result of the ban on the sale of coal in Dublin in 1990. The drop in exposure was associated with decreases in cardiovascular and respiratory mortality (13% and 7%, respectively). Similar improvements both in exposure to PM (but not SO_2) and health statistics were observed in other Irish cities over the years 1995–1998 (Goodman et al. 2009). However, further analyses taking into account the general decreasing trend in cardiovascular mortality due to other factors, did not confirm the observed health benefits (Dockery et al. 2013).

Low emission zones in Rome, Italy (2001–2005)

The effect of the low emission zones, implemented in two city areas in Rome on traffic-related PM10 and NO_2 concentrations and on mortality in 2001–2005 were examined by Cesaroni et al. (2012). A gain of 3.4 days per person (921 years of life gained per 100,000 population) due to reductions in NO_2 associated with the interventions was observed.

The London congestion charging zone (2003), and Stockholm congestion charging trial (2006)

As a result of the introduction of traffic congestion charging zone in London in 2003, after one year, 30% less congestion was observed (TfL 2004). Etimated gain in life was found to be 183 years per 100,000 of residents of congestion charging zone (Tonne et al. 2010).

In Stockholm, in the case of a seven month—congestion charging trial implemented in 2006 (Eliasson 2008), an estimate of life gain was 206 years per 100,000 population for the area of Greater Stockholm. An assumption was made that introduced changes in traffic organization would be continued over a 10-year period.

Future research needs

In an EPA criteria document, an important statetment has been made, namely that there are many components contributing to the health effects, but there is not sufficient evidence to differentiate those constituents (or sources) that are more closely related to specific health outcomes (EPA 2009).

Greater understanding of the most 'responsible' pollutants for causing the greatest diseases burden should be developed in relation to the sources (traffic vs. low emission), chemicals, and susceptible groups. As a result, the most hazardous pollutants could be identified and effective control measures introduced (WHO 2013).

As far as biological mechanism is concerned, major uncertainty exists about the pathway(s) whereby air pollution in the lungs affects the circulatory system. Identification of the main mediators will provide the opportunity to start a process of developing measures to mitigate the cardiovascular response effectively. In the meantime, continuous efforts should be made to keep the concentrations of air pollution as low as possible (WHO 2013).

Outdoor air pollution and newborns outcomes

Introduction

Nowadays, special attention is paid to the impact of air pollution on birth outcomes. An important advantage of these studies is the shorter exposure window, which is typically nine months, compared to the time spans from the studies of chronic exposure to air pollution in children and adults, which can even be years (Woodruff et al. 2009). This can make it easier to evaluate the exposures of interest more effectively and to decrease the possibility of other risk factors influencing the outcomes. Although a growing number of studies that have evaluated potential links between air pollution and birth outcomes exist, their results are not consistent. The differences in methodologies used among these epidemiologic studies and important areas for future research were discussed in two papers (Slama et al. 2008; Woodruff et al. 2009).

The heterogeneity in the published studies may arise from differences in many aspects of the study design and available data.

Study design

A commonly used method for the assessment of association between air pollution and birth outcomes is a linkage of outcome and covariate data from routinely collected records (such as birth certificates) with ambient air quality monitoring data (Ha et al. 2001, 2004; Liu 2003, 2007; Jalaludin et al. 2007; Bell et al. 2007; Suh et al. 2009; Morello-Frosch et al. 2010; Darrow et al. 2009a,b, 2011). A large sample size available for the analysis at low costs is the advantage of this approach but it has also several limitations. These include the possibility of exposure misclassification and the lack of data about confounding factors. For this reason, nowadays prospective cohort studies with recruitment early or even before pregnancy are more frequently conducted (Perera et al. 2003, 2004; Jedrychowski et al. 2003, 2004; Choi et al. 2006, 2008, 2012; van den Hooven et al. 2009; Estarlich et al. 2011). The prospective cohort study design enables a more reliable identification of exposures and outcomes with their verification by biomarkers or outcome measurements and notification of any changes in exposure levels. Such a design also allows for better estimation of confounders and co-exposures. On the other hand, it is more costly and time consuming. These two designs can be connected by conducting case-control studies with a collection of additional information at the individual level for a sample nested within a cohort constituted from birth records (Ritz et al. 2007). This allows researchers to achieve more detailed information for the subset of population (Slama et al. 2008). As another example, in the analysis performed by Huynh et al. (2006), cases (births between 24 and 36 weeks of gestation) were matched to controls (gestational age 39–44 weeks).

Outcomes of interest

Majority of the studies evaluating the impact of air pollution on birth outcomes have focused on measures of fetal growth and preterm delivery (Maisonet et al. 2004; Lacasaña et al. 2005; Sram et al. 2005; Ghosh et al. 2007; Woodruff et al. 2009; Bosetti et al. 2010; Proietti et al. 2013). Frequently used biometric measures at birth are: reduction in birth weight as a continuous variable, low birth weight (LBW) as birth weight below 2,500 g, and very low birth weight (VLBW), which is birth weight below 1500 g. Preterm birth (PB) is defined as a birth below 37 completed weeks of gestation. It is important to be aware of the fact that for evaluating the impact of air pollution on LBW and VLBW usually a bigger sample size is required. Other researchers use intrauterine growth restriction (IUGR) defined as low birth weight in full-term infants or small for gestational age (SGA) as the birth weight below the 10th percentile of birth weight distribution for a specific gestational age and sex based on national standards for life births as the outcome. To distinguish between reduced birth weight resulting from growth restriction from that of preterm delivery, most researchers assess birth weight at term or account for the gestational age in the model. In the analysis looking at preterm delivery, the problem can occur in accuracy of assessment of pregnancy duration, which can be based on a woman's recall of last menstrual period and/or ultrasound data.

Some of the studies, although less frequently, have also looked at other poor pregnancy outcomes such as: birth defects, miscarriages, stillbirth and infant mortality.

Gestational duration

Among studies evaluating the association between air pollution and preterm birth (PB) published since 2000, most have been performed in US (Wilhelm and Ritz 2003, 2005; Sagiv et al. 2005; Parker et al. 2005; Huynh et al. 2006; Ritz et al. 2000, 2007; Darrow et al. 2009a; Chang et al. 2012) and Europe: Czech Republic, Poland, Netherlands, Sweden (Bobak 2000; van den Hooven et al. 2009; Gehring et al. 2011a,b; Malmqvist et al. 2011), and the other studies have been performed in Canada (Brauer et al. 2008; Liu et al. 2003), China (Zhao et al. 2011), South Korea (Leem et al. 2006; Kim et al. 2007; Suh et al. 2009) and Australia (Hansen et al. 2006).

In the analysis performed by Bobak (2000), on 78,148 singleton live births registered in 1990–1991, PB were associated with total suspended particles (TSP) exposure during the first trimester of pregnancy but no significant associations were found for the second and third trimester. For SO_2 exposure, the association was stronger than the one observed for TSP and more pronounced for the first trimester but for other trimesters it was also statistically significant. Analysis by Leem et al. (2006) in Korea on more than 52,000 singleton births registered in 2001–2002 indicated an increased risk of preterm delivery for the highest quartile of exposure to PM10 during the first trimester. No significant association for the third trimester was noticed. The same pattern was observed in the study performed in Australia (Hansen et al. 2006). On the contrary, Suh et al. (2009) indicated an increased risk of PB for the third trimester exposure to PM10. Analysis performed in South California indicated a risk of PB increased by 20% for a 50 $\mu g\ m^{-3}$ increase in the average level of PM10 during the 6 weeks before birth and 16% increase during the first month of pregnancy (Ritz et al. 2000). PM10 effects showed no regional pattern but they were slightly reduced when several covariates were taken into account. The same authors performed an assessment within the cohort of 58,316 births and in case-control study nested in a birth cohort (N = 2543) (Ritz et al. 2007). They assessed the effect of PM2.5, NO_2, O_3, and CO exposure (during the first trimester of pregnancy and six weeks before birth) on PB. This nested case-control study allowed for inclusion of additional confounders such as active and passive smoking and alcohol consumption during pregnancy in the analysis. The highest level of exposure (PM2.5 > 21.4 $\mu g\ m^{-3}$) during the first trimester of pregnancy was associated with an increased risk of preterm delivery for the whole cohort and for the subsample. In the subsample, such a risk was not modified by adjustment for additional factors. Clear dose-response pattern of PB with distance weighted traffic density primarily for women who delivered in winter or fall was documented in another analysis of this group (Wilhelm and Ritz 2003). Wilhelm and Ritz (2005) also examined whether varying residential distance from monitoring stations (for CO, PM2.5, and PM10) affects risk estimates. They found that women with high first trimester CO, residing within a 1-mile distance had a risk of PB increased by 27%. Similar size effect was observed for exposure to particulates. In contrast, smaller or no effects were observed beyond a 1-mile distance of residence from the station. In the study on 10,673 preterm births and 32,019 matched controls born in California, an increased risk of 15% was observed for every 10 $\mu g\ m^{-3}$ of PM2.5 (Huynh et al. 2006). In the analysis performed in Canada, Brauer et al. (2008) estimated residential exposure to air pollution using various methods including nearest, inverse-distance weighting and temporally adjusted land use regression (LUR). They found no consistent association for birth before 37 weeks of gestation with any of the exposure metrics. The analysis performed by Gehring et al. (2011a,b) based on ABCD and PIAMA study with a land-use regression model did not indicate statistically significant association with any of air pollution measures and PB (Gehring 2011a,b). On the other hand, in China high pollution levels of NO_2, SO_2, and PM10 had a consistent cumulative effect on the risk of PB (Zhao et al. 2011).

To summarize, the results of the studies evaluating the impact of air pollution (with the measurement of different pollutants, by means of different methods of exposure assessment and using different study designs) on preterm birth are mixed and inconclusive, although some of them indicate that such an association exists.

Fetal growth

Majority of the studies evaluating the impact of air pollution on pregnancy outcomes have focused on fetal growth parameters (Bobak 2000; Rogers et al. 2000; Ha et al. 2001; Maisonet et al. 2001; Chen et al. 2002; Liu et al. 2003, 2007; Perera et al. 2003, 2004; Jedrychowski et al. 2003, 2004, 2009; Gouveia et al. 2004; Mannes et al. 2005; Parker et al. 2005; Salam et al. 2005; Choi et al. 2006, 2008; Rogers and Dunlop 2006; Bell et al. 2007; Hansen et al. 2007; Slama et al. 2007; Brauer et al. 2008; Aguilera et al. 2009; Rich et al. 2009; Ballester et al. 2010; Morello-Frosch et al. 2010; Madsen et al. 2010; Seo et al. 2010; Darrow et al. 2011; Gehring et al. 2011a,b; Malmqvist et al. 2011; Xu 2011). Most studies in this field use birth weight as an outcome variable or LBW in term neonates. Birth weight is easily obtainable from existing records and it is also a good marker of intrauterine development and a predictor of children's health. Many studies have found evidence for LBW at term associated with higher exposure to air pollution, whereas other studies have not found any association. In order to understand how differences in research methods contribute to variations in the findings, The International Collaboration on Air Pollution and Pregnancy Outcomes (ICAPPO) was formed (Woodruff et al. 2010; Parker et al. 2011). Based on the data from the individual studies (Ha et al. 2004; Gouveia et al. 2004; Bell et al. 2007, 2008; Jalaludin et al. 2007; Pesatori et al. 2008; Brauer et al. 2008; Glinianaia et al. 2008; Darrow et al. 2009a,b; Rich et al. 2009; van den Hooven et al. 2009; Morello-Frosch et al. 2010; Lapeule et al. 2010; Pearce et al. 2010; Gehring et al. 2011a) they estimated odds ratio for the association between PM10 (averaged over pregnancy) and LBW at term (37–42 complete weeks of gestation) among life-born singleton births (Parker et al. 2010). The risk of term LBW associated with 10 $\mu g\ m^{-3}$ increase in average PM10 concentration during pregnancy, adjusted for socio-economic status (SES), was calculated. It ranged from 0.63 (95% CI 0.3–1.4) for Netherlands to 1.2 (95% CI 0.6–2.2) for Vancouver, with 6 research groups (out of 13) reporting statistically significant adverse associations (Parker et al. 2011). Analysis with additional confounders indicated slightly stronger associations between air pollution and LBW compared with the analysis with inclusion of only SES as a confounder. After controlling for SES, the reduction in mean birth weight associated with a PM10 increase of 10 $\mu g\ m^{-3}$ ranged from 2 to 20 g for most of locations. An interesting analysis was also performed for four areas within the Spanish Children's Health and Environment (INMA) mother and child cohort study. The analysis was performed to evaluate the possible effect of exposure to NO_2 on anthropometric measures at birth. In the combined analysis, an increase of 10 $\mu g\ m^{-3}$ in NO_2 exposure during pregnancy was associated with a decrease in birth length of –0.9 mm. For the subset of women who spent $\geq$ 15 hr/day at home, the association was even stronger (Estarlich et al. 2011). For the same subset of women, a reduction of 22 g in birth weight was associated with each 10 $\mu g/m^3$ increase in NO_2 exposure in the second trimester. A series of analyses of the relationship between in utero exposure to airborne polycyclic aromatic hydrocarbons (PAH) and anthropometric parameters at birth were performed based on prospective studies in New York, US and Krakow, Poland

(Perera et al. 2003, 2004; Choi et al. 2006, 2008; Jedrychowski et al. 2003, 2004, 2009). In these studies, PAH exposure was based on personal air monitoring or determinations of PAH-DNA adduct levels. PAH exposure was associated with a significantly reduced birth weight in Krakow Caucasians and in NYC African Americans but not in NYC Dominicans (Choi et al. 2006). Additional analysis performed on African-Americans and Dominicans residing in NYC indicated that 1-ln-unit increase in prenatal PAH exposure was associated with a 2-fold increase in the risk of symmetric intrauterine growth restriction and 0.04% increase in cephalization index in African-Americans (Choi et al. 2008). These effects were not observed in Dominicans, which as postulated by authors, might reflect reduction of the risk by healthy lifestyle of this ethnic group.

To conclude, most of the existing studies indicate association between exposure to air pollution and some restrictions in fetal growth.

Exposure assessment

Studies vary in terms of the sets of pollutants they consider and the methods used for assessing them. Most studies have examined particulate matter (PM), which can be measured as total suspended particles (TSP), particulate matter with an aerodynamic diameter smaller than 10 μm (PM10), and finer particles, smaller than 2.5 μm (PM2.5). It is important to notice that despite similar gravimetric values of PM, depending on the source, different particulates composition may arise (Woodruff et al. 2003, 2009; Ritz and Wilhelm 2008; Proietti et al. 2013). As an example, particles measured at the seaside derive mostly from natural sources, whereas urban airborne particulates derive mostly from vehicle resuspension from the road and abrasion processes from wheels and brakes. Biological effects can also depend on the content of organic and elemental carbon and PAH. As an example, the analysis of data from the prospective cohort studies in New York, US and Krakow, Poland (performed using the same methodology) indicated that personal PAH exposure was 10-fold higher in Krakow than in the NYC cohort, with the major source of PAH in Krakow being coal burning, whereas diesel fuel combustion was the major source in NYC (Jedrychowski et al. 2003; Tonne et al. 2004; Choi et al. 2006). In addition, proportions of specific compounds in the total PAH mixture differed widely across these two locations. Among the Krakow women, the PAHs contributing the highest proportion to the total mixture were: benzo(b)fluoranthene (23.6%) and benz(a)anthracene (15.8%). In NYC, the highest proportions were contributed by: benzo(g,h,i)perylene (32.5%), benzo(b)fluoranthene (15.0%), and indeno(1,2,3-cd)pyrene (15.0%) (Jedrychowski et al. 2003).

Other pollutants examined in the studies evaluating the impact of air pollution on newborns parameters include: CO, O_3, SO_2 and NO_2, and PAH.

The chosen biomarkers of exposure have differed between the studies. Some studies consider exposure to pollutants separately, while other consider the exposures simultaneously. In a multi-pollutant analysis, the problem can arise from strong between-pollutant correlations, accuracy, and availability of data for specific air pollutant and heterogeneous degrees of spatial resolution for various pollutants. It is important to notice that regional and demographic differences in the studies' populations may contribute to variations is the studies conclusions (Woodruff et al. 2009; Parker and Woodruff 2008).

Selection of a pollutant for analysis may also depend on a chosen reproductive endpoint.

The precision of estimating individual exposure affects the power of the study and the detected effect of air pollution on health. The most common approach is to assess residential exposure at the subject's address with data concerning the district, postal code, or street number. The data from the existing networks of ambient monitoring stations can be used in different ways depending on the study. Some researchers use average over geographic areas, other nearest monitor measurement, or inverse distance–weighted (IDW) averages from multiple monitors from residence (Parker et al. 2011). Data from monitoring stations allow for including large numbers of births but have several limitations. In such an approach, it is assumed that in the considered area the pollution level is homogenous, that the subject did not move during pregnancy, and that they stayed most of the time at home, or the exposure in the workplace is comparable to that at home (Proiertti et al. 2013). It is also important to be aware of the fact that monitors are not always located in all the places people live in and do not always provide continuous measurement data. Monitoring network locations are sited for policy and regulatory purposes not for health studies, and for this reason they can be useful but not ideal for epidemiological analysis (Woodruff et al. 2009).

Several approaches allow taking into account area variations in pollution. Information about the distance from the road or distance-weighted traffic density constitutes a simple source model available in many locations (Slama et al. 2008). The researchers also have used other models to estimate exposure such as: dispersion modelling (mathematical simulation of how air pollutants disperse in the ambient atmosphere) (Generation R study—Wasseling et al. 2002; Norway—Madsen et al. 2010), land use regression (LUR) (PIAMA—Gethring et al. 2011a; LISA study—Slama et al. 2007) or two-stage geostatistical approaches incorporating monitoring station data and information on temporally or spatially varying covariates (Slama et al. 2008). LUR models may provide an important complement to personal and biomonitoring for assessing exposure to pollutants with spatially heterogeneous concentrations in large study populations (Ritz and Wilhelm 2008). In the LUR modelling approach, concentrations of exhaust markers (such as NOx) are measured simultaneously at any locations throughout an urban area using relatively inexpensive passive monitors. Different geographic information system (GIS) parameters (including traffic and population density) are used to predict the measurement concentration. The model can then be used to estimate concentration at home and at work locations based on the GIS parameter values at those locations (Ritz and Wilhelm 2008). LUR models typically yield yearly exposure assessment so they are meant to characterize special rather than temporal variability in air pollution levels (Ritz and Wilhelm 2008; Slama et al. 2008). This may not work for pregnancy where shorter term of exposures are of interest. One option is to incorporate temporal variability into LUR models based on measurements from background monitoring stations (Slama et al. 2008).

Personal monitoring may provide an estimate of exposure less prone to misclassification than an ecologic approach. A detailed analysis of personal PAH exposure (based on 48-hr personal air monitoring), derives from US and Polish cohorts of pregnant women (Jedrychowski et al. 2003, 2004, 2009; Perera et al. 2003, 2004;

Choi et al. 2006, 2008). The advantage of assessment of exposure based on personal monitoring in comparison with the data from monitoring stations (based on address of a place of residence) is a more accurate assessment of exposure (as the monitoring stations may be located at a significant distance from the place of residence and thus, do not reflect real exposure levels) and the inclusion of the whole exposure spectrum taking into account activity patterns (such as exposure in the workplace or transport). The costs and changing activity pattern during pregnancy which will not be adhered by a single measurement may constitute limitations of this approach.

To date, there is little consensus on appropriate biomarkers for ambient air pollutants except for PAH and their metabolites. Exposure to PAH can be assessed based on their metabolites analysed in urine or measurement of PAH-DNA adducts level in maternal or cord blood (Jedrychowski et al. 2003, 2004, 2009; Perera et al. 2003, 2004; Choi et al. 2006, 2008). The issue, which needs to be addressed when selecting biomarkers of exposure is the ability to discern source contributions like diet vs. traffic and appropriate timing of measurement (Ritz and Wilhelm 2008). As an example, urinary PAH metabolites measured at a single point in time during pregnancy will reflect exposures that occurred in the previous few days unless exposure is continuous, while PAH-DNA adducts will indicate a longer time period of exposure, which is approximately one month. Taking this into account, researchers may need to target the relevant exposure window or perform repeated measurements. On the other hand, in the studies performed by Perera et al. 2006, 2009; Edwards et al. 2010; and Polanska et al. 2011, the authors considered a single monitoring time point to be a reasonable indicator of prenatal exposure over the last two trimesters of pregnancy because measurements during the second and third trimesters were correlated.

Critical window of exposure

The issue of concern and the challenge of epidemiology is to identify whether there are particular periods of susceptibility during pregnancy when air pollution exposure is particularly harmful to the fetus and its development. For LBW and preterm delivery, the first (when placental attachment and development occurs) and third trimesters (when fetal growth velocity is highest) exposure have been implicated as having the most relevance, while for birth defects the development time of specific organs has to be considered. Most published studies, in addition to the whole pregnancy assessment, have focused on evaluating exposure by trimester or gestational month. The results of those studies are not consistent and do not indicate a specific time window of susceptibility. Some of them have reported effects due to the first trimester exposures others only for the third trimester (Proietti et al. 2013; Woodruff et al. 2009; Slama et al. 2008). There are also studies that have indicated effects from the second trimester exposure or exposure during more than one trimester (Woodruff et al. 2009). Such differences in the study results may be due to the various methods used to consider correlated exposures among trimesters and pollutants. In addition, it is difficult to distinguish one trimester from other time periods as being important because exposures among trimesters are correlated (Woodruff et al. 2009). The method used by Bell et al. (2007) can be a useful approach in the case of simultaneous adjustment for all trimester-specific exposure variables. In that study, exposure during each trimester was modeled as a function of exposure in other trimesters, and the results from these trimester-specific models were included in the subsequent trimester-specific regression models to control for other trimester's exposure (Bell et al. 2007).

Confounding

Taking into account the fact that pollution levels vary in time and space, any factor influencing pregnancy outcomes and varying with time or space in a way similar to air pollutants is a potential confounder. Socio-economic status (SES) and related factors are thought to be important confounders because they (such as low SES) are associated with the occurrence of adverse reproductive outcomes and air pollution levels in neighborhoods (Woodruff et al. 2009; Slama et al. 2008). As an example, higher levels of traffic air pollutants are often observed in the city center than in the suburbs and in the US cities; people with lower SES more often live in the city centers than in the suburbs and thus, are exposed to higher levels of these pollutants (Slama et al. 2008). An opposite pattern can be observed in Europe where in the same cities the city centers are more often inhabited by residents with higher SES. The following variables can be used as determinants of SES: ethnicity, educational and employment level, income, occupation, and health insurance. There is also a question which factors best describe SES and how they need to be measured to control for them. Selection of accurate measure of SES can be done at a country level taking into account their availability and accuracy. The other confounders pointed in most studies are: maternal age, marital status, parity, body mass index, weight gain during pregnancy and maternal behaviors, particularly smoking. While the birth records usually contain information such as maternal age, education, and parity, there is no information about other variables, so this constitutes a concern if the unmeasured variables may confound the observed association.

Ritz et al. (2007) conducted a case-control survey nested within a birth cohort and collected detailed risk factor information in order to assess the extent to which residual confounding and exposure misclassification may impact air pollution effect estimates. The researchers indicated that many confounders recommended to be included in the analysis, including smoking, ETS, alcohol drinking income, occupation, or body mass index did not have a large effect on the relationship between air pollution and preterm delivery, and that variables obtained from birth certificate are sufficient to control for potential confounding by these factors. In addition, they found that accounting for time-activity patterns of pregnant women in an effort to reduce exposure misclassification tended to strengthen effect estimates (Ritz et al. 2007; Ritz and Wilhelm 2008). The other study indicated that maternal high education and gestational age had the largest effects on the relationship between air pollution and birth weight (Slama et al. 2007). A separate problem can be identified when mothers are excluded from the study because they do not live near monitoring locations, which can result in selection bias (Woodfruff et al. 2009). Mothers living near monitors may differ from those living far from them. On the other hand, some researchers conclude that inclusion of mothers living near monitors affects generalizability of the study rather than bias. The other variable, which needs to be considered, is season, which is associated with air pollution levels and premature delivery although part of this association may be attributable to seasonality in air pollution levels (Slama 2008). The association between season and PB could also result from exposure to other factors such as infectious diseases. They vary depending on the season and for that reason, season should be considered as a potential confounder. Taking into account that season of birth is influenced by duration of pregnancy, which may be shortened by exposure to air pollution (and that confounders based on their definition should not be affected by exposure), season of conception (season of last menstrual period) seems like a more appropriate adjustment rather than season of birth. In some settings, the association of a season with air pollution might be very strong. In such a case, controlling for season might produce over adjustment or make the estimates associated with air pollution unstable so it would be more appropriate to adjust for seasonally varying factors (Slama et al. 2008).

The other concern is to distinguish between a confounder and an effect modifier, which could allow for identification of subgroups more vulnerable to the effects of air pollution (Woodruff et al. 2009). The same variables can be confounders or effect modifiers depending on the characteristics of the study population or selected hypothesis. Some speculations exist that the association may be stronger among male than female infants or in mothers from poorer/richer neighborhoods. A stronger effect of air pollution on birth weight was also reported for parous than nulliparous women in the study in which exposure was estimated from the home address, which as interpreted by the authors could result from the fact that home address based exposure estimate is more accurate for the parous pregnant women because they are more likely to stay at home to take care of their other children than for the nulliparous women (Ritz and Yu 1999; Slama et al. 2008).

Biological mechanisms

Fetal growth is influenced by alterations of utero-placental and umbilical blood flow, and transplacental glucose and oxygen transport (Slama et al. 2008). Studies performed on nonpregnant adults indicate that PM levels have been associated with plasma viscosity and endothelial function (Pope and Dockery 2006). If such an association exists also in pregnant women, air pollution induced changes in plasma viscosity and artery vasoconstriction may in turn influence maternal-placental exchanges and hence, fetal growth. Some studies also indicate linkage between short-term changes in air pollutants and endothelial function or inflammatory response (Slama et al. 2008). The postulated mechanisms of the foetal toxicity of PAH may involve the induction of apoptosis after DNA damage from PAH, antiestrogenic effects of PAH, binding to the human aryl hydrocarbon receptor to induce P450 enzymes, or to receptors for placental growth factors, which can result in a decreased exchange of oxygen and nutrients (Dejmek et al. 2000; Choi et al. 2006).

Future research needs

The existing evidence suggests that air pollution may play a role in adverse pregnancy outcomes. The majority of research in this field gives the opportunity for international collaboration between researchers, i.e., to apply the same or similar methods to analyze the existing data (Woodruff et al. 2009). Applying a consistent analytic strategy would make it possible to reconcile some of the apparent inconsistences in the effect estimates observed across the studies (Woodruff et al. 2009). Apart from the broadly studied pregnancy outcomes (such as birth weight and preterm delivery), other perinatal end points may be sensitive to air pollutant exposure and could be considered in future studies (Slama et al. 2008). In addition, evaluating the outcomes such as fetal loss (both as an end point or as a potential bias), pregnancy related hypertension or preeclampsia, and impact on endothelial function or inflammatory response may provide insights into biological mechanisms. Animal studies can be also useful when it comes to identification of the windows of exposure to air pollution, and reproductive outcomes which are difficult to ascertain with available epidemiological data and specific pollutants or pollutant mixtures (Woodruff et al. 2009). There is a need to improve methods of exposure assessment as spatial resolution is often inadequate (Slama et al. 2008). The dispersion and LUR models with temporal component and with information about time-activity patterns have potential in future studies. In addition, development of biomarkers of exposure to traffic-related air pollution could be also useful. Apart from the existing large studies evaluating the data from monitoring stations and birth certificate records, studies that collect detailed exposure and covariate information with the collected biological samples in nested subgroups of larger populations or cohort studies, should be further encouraged.

Final comments

Different environmental epidemiological studies give the possibility to assess the relationship between air pollution and the wide-range of health conditions in humans. While interpreting their results, it is important to be aware of their limitations, which can relate to exposure and outcome assessment and control for confounding factors and co-exposures. Commonly used methods for the assessment of association between air pollution and health outcomes is a linkage of outcome and covariate data from the routinely collected records (hospital records, birth, or death certificates) with the information about air pollution level (from monitoring data). Although this kind of analysis allows researchers to achieve a big sample size and can be done relatively quickly and at low costs, it has several limitations such as exposure misclassification and the lack of data about potential confounding factors. When using data from monitoring stations, researchers assume that in a specific area the pollution level is homogenous and that the subject spent most of the time in such an area. It is important to be aware of the fact that monitoring network locations are cited for policy and regulatory purposes and not for health studies, so that they may be useful but not ideal for an epidemiological analysis. Additionally, it is not clear from the existing studies which pollutant component is responsible for causing the greatest disease burden in relation to sources, chemicals and susceptible groups. The improved methods of exposure assessment together with the development of biomarkers of exposure seem like one of the important advantages of future studies. The analytical epidemiological studies (especially prospective cohort studies) enable a more reliable identification of exposure and outcomes, give possibility for their verification by a biomarker or outcome measurements, notification of any changes in exposure level, and assessment of confounding factors. On the other hand, those studies are usually based on a smaller sample size and they are more costly and time consuming. Majority of the research, especially that focusing on the impact of air pollution on birth outcomes, gives the opportunity for international collaboration and reconciliation of some of the apparent inconsistencies in the effect estimates observed across the studies. Animal studies can be also crucial and together with environmental epidemiology, can be useful for identification of biological mechanisms of the impact of air pollution on health outcomes.

References

Aguilera, I., M. Guxens, R. Garcia-Esteban, T. Corbella, M.J. Nieuwenhuijsen, C.M. Foradada and J. Sunyer. 2009. Association between GIS-based exposure to urban air pollution during pregnancy and birth weight in the INMA Sabadell Cohort. Environ. Health Perspect. 117: 1322–1327.

Allen, R.W., C. Carlsten, B. Karlen, S. Leckie, S. van Eeden, S. Vedal, I. Wong and M. Brauer. 2011. An air filter intervention study of endothelial function among healthy adults in a woodsmoke-impacted community. Am. J. Respir. Crit. Care Med. 183(9): 1222–1230.

Andersen, Z.J., O. Raaschou-Nielsen, M. Ketzel, S.S. Jensen, M. Hvidberg, S. Loft, A. Tjønneland, K. Overvad and M. Sørensen. 2012a. Diabetes incidence and long-term exposure to air pollution: a cohort study. Diabetes Care 35(1): 92–98.

Anderson, H.R., R.W. Atkinson, S.A. Bremner, J. Carrington and J. Peacock. 2007. Quantitative systematic review of short term associations between ambient air pollution (particulate matter, ozone, nitrogen dioxide, sulphur dioxide and carbon monoxide), and mortality and morbidity. London, Department of Health (https://www.gov.uk/government/publications/quantitative-systematic-review-of-short-term-associations-between-ambient-air-pollution-particulate-matter-ozone-nitrogen-dioxide-sulphur-dioxide-and-carbon-monoxide-and-mortality-and-morbidity).

Anderson, H.R., B. Armstrong, S. Hajat, R. Harrison, V. Monk, J. Poloniecki, A. Timmis and P. Wilkinson. 2010. Air pollution and activation of implantable cardioverter defibrillators in London. Epidemiology 21(3): 405–413.

Anderson, H.R., G. Favarato and R.W. Atkinson. 2013a. Long-term exposure to outdoor air pollution and the prevalence of asthma: meta-analysis of multi-community prevalence studies. Air Quality, Atmosphere and Health 6(1): 57–68.

Anderson, H.R., G. Favarato and R.W. Atkinson. 2013b. Long-term exposure to air pollution and the incidence of asthma: meta-analysis of cohort studies. Air Quality, Atmosphere and Health 6(1): 47–56.

Atkinson, R.W., G.W. Fuller, H.R. Anderson, R.M. Harrison and B. Armstrong. 2010. Urban ambient particle metrics and health: a time-series analysis. Epidemiology 21(4): 501–511.

Baccini, M., A. Biggeri, G. Accetta, T. Kosatsky, K. Katsouyanni, A. Analitis, H.R. Anderson, L. Bisanti, D. D'Ippoliti, J. Danova, B. Forsberg, S. Medina, A. Paldy, D. Rabczenko, C. Schindler and P. Michelozzi. 2008. Heat effects on mortality in 15 European cities. Epidemiology 19(5): 711–719.

Ballester, F., M. Estarlich, C. Iniguez, S. Llop, R. Ramon, A. Esplugues, M. Lacasana and M. Rebagliato. 2010. Air pollution exposure during pregnancy and reduced birth size: a prospective birth cohort study in Valencia, Spain. Environ. Health (9): 6: 1–11.

Barrett, E.G., K.C. Day, A.P. Gigliotti, M.D. Reed, J.D. McDonald, J.L. Mauderly and S.K. Seilkop. 2011. Effects of simulated downwind coal combustion emissions on pre-existing allergic airway responses in mice. Inhal. Toxicol. 23(13): 792–804.

Bauer, M., S. Moebus, S. Möhlenkamp, N. Dragano, M. Nonnemacher, M. Fuchsluger, C. Kessler, H. Jakobs, M. Memmesheimer, R. Erbel, K.H. Jöckel and B. Hoffmann. 2010. Urban particulate matter air pollution is associated with subclinical atherosclerosis: results from the HNR (Heinz Nixdorf Recall) study. J. Am. Coll. Cardiol. 56(22): 1803–1808.

Beelen, R., G. Hoek, P.A. van den Brandt, R.A. Goldbohm, P. Fischer, L.J. Schouten, M. Jerrett, E. Hughes, B. Armstrong and B. Brunekreef. 2008. Long-term effects of traffic-related air pollution on mortality in a Dutch cohort (NLCS-AIR study). Environ. Health Perspect. 116(2): 196–202.

Bell, M.L., K. Ebisu and K. Belanger. 2007. Ambient air pollution and low birth weight in Connecticut and Massachusetts. Environ. Health Perspect. 115: 1118–1124.

Bell, M.L., K. Ebisu and K. Belanger. 2008. The relationship between air pollution and low birth weight: effects by mother's age, infant sex, co-pollutants, and pre-term births. Environ. Res. Lett. 3(4): 44003.

Bell, M.L., K. Belanger, K. Ebisu, J.F. Gent, H.J. Lee, P. Koutrakis and B.P. Leaderer. 2010. Prenatal exposure to fine particulate matter and birth weight: variations by particulate constituents and sources. Epidemiology 21(6): 884–891.

Bind, M.A., A. Baccarelli, A. Zanobetti, L. Tarantini, H. Suh, P. Vokonas and J. Schwartz. 2012. Air pollution and markers of coagulation, inflammation, and endothelial function: associations and epigene-environment interactions in an elderly cohort. Epidemiology 23(2): 332–340.

Bobak, M. 2000. Outdoor air pollution, low birth weight, and prematurity. Environ. Health Perspect. 108: 173–176.

Bosetti, C., M.J. Nieuwenhuijsen, S. Gallus, S. Cipriani, C. La Vecchia and F. Parazzini. 2010. Ambient particulate matter and preterm birth or birth weight: a review of the literature. Arch. Toxicol. 84(6): 447–460.

Brauer, M., C. Lencar, L. Tamburic, M. Koehoorn, P. Demers and C. Karr. 2008. A cohort study of traffic-related air pollution impacts on birth outcomes. Environ. Health Perspect. 116: 680–686.

Brook, R.D. 2008. Cardiovascular effects of air pollution. Clinical Science 115(6): 175–187.

Brook, R.D., M. Jerrett, J.R. Brook and M. Finkelstein. 2008. The relationship between diabetes mellitus and traffic-related air pollution. J. Occup. Environ. Med. 50: 32–38.

Brook, R.D., S. Rajagopalan, C.A. Pope 3rd, J.R. Brook, A. Bhatnagar, A.V. Diez-Roux, F. Holguin, Y. Hong, R.V. Luepker, M.A. Mittleman, A. Peters, D. Siscovick, S.C. Smith, Jr., L. Whitsel and J.D. Kaufman. 2010. Particulate matter air pollution and cardiovascular disease: an update to the scientific statement from the American Heart Association. Circulation 121(21): 2331–2378.

Burnett, R.T., D. Stieb, J.R. Brook, S. Cakmak, R. Dales, M. Raizenne, R. Vincent and T. Dann. 2004. Associations between short-term changes in nitrogen dioxide and mortality in Canadian cities. Arch. Environ. Health 59(5): 228–236.

Cesaroni, G., H. Boogaard, S. Jonkers, D. Porta, C. Badaloni, G. Cattani, F. Forastiere and G. Hoek. 2012. Health benefits of traffic-related air pollution reduction in different socioeconomic groups: the effect of low-emission zoning in Rome. Occup. Environ. Med. 69(2): 133–139.

Chang, H.H., B.J. Reich and M.L. Miranda. 2012. Time-to-event analysis of fine particle air pollution and preterm birth: results from North Carolina, 2001–2005. Am. J. Epidemiol. 175: 91–98.

Chen, H., M.S. Goldberg and P.J. Villeneuve. 2008. A systematic review of the relation between long-term exposure to ambient air pollution and chronic diseases. Rev. Environ. Health 23(4): 243–297.

Chen, L., W. Yang, B.L. Jennison, A. Goodrich and S.T. Omaye. 2002. Air pollution and birth weight in northern Nevada, 1991–1999. Inhal. Toxicol. 14: 141–157.

Choi, H., W. Jedrychowski, J. Spengler, D.E. Camann, R.M. Whyatt, V. Rauh, W.Y. Tsai and F.P. Perera. 2006. International studies of prenatal exposure to polycyclic aromatic hydrocarbons and fetal growth. Environ. Health Perspect. 114(11): 1744–1750.

Choi, H., V. Rauh, R. Garfinkel, Y. Tu and F.P. Perera. 2008. Prenatal exposure to airborne polycyclic aromatic hydrocarbons and risk of intrauterine growth restriction. Environ. Health Perspect. 116(5): 658–665.

Choi, H., L. Wang, X. Lin, J.D. Spengler and F.P. Perera. 2012. Fetal window of vulnerability to airborne polycyclic aromatic hydrocarbons on proportional intrauterine growth restriction. PloS ONE 7(4): e35464.

Clancy, L., P. Goodman, H. Sinclair and D.W. Dockery. 2002. Effect of air-pollution control on death rates in Dublin, Ireland: an intervention study. Lancet 360(9341): 1210–1214.

Crouse, D.L., P.A. Peters, A. van Donkelaar, M.S. Goldberg, P.J. Villeneuve, O. Brion, S. Khan, D.O. Atari, M. Jerrett, C.A. Pope 3rd, M. Brauer, J.R. Brook, R.V. Martin, D. Stieb and R.T. Burnett. 2012. Risk of nonaccidental and cardiovascular mortality in relation to long-term exposure to low concentrations of fine particulate matter: a Canadian national-level cohort study. Environ. Health Perspect. 120(5): 708–714.

Darrow, L.A., M. Klein, W.D. Flanders, L.A. Waller, A. Correa, M. Marcus, J.A. Mulholland, A.G. Russell and P.E. Tolbert. 2009a. Ambient air pollution and preterm birth: a time-series analysis. Epidemiology 20(5): 689–698.

Darrow, L.A., M.J. Strickland, M. Klein, L.A. Waller, W.D. Flanders, A. Correa, M. Marcus and P.E. Tolbert. 2009b. Seasonality of birth and implications for temporal studies of preterm birth. Epidemiology 20(5): 699–706.

Darrow, L.A., M. Klein, M.J. Strickland, J.A. Mulholland and P.E. Tolbert. 2011. Ambient air pollution and birth weight in full-term infants in Atlanta, 1994–2004. Environ. Health Perspect. 119: 731–737.

De Hartog, J.J., T. Lanki, K.L. Timonen, G. Hoek, N.A. Janssen, A. Ibald-Mulli, A. Peters, J. Heinrich, T.H. Tarkiainen, R. van Grieken, J.H. van Wijnen, B. Brunekreef and J. Pekkanen. 2009. Associations between PM2.5 and heart rate variability are modified by particle composition and beta-blocker use in patients with coronary heart disease. Environ. Health Perspect. 117(1): 105–111.

Dejmek, J., I. Solanský, I. Benes, J. Lenícek and R.J. Šrám. 2000. The impact of polycyclic aromatic hydrocarbons and fine particles on pregnancy outcome. Environ. Health Perspect. 108(12): 1159–1164.

Delfino, R.J., T. Tjoa, D.L. Gillen, N. Staimer, A. Polidori, M. Arhami, L. Jamner, C. Sioutas and J. Longhurst. 2010a. Traffic-related air pollution and blood pressure in elderly subjects with coronary artery disease. Epidemiology 21(3): 396–404.

Dockery, D.W., C.A. Pope 3rd, X. Xu, J.D. Spengler, J.H. Ware, M.E. Fay, B.G. Ferris, Jr. and F.E. Speizer. 1993. An association between air pollution and mortality in six U.S. cities. N. Engl. J. Med. 329(24): 1753–1759.

Dockery, D.W., D.Q. Rich, P.G. Goodman, L. Clancy, P. Ohman-Strickland, P. George and T. Kotlov; HEI Health Review Committee. 2013. Effect of air pollution control on mortality and hospital admissions in Ireland. Res. Rep. Health Eff. Inst. (176): 3–109.

Dominici, F., M. Daniels, A. McDermott, S.L. Zeger and J.M. Samet. 2003. Shape of the exposure-response relation and mortality displacement in the NMMAPS database. pp. 91–96. *In*: Revised Analyses of Time-Series Studies of Air Pollution and Health. Health Effects Institute, Charlestown, MA, USA.

Dominici, F., R.D. Peng, K. Ebisu, S.L. Zeger, J.M. Samet and M.L. Bell. 2007. Does the effect of PM10 on mortality depend on PM nickel and vanadium content? A reanalysis of the NMMAPS data. Environ. Health Perspect. 115(12): 1701–1703.

Downs, S.H., Ch. Schindler, L.-J.S. Liu, D. Keidel, L. Bayer-Oglesby, M.H. Brutsche, M.W. Gerbase, R. Keller, N. Künzli, P. Leuenberger, N.M. Probst-Hensch, J.-M. Tschopp, J.-P. Zellweger, T. Rochat, J. Schwartz, U. Ackermann-Liebrich and the SAPALDIA Team. 2007. Reduced exposure to PM10 and attenuated age-related decline in lung function. N. Engl. J. Med. 357(23): 2338–2347.

Edwards, S.C., W. Jedrychowski, M. Butscher, D. Camann, A. Kieltyka, E. Mroz, E. Flak, Z. Li, S. Wang, V. Rauh and F. Perera. 2010. Prenatal exposure to airborne polycyclic aromatic hydrocarbons and children's intelligence at 5 years of age in a prospective cohort study in Poland. Environ. Health Perspect. 118(9): 1326–1331.

Eliasson, J. 2008. Lessons from the Stockholm congestion charging trial. Transport Policy 15(6): 395–404.

EPA. 2009. Integrated science assessment for particulate matter (final report). Washington, DC, United States Environmental Protection Agency (http://cfpub.epa.gov/ncea/cfm/recordisplay.cfm?deid=216546#Download).

Estarlich, M., F. Ballester, I. Aguilera, A. Fernandez-Somoano, A. Lertxundi, S. Llop, C. Freire, A. Tardon, M. Basterrechea, J. Sunyer and C. Iniguez. 2011. Residential exposure to outdoor air pollution during pregnancy and anthropometric measures at birth in a multicenter cohort in Spain. Environ. Health Perspect. 119: 1333–1338.

Gehring, U., J. Heinrich, U. Krämer, V. Grote, M. Hochadel, D. Sugiri, M. Kraft, K. Rauchfuss, H.G. Eberwein and H.E. Wichmann. 2006. Long-term exposure to ambient air pollution and cardiopulmonary mortality in women. Epidemiology 17(5): 545–551.

Gehring, U., A.H. Wijga, P. Fischer, J.C. de Jongste, M. Kerkhof, G.H. Koppelman, H.A. Smit and B. Brunekreef. 2011a. Traffic-related air pollution, preterm birth and term birth weight in the PIAMA birth cohort study. Environ. Res. 111: 125–135.

Gehring, U., M. van Eijsden, M.B. Dijkema, M.F. van der Wal, P. Fischer and B. Brunekreef. 2011b. Traffic-related air pollution and pregnancy outcomes in the Dutch ABCD birth cohort study. Occup. Environ. Med. 68: 36–43.

Ghosh, R., J. Rankin, T. Pless-Mulloli and S. Glinianaia. 2007. Does the effect of air pollution on pregnancy outcomes differ by gender? A systematic review. Environ. Res. 105(3): 400–408.

Glinianaia, S.V., J. Rankin, T. Pless-Mulloli, M.S. Pearce, M. Charlton and L. Parker. 2008. Temporal changes in key maternal and fetal factors affecting birth outcomes: a 32-year population-based study in an industrial city. BMC Pregnancy Childbirth 8: 39.

Goodman, P.G., D.Q. Rich, A. Zeka, L. Clancy and D.W. Dockery. 2009. Effect of air pollution controls on black smoke and sulfur dioxide concentrations across Ireland. Air Waste Manage. 59(2): 207–213.

Gouveia, N., S.A. Bremner and H.M. Novaes. 2004. Association between ambient air pollution and birth weight in Sao Paulo, Brazil. J. Epidemiol. Community Health 58: 11–17.

Gowers, A.M., P. Cullinan, J.G. Ayres, H.R. Anderson, D.P. Strachan, S.T. Holgate, I.C. Mills and R.L. Maynard. 2012. Does outdoor air pollution induce new cases of asthma? Biological plausibility and evidence; a review. Respirology 17(6): 887–898.

Ha, E.H., Y.C. Hong, B.E. Lee, B.H. Woo, J. Schwartz and D.C. Christiani. 2001. Is air pollution a risk factor for low birth weight in Seoul? Epidemiology 12: 643–648.

Ha, E.H., B.E. Lee, H.S. Park, Y.S. Kim, H. Kim and Y.J. Kim. 2004. Prenatal exposure to PM10 and preterm birth between 1998 and 2000 in Seoul, Korea. J. Prev. Med. Public Health 37: 300–305.

Hansen, C., A. Neller, G. Williams and R. Simpson. 2006. Maternal exposure to low levels of ambient air pollution and preterm birth in Brisbane, Australia. BJOG. 113(8): 935–941.

Hansen, C., A. Neller, G. Williams and R. Simpson. 2007. Low levels of ambient air pollution during pregnancy and fetal growth among term neonates in Brisbane, Australia. Environ. Res. 103(3): 383–389.

Hildebrandt, K., R. Rückerl, W. Koenig, A. Schneider, M. Pitz, J. Heinrich, V. Marder, M. Frampton, G. Oberdörster, H.E. Wichmann and A. Peters. 2009. Short-term effects of air pollution: a panel study of blood markers in patients with chronic pulmonary disease. Part. Fibre Toxicol. 6: 25.

Hoffmann, B., S. Moebus, A. Stang, E.-M. Beck, N. Dragano, S. Möhlenkamp, A. Schmermund, M. Memmesheimer, K. Mann, R. Erbel and K.H. Jöckel. 2006. Residence close to high traffic and prevalence of coronary heart disease. European Heart J. 27(22): 2696–2702.

Hoffmann, B., S. Moebus, S. Möhlenkamp, A. Stang, N. Lehmann, N. Dragano, A. Schmermund, M. Memmesheimer, K. Mann, R. Erbel and K.H. Jöckel. 2007. Residential exposure to traffic is associated with coronary atherosclerosis. Circulation 116(5): 489–496. REVIHAAP Project: Technical Report Pages 249.

Huynh, M., T.J. Woodruff, J.D. Parker and K.C. Schoendorf. 2006. Relationships between air pollution and preterm birth in California. Paediatr. Perinat. Epidemiol. 20: 454–461.

Ito, K., W.F. Christensen, D.J. Eatough, R.C. Henry, E. Kim, F. Laden, R. Lall, T.V. Larson, L. Neas, P.K. Hopke and G.D. Thurston. 2006. PM source apportionment and health effects: 2. An investigation of intermethod variability in associations between source-apportioned fine particle mass and daily mortality in Washington, DC. J. Expo. Sci. Environ. Epidemiol. 16(4): 300–310.

Ito, K., R. Mathes, Z. Ross, A. Nádas, G. Thurston and T. Matte. 2011. Fine particulate matter constituents associated with cardiovascular hospitalizations and mortality in New York City. Environ. Health Perspect. 119(4): 467–473.

Jacquemin, B., F. Kauffmann, I. Pin, N. Le Moual, J. Bousquet, F. Gormand, J. Just, R. Nadif, C. Pison, D. Vervloet, N. Künzli and V. Siroux. 2012. Air pollution and asthma control in the epidemiological study on the genetics and environment of asthma. J. Epidemiol. Community Health 66(9): 796–802.

Jalaludin, B, T. Mannes, G. Morgan, D. Lincoln, V. Sheppeard and S. Corbett. 2007. Impact of ambient air pollution on gestational age is modified by season in Sydney, Australia. Environ. Health 6: 16.

Janssen, N.A.H., M.E. Gerlofs-Nijland, T. Lanki, R.O. Salonen, F. Cassee, G. Hoek, P. Fischer, B. Brunekreef and M. Krzyzanowski. 2012. Health effects of black carbon. Copenhagen, WHO Regional Office for Europe (http://www.euro.who.int/__data/assets/pdf_file/0004/162535/e96541.pdf).

Jędrychowski, W., R.M. Whyatt, D.E. Camann, U.V. Bawle, K. Peki, J.D. Spengler, T.S. Dumyahn, A. Penar and F.P. Perera. 2003. Effect of prenatal PAH exposure on birth outcomes and neurocognitive development in a cohort of newborns in Poland. Study design and preliminary ambient data. Int. J. Occup. Med. Environ. Health 16(1): 21–29.

Jedrychowski, W., I. Bendkowska, E. Flak, A. Penar, R. Jacek, I. Kaim, J.D. Spengler, D. Camann and F.P. Perera. 2004. Estimated risk for altered fetal growth resulting from exposure to fine particles during pregnancy: an epidemiologic prospective cohort study in Poland. Environ. Health Perspect. 112(14): 1398–1402.

Jedrychowski, W., F. Perera, D. Mrozek-Budzyn, E. Mroz, E. Flak, J.D. Spengler, S. Edwards, R. Jacek, L. Kaim and Z. Skolicki. 2009. Gender differences in fetal growth of newborns exposed prenatally to airborne fine particulate matter. Environ. Res. 109: 447–456.

Jerrett, M., R.T. Burnett, R. Ma, C.A. Pope 3rd, D. Krewski, K.B. Newbold, G. Thurston, Y. Shi, N. Finkelstein, E.E. Calle and M.J. Thun. 2005. Spatial analysis of air pollution and mortality in Los Angeles. Epidemiology 16(6): 727–736.

Jerrett, M., R.T. Burnett, C.A. Pope 3rd, K. Ito, G. Thurston, D. Krewski, Y. Shi, E. Calle and M. Thun. 2009. Long-term ozone exposure and mortality. N. Engl. J. Med. 360(11): 1085–1095.

Katsouyanni, K., J.M. Samet, H.R. Anderson, R. Atkinson, A. Le Tertre, S. Medina, E. Samoli, G. Touloumi, R.T. Burnett, D. Krewski, T. Ramsay, F. Dominici, R.D. Peng, J. Schwartz and A. Zanobetti. 2009. Air pollution and health: a European and North American approach (APHENA). Boston. Health Effects Institute (http://pubs.healtheffects.org/getfile.php?u=518, accessed 14 March 2013).

Kim, J.J., K. Huen, S. Adams, S. Smorodinsky, A. Hoats, B. Malig, M. Lipsett and B. Ostro. 2008. Residential traffic and children's respiratory health. Environ. Health Perspect. 116(9): 1274–1279.

Kim, O.J., E.H. Ha, B.M. Kim, J.H. Seo, H.S. Park, W.J. Jung, B.E. Lee, Y.J. Suh, Y.J. Kim, J.T. Lee, H. Kim and Y.C. Hong. 2007. PM10 and pregnancy outcomes: a hospital-based cohort study of pregnant women in Seoul. J. Occup. Environ. Med. 49(12): 1394–1402.

Kim, S.Y., J.L. Peel, M.P. Hannigan, S.J. Dutton, L. Sheppard, M.L. Clark and S. Vedal. 2012. The temporal lag structure of short-term associations of fine particulate matter chemical constituents and cardiovascular and respiratory hospitalizations. Environ. Health Perspect. 120(8): 1094–1099.

Kovats, R.S. and S. Hajat. 2008. Heat stress and public health: a critical review. Ann. Rev. Public. Health 29: 41–55.

Krämer, U., C. Herder, D. Sugiri, K. Strassburger, T. Schikowski, U. Ranft and W. Rathmann. 2010. Traffic-related air pollution and incident type 2 diabetes: results from the SALIA cohort study. Environ. Health Perspect. 118(9): 1273–1279.

Krewski, D., M. Jerrett, R.T. Burnett, R. Ma, E. Hughes, Y. Shi, M.C. Turner, C.A. Pope 3rd, G. Thurston, E.E. Calle, M.J. Thun, B. Beckerman, P. DeLuca, N. Finkelstein, K. Ito, D.K. Moore, K.B. Newbold, T. Ramsay, Z. Ross, H. Shin and B. Tempalski. 2009. Extended follow-up and spatial analysis of the American Cancer Society study linking particulate air pollution and mortality. Res. Rep. Health Eff. Inst. 140: 5–114.

Lacasaña, M., A. Esplugues and F. Ballester. 2005. Exposure to ambient air pollution and prenatal and early childhood health effects. Eur. J. Epidemiol. 20(2): 183–199.

Leem, J.H., B.M. Kaplan, Y.K. Shim, H.R. Pohl, C.A. Gotway, S.M. Bullard, J.F. Rogers, M.M. Smith and C.A. Tylenda. 2006. Exposures to air pollutants during pregnancy and preterm delivery. Environ. Health Perspect. 114(6): 905–910.

Lepeule, J., F. Caini, S. Bottagisi, J. Galineau, A. Hulin, N. Marquis, A. Bohet, V. Siroux, M. Kaminski, M.A. Charles, R. Slama and EDEN Mother–Child Cohort Study Group. 2010. Maternal exposure to nitrogen dioxide during pregnancy and offspring birth weight: comparison of two exposure models. Environ. Health Perspect. 118: 1483–1489.

Lipfert, F.W., R.E. Wyzgab, J.D. Batyc and J.P. Millerc. 2006. Traffic density as a surrogate measure of environmental exposures in studies of air pollution health effects: long-term mortality in a cohort of US veterans. Atmos. Environ. 40(1): 154–169.

Liu, S., D. Krewski, Y. Shi, Y. Chen and R.T. Burnett. 2003. Association between gaseous ambient air pollutants and adverse pregnancy outcomes in Vancouver, Canada. Environ. Health Perspect. 111: 1773–1778.

Liu, S., D. Krewski, Y. Shi, Y. Chen and R.T. Burnett. 2007. Association between maternal exposure to ambient air pollutants during pregnancy and fetal growth restriction. J. Expo. Sci. Environ. Epidemiol. 17: 426–432.

Logan, W.P. 1953. Mortality in the London fog incident, 1952. Lancet 261(6755): 336–338.

Madsen, C., U. Gehring, S.E. Walker, B. Brunekreef, H. Stigum, O. Naess and P. Nafstad. 2010. Ambient air pollution exposure, residential mobility and term birth weight in Oslo, Norway. Environ. Res. 110: 363–371.

Maisonet, M., T.J. Bush, A. Correa and J.J. Jaakkola. 2001. Relation between ambient air pollution and low birth weight in the Northeastern United States. Environ. Health Perspect. 109(3): 351–356.

Maisonet, M., A. Correa, D. Misra and J.J. Jaakkola. 2004. A review of the literature on the effects of ambient air pollution on fetal growth. Environ. Res. 95(1): 106–115.

Malmqvist, E., A. Rignell-Hydbom, H. Tinnerberg, J. Bjork, E. Stroh, K. Jakobsson, R. Rittner and L. Rylander. 2011. Maternal exposure to air pollution and birth outcomes. Environ. Health Perspect. 119: 553–558.

Mannes, T., B. Jalaludin, G. Morgan, D. Lincoln, V. Sheppeard and S. Corbett. 2005. Impact of ambient air pollution on birth weight in Sydney, Australia. Occup. Environ. Med. 62: 524–530.

Morello-Frosch, R., B.M. Jesdale, J.L. Sadd and M. Pastor. 2010. Ambient air pollution exposure and full-term birth weight in California. Environ. Health 9: 44.

Naeher, L.P., M. Brauer, M. Lipsett, J.T. Zelikoff, C.D. Simpson, J.Q. Koenig and K.R. Smith. 2007. Woodsmoke health effects: a review. Inhal. Toxicol. 19(1): 67–106.

Ostro, B., R. Broadwin, S. Green, W.Y. Feng and M. Lipsett. 2006. Fine particulate air pollution and mortality in nine California counties: results from CALFINE. Environ. Health Perspect. 114(1): 29–33.

Ostro, B., L. Roth, B. Malig and M. Marty. 2009. The effects of fine particle components on respiratory hospital admissions in children. Environ. Health Perspect. 117(3): 475–480.

Ostro, B., M. Lipsett, P. Reynolds, D. Goldberg, A. Hertz, C. Garcia, K.D. Henderson and L. Bernstein. 2010. Long-term exposure to constituents of fine particulate air pollution and mortality: results from the California Teachers Study. Environ. Health Perspect. 118(3): 363–369.

Parker, J.D. and T.J. Woodruff. 2008. Influences of study design and location on the relationship between particulate matter air pollution and birth weight. Paediatr. Perinat. Epidemiol. 22(3): 214–227.

Parker, J.D., T.J. Woodruff, R. Basu and K.C. Schoendorf. 2005. Air pollution and birth weight among term infants in California. Pediatrics 115: 121–128.

Parker, J.D., D.Q. Rich, S.V. Glinianaia, J.H. Leem, D. Wartenberg, M.L. Bell, M. Bonzini, M. Brauer, L. Darrow, U. Gehring, N. Gouveia, P. Grillo, E. Ha, E.H. van den Hooven, B. Jalaludin, B.M. Jesdale, J. Lepeule, R. Morello-Frosch, G.G. Morgan, R. Slama, F.H. Pierik, A.C. Pesatori, S. Sathyanarayana, J. Seo, M. Strickland, L. Tamburic and T.J. Woodruff. 2011. The international collaboration on air pollution and pregnancy outcomes: initial results. Environ. Health Perspect. 119(7): 1023–1028.

Pearce, M.S., S.V. Glinianaia, J. Rankin, S. Rushton, M. Charlton, L. Parker and T. Pless-Mulloli. 2010. No association between ambient particulate matter exposure during pregnancy and still birth risk in the north of England, 1962–1992. Environ. Res. 110: 118–122.

Perera, F.P., V. Rauh, W.Y. Tsai, P. Kinney, D. Camann, D. Barr, T. Bernert, R. Garfinkel, Y.H. Tu, D. Diaz, J. Dietrich and R.M. Whyatt. 2003. Effects of transplacental exposure to environmental pollutants on birth outcomes in a multiethnic population. Environ. Health Perspect. 111(2): 201–205.

Perera, F.P., V. Rauh, R.M. Whyatt, W.Y. Tsai, J.T. Bernert, Y.H. Tu, H. Andrews, J. Ramirez, L. Qu and D. Tang. 2004. Molecular evidence of an interaction between prenatal environmental exposures and birth outcomes in a multiethnic population. Environ. Health Perspect. 112(5): 626–630.

Perera, F.P., V. Rauh, R.M. Whyatt, W.Y. Tsai, D. Tang, D. Diaz, L. Hoepner, D. Barr, Y.H. Tu, D. Camann and P. Kinney. 2006. Effect of prenatal exposure to airborne polycyclic aromatic hydrocarbons on neurodevelopment in the first 3 years of life among inner-city children. Environ. Health Perspect. 114(8): 1287–1292.

Perera, F.P., Z. Li, R. Whyatt, L. Hoepner, S. Wang, D. Camann and V. Rauh. 2009. Prenatal airborne polycyclic aromatic hydrocarbon exposure and child IQ at age 5 years. Pediatrics 124(2): 195–202.

Pesatori, A.C., M. Bonzini, M. Carugno, N. Giovannini, V. Signorelli and A. Baccarelli. 2008. Ambient air pollution affects birth and placental weight. A study from Lombardy (Italy) region. Epidemiology 19(suppl.): 178–179.

Peters, A., D.W. Dockery, J.E. Muller and M.A. Mittleman. 2001. Increased particulate air pollution and the triggering of myocardial infarction. Circulation 103: 2810–2815.

Polanska, K., W. Hanke, W. Sobala, S. Brzeznicki and D. Ligocka. 2011. Predictors of environmental exposure to polycyclic aromatic hydrocarbons among pregnant women—prospective cohort study in Poland. Int. J. Occup. Med. Environ. Health 24(1): 8–17.

Pope, C.A. and D.W. Dockery. 2006. Health effects of fine particulate air pollution: lines that connect. Air Waste Manag. Assoc. 56: 709–742.

Pope, C.A. 3rd, R.T. Burnett, M.J. Thun, E.E. Calle, D. Krewski, K. Ito and G.D. Thurston. 2002. Lung cancer, cardiopulmonary mortality, and long-term exposure to fine particulate air pollution. J. Am. Med. Assoc. 287(9): 1132–1141.

Pope, C.A. 3rd, M. Ezzati and D.W. Dockery. 2009. Fine-particulate air pollution and life expectancy in the United States. N. Engl. J. Med. 360(4): 376–386.

Proietti, E., M. Röösli, U. Frey and P. Latzin. 2013. Air pollution during pregnancy and neonatal outcome: a review. J. Aerosol Med. Pulm. Drug Deliv. 26(1): 9–23.

Puett, R.C., J.E. Hart, J.D. Yanosky, C. Paciorek, J. Schwartz, H. Suh, F.E. Speizer and F. Laden. 2009. Chronic fine and coarse particulate exposure, mortality, and coronary heart disease in the Nurses' Health Study. Environ. Health Perspect. 117(11): 1697–1701.

Puett, R.C., J.E. Hart, H. Suh, M. Mittleman and F. Laden. 2011. Particulate matter exposures, mortality, and cardiovascular disease in the health professionals follow-up study. Environ. Health Perspect. 119(8): 1130–1135.

Raaschou-Nielsen, O., M. Sørensen, M. Ketzel, O. Hertel, S. Loft, A. Tjønneland, K. Overvad and Z.J. Andersen. 2013. Long-term exposure to traffic-related air pollution and diabetes-associated mortality: a cohort study. Diabetologia 56(1): 36–46.

Rich, D.Q., K. Demissie, S.E. Lu, L. Kamat, D. Wartenberg and G.G. Rhoads. 2009. Ambient air pollutant concentrations during pregnancy and the risk of fetal growth restriction. J. Epidemiol. Community Health 63: 488–496.

Rich, D.Q., W. Zareba, W. Beckett, P.K. Hopke, D. Oakes, M.W. Frampton, J. Bisognano, D. Chalupa, J. Bausch, K. O'Shea, Y. Wang and M.J. Utell. 2012. Are ambient ultrafine, accumulation mode, and fine particles associated with adverse cardiac responses in patients undergoing cardiac rehabilitation? Environ. Health Perspect. 120(8): 1162–1169.

Ritz, B. and F. Yu. 1999. The effect of ambient carbon monoxide on low birth weight among children born in southern California between 1989 and 1993. Environ. Health Perspect. 107(1): 17–25.

Ritz, B. and M. Wilhelm. 2008. Ambient air pollution and adverse birth outcomes: methodologic issues in an emerging field. Basic Clin. Pharmacol. Toxicol. 102(2): 182–190.

Ritz, B., F. Yu, G. Chapa and S. Fruin. 2000. Effect of air pollution on preterm birth among children born in Southern California between 1989 and 1993. Epidemiology 11(5): 502–511.

Ritz, B., M. Wilhelm, K.J. Hoggatt and J.K. Ghosh. 2007. Ambient air pollution and preterm birth in the environment and pregnancy outcomes study at the University of California, Los Angeles. Am. J. Epidemiol. 166: 1045–1052.

Roemer, W.H. and J.H. van Wijnen. 2001. Daily mortality and air pollution along busy streets in Amsterdam, 1987–1998. Epidemiology 12(6): 649–653.

Rogers, J.F. and A.L. Dunlop. 2006. Air pollution and very low birth weight infants: a target population? Pediatrics 118(1): 156–164.

Rogers, J.F., S.J. Thompson, C.L. Addy, R.E. McKeown, D.J. Cowen and P. Decouflé. 2000. Association of very low birth weight with exposures to environmental sulfur dioxide and total suspended particulates. Am. J. Epidemiol. 151(6): 602–613.

Rückerl, R., A. Schneider, S. Breitner, J. Cyrys and A. Peters. 2011. Health effects of particulate air pollution: a review of epidemiological evidence. Inhal. Toxicol. 23(10): 555–592.

Sagiv, S.K., P. Mendola, D. Loomis, A.H. Herring, L.M. Neas, D.A. Savitz and C.A. Poole. 2005. A time-series analysis of air pollution and preterm birth in Pennsylvania, 1997–2001. Environ. Health Perspect. 113(5): 602–606.

Salam, M.T., J. Millstein, Y.F. Li, F.W. Lurmann, H.G. Margolis and F.D. Gilliland. 2005. Birth outcomes and prenatal exposure to ozone, carbon monoxide, and particulate matter: results from the Children's Health Study. Environ. Health Perspect. 113: 1638–1644.

Schwartz, J., F. Laden and A. Zanobetti. 2002. The concentration-response relation between PM and daily deaths. Environmental Health Perspectives Environ. Health Perspect. 110: 1025–1029.

Seo, J.H., J.H. Leem, E.H. Ha, O.J. Kim, B.M. Kim, J.Y. Lee, H.S. Park, H.C. Kim, Y.C. Hong and Y.J. Kim. 2010. Population attributable risk of low birth weight related to PM10 pollution in seven Korean cities. Paediatr. Perinat. Epidemiol. 24: 140–148.

Slama, R., V. Morgenstern, J. Cyrys, A. Zutavern, O. Herbarth, H.E. Wichmann and J. Heinrich. 2007. Traffic-related atmospheric pollutants levels during pregnancy and offspring's term birth weight: a study relying on a land-use regression exposure model. Environ. Health Perspect. 115: 1283–1292.

Slama, R., L. Darrow, J. Parker, T.J. Woodruff, M. Strickland, M. Nieuwenhuijsen, S. Glinianaia, K.J. Hoggatt, S. Kannan, F. Hurley, J. Kalinka, R. Srám, M. Brauer, M. Wilhelm, J. Heinrich and B. Ritz. 2008. Meeting report: atmospheric pollution and human reproduction. Environ. Health Perspect. 116(6): 791–798.

Smith, K.R., M. Jerrett, H.R. Anderson, R.T. Burnett, V. Stone, R. Derwent, R.W. Atkinson, A. Cohen, S.B. Shonkoff, D. Krewski, C.A. Pope 3rd, M.J. Thun and G. Thurston. 2009. Public health benefits of strategies to reduce greenhouse-gas emissions: health implications of short-lived greenhouse pollutants. Lancet 374(9707): 2091–2103.

Son, J.Y., J.T. Lee, K.H. Kim, K. Jung and M.L. Bell. 2012. Characterization of fine particulate matter and associations between particulate chemical constituents and mortality in Seoul, Korea. Environ. Health Perspect. 120(6): 872–878.

Srám, R.J., B. Binková, J. Dejmek and M. Bobak. 2005. Ambient air pollution and pregnancy outcomes: a review of the literature. Environ. Health Perspect. 113(4): 375–382.

Suh, Y.J., H. Kim, J.H. Seo, H. Park, Y.J. Kim, Y.C. Hong and E.H. Ha. 2009. Different effects of PM10 exposure on preterm birth by gestational period estimated from time-dependent survival analyses. Int. Arch. Occup. Environ. Health 82: 613–621.

TfL. 2004. Congestion charging central London: impacts monitoring—second annual report. London, Transport for London (http://www.tfl.gov.uk/assets/downloads/Impacts-monitoring-report-2.pdf).

Tonne, C., R.M. Whyatt, D.E. Camann, F.P. Perera and P.L. Kinney. 2004. Predictors of personal polycyclic aromatic hydrocarbon exposures among pregnant minority women in New York City. Environ. Health Perspect. 112: 754–759.

Tonne, C., S. Beevers, F.J. Kelly, L. Jarup, P. Wilkinson and B. Armstrong. 2010. An approach for estimating the health effects of changes over time in air pollution: an illustration using cardio-respiratory hospital admissions in London. Occup. Environ. Med. 67(6): 422–427.

van den Hooven, E.H., V.W. Jaddoe, Y. de Kluizenaar, A. Hofman, J.P. Mackenbach, E.A. Steegers, H.M. Miedema and F.H. Pierik. 2009. Residential traffic exposure and pregnancy-related outcomes: a prospective birth cohort study. Environ. Health 8: 59.

Weichenthal, S. 2012. Selected physiological effects of ultrafine particles in acute cardiovascular morbidity. Environ. Res. 115: 26–36.

Wesseling, J., J. den Boeft, G.A.C. Boersen, K. Hollander, K.D. van den Hout and M.P. Keuken. 2002. Development and validation of the new TNO model for the dispersion of traffic emissions. pp. 456–460. *In*: 8th International Conference on Harmonisation within Atmospheric Dispersion Modelling for Regulatory Purposes. Demetra Ltd., Sofia, Bulgaria.

WHO Regional Office for Europe. 2006. Air quality guidelines global update 2005: particulate matter, ozone, nitrogen dioxide and sulfur dioxide. WHO Regional Office for Europe. Copenhagen, Denmark (http://www.euro.who.int/__data/assets/pdf_file/0005/78638/E90038.pdf).

WHO Regional Office for Europe. 2013. Review of evidence on health aspects of air pollution—REVIHAAP Project Technical Report. http://www.euro.who.int/data/assets/pdf_file/0020/182432/e96762-final.pdf.

Wilhelm, M. and B. Ritz. 2003. Residential proximity to traffic and adverse birth outcomes in Los Angeles county, California, 1994–1996. Environ. Health Perspect. 111(2): 207–216.

Wilhelm, M. and B. Ritz. 2005. Local variations in CO and particulate air pollution and adverse birth outcomes in Los Angeles County, California, USA. Environ. Health Perspect. 113(9): 1212–1221.

Woodruff, T.J., J.D. Parker, A.D. Kyle and K.C. Schoendorf. 2003. Disparities in exposure to air pollution during pregnancy. Environ. Health Perspect. 111(7): 942–946.

Woodruff, T.J., J.D. Parker, L.A. Darrow, R. Slama, M.L. Bell, H. Choi, S. Glinianaia, K.J. Hoggatt, C.J. Karr, D.T. Lobdell and M. Wilhelm. 2009. Methodological issues in studies of air pollution and reproductive health. Environ. Res. 109(3): 311–320.

Woodruff, T.J., J.D. Parker, K. Adams, M.L. Bell, U. Gehring, S. Glinianaia, E.H. Ha, B. Jalaludin and R. Slama. 2010. International Collaboration on Air Pollution and Pregnancy Outcomes (ICAPPO). Int. J. Environ. Res. Public Health 7(6): 2638–2652.

Xu, X., R.K. Sharma, E.O. Talbott, J.V. Zborowski, J. Rager, V.C. Arena and C.D. Volz. 2011. PM10 air pollution exposure during pregnancy and term low birth weight in Allegheny County, PA, 1994–2000. Int. Arch. Occup. Environ. Health 84: 251–257.

Zanobetti, A. and J. Schwartz. 2009. The effect of fine and coarse particulate air pollution on mortality: a national analysis. Environ. Health Perspect. 117(6): 898–903.

Zanobetti, A. and J. Schwartz. 2011. Ozone and survival in four cohorts with potentially predisposing diseases. Am. J. Respir. Crit. Care Med. 184(7): 836–841.

Zhao, Q., Z. Liang, S. Tao, J. Zhu and Y. Du. 2011. Effects of air pollution on neonatal prematurity in Guangzhou of China: a time-series study. Environ. Health 10: 2.

11

Dose-Effect and Dose-Risk Relationships as the Tools for the Quantitative Description of the Adverse Health Effects of Inhaled Pollutants: Chances and Limitations

Jozef S. Pastuszka

Dose as a key factor determining adverse health effects of environmental origin

There is vast knowledge about the adverse health effects appearing inside the population exposed to air pollutants (prognosis of the health effects for a single person needs additional information, first of all resulting from the medical investigation of this person before exposure). Most of the volume of this book is also devoted to the problems of characteristics of air pollutants and their health impacts.

Although the qualitative prognosis of the health effects resulting from exposure to air pollutants is rather easy, the quantitative prognosis requires, as a rule, advanced analysis of existing data and often also additional studies. Trying to precisely predict the adverse health effect of the specific air pollutant inhaled by the exposed population, it should be clearly stated that the factor responsible for the kind of effect and its intensity is not the concentration (C) but dose (D) being, as a rule, the mass or amount of absorbed pollutant.

Silesian University of Technology, Division of Energy and Environmental Engineering, Department of Air Protection, 22B Konarskiego St., 44-100 Gliwice, Poland.
E-mail: Jozef.Pastuszka@polsl.pl

The idea of the importance of the dose has been formulated unusually slowly. Although the toxicity of various substances, including gases, has been studied for ages, it was only in the early 1900s Fritz Haber, who was a German scientist (from Silesia), studied acute lethality of war gases and contended that there was a constant relationship between the concentration (C, mg/m^3) of the gas and the length of time (t, min) leading to the death of an animal (Haber 1924). This relationship between C and t was expressed as $C \cdot t = k$ and became known as Haber's rule (Witschi 1999; Miller et al. 2000). While Warren (1900), who used a basically similar equation for the strength of the salt solution that killed *Daphnia magna*, preceded Haber, the concept of $C \cdot t = k$ has been ascribed to Haber (Miller et al. 2000). Many years later, various aspects of Haber's rule have served as the kernel for setting short-term exposure limits for workplace exposures (Miller et al. 2000). At last the fundamental significance of dose in predicting adverse health effects on people present in the polluted environment became clear because dose is the product of the concentration, time of exposure, and the coefficient of absorption of the pollutant.

In the meantime, a young Croatian scientist Mirko Fugaš (Fugaš et al. 1972; Fugaš 1976, 1986) found that the received dose D of air pollutant depends on the concentration C_k of the pollutant in the air of the microenvironment (k), time of exposure to this pollutant in the k-microenvironment (t_k), and on the inhalation rate, or generally, the contact rate w. Hence, the dose can be calculated using the simple equation:

$$D = w\sum_k C_k t_k \qquad (11\text{-}1)$$

or

$$D = w\sum_k E_k \qquad (11\text{-}2)$$

where:

$$E_k = C_k t_k \qquad (11\text{-}3)$$

is the average exposure in the kth microenvironment; t_k is time spent in the kth microenvironment; and C_k is the concentration of the analyzed pollutant in the kth microenvironment.

It should be noted that the term "exposure" means both a product of concentration and time of exposure and, more generally, a contact between an airborne contaminant and a surface of the human body, either outer (for example, the skin) or inner (for example, respiratory tract epithelium).

Making use of concentration levels in the so-called microenvironments (room, car, train, outdoor environment, occupational site, etc.) coupled with activity pattern data (typically the time spent in each microenvironment) yields more accurate exposure information and is increasingly used for estimating population exposure distributions (Kruize et al. 2003; Mekel et al. 2014). Although the definition of the exposure as the sum of the products of concentration and time of exposure in all the analyzed microenvironments is very useful in practice, the basic theoretical approach requires the integrated exposure to be characterized as follows:

$$E_j = \int_{t1}^{t2} C_j(t)dt \tag{11-4}$$

where E_j is the integrated exposure of the jth individual to a concentration C of an agent for the time period from t_1 to t_2 (associated with a biological response).

Exposure assessment has been developed as the distinctly separated branch of science during the last thirty years (Lioy 2010; Ryswyk et al. 2014). Generally, depending on the purpose of the exposure assessment, exposure estimates may reflect the average exposure, peak exposure, or high-end exposure experienced by a small sample of the population (Mekel et al. 2014). As a rule, point values for each of the exposure variables in the model are used: central tendency estimates like mean or median values for estimation of the average exposure for the studied population. For estimating high-end exposure, unfavorable values (e.g., > 90th percentile, maximum value) for the exposure variables might be used as input (Mekel et al. 2014). Unfortunately, in this case, the estimation of the exposure is difficult. According to Mekel et al. (2014), to overcome this problem, the employment of distributions for exposure variables in the so-called probabilistic exposure assessments or distribution-based assessment may be used.

Basically, there are two general approaches to air pollution exposure assessment: (1) air monitoring, which depends on either direct measurements (personal monitoring) or indirect measurements (fixed-site monitors combined with data on time-activity patterns), and (2) biological measurements that use biological markers to assess exposure (Sexton and Ryan 1988).

Ambient concentration measurements are carried out using central-site-monitors most commonly. However, this approach may introduce exposure prediction errors and misclassification of exposures for pollutants that are spatially heterogeneous (Özkaynak et al. 2013), such as those associated with traffic emissions, for example, carbon monoxide, elemental carbon, nitrogen oxides, and particulate matter, especially PM10 and PM2.5 (Pastuszka et al. 2010; Kozielska et al. 2013). Therefore, alternative air quality and human exposure metrics were applied in recent air pollution health effect studies (Özkaynak et al. 2013). These metrics include: central site or interpolated monitoring data, regional pollution levels predicted using the national scale air quality models, or from measurements combined with local-scale air quality models, hybrid models that include satellite data, statistically blended modeling, and measurement data concentrations adjusted by home infiltration rates and others. All these alternative exposure metrics were applied in epidemiological applications to health outcomes, including daily mortality and respiratory hospital admissions, daily hospital emergency department visits, daily myocardial infarction, and daily adverse birth outcomes.

Traditionally, indirect methods such as ambient environmental monitoring or self-administered questionnaires were used for the assessment of chemical exposures. However, with the rapid advancement of sensitive analytical techniques, biomonitoring has become very popular for the characterization of chemical exposure (Sexton et al. 2004; Heffernan et al. 2014). Referring biomonitoring to chemical biomonitoring only, it can be defined as the systematic collection of human tissues and fluids, such as blood, urine, hair, and transformation products to facilitate exposure assessment (Angerer et al. 2007; Heffernan et al. 2014).

Dose-effect relationship

On the basis of a number of studies on adverse health effects caused by exposure to air pollutants, it was possible to prepare the dose-effect relationship for specific pollutants. Dose-effect assessment is the quantitative relationship between the magnitude of dose (or exposure) and the occurrence of human health effects.

Such a relationship is presented in Fig. 11-1. While on the horizontal axis there are the values of dose (for example, in mg), the vertical axis contains the description of various health effects, coded as: *a*, *b*, *c*, etc. Since these “values” on the vertical axis are not continuous in character, the graph in Fig. 11-1 should be treated as a general illustration of the dose-effect relationship prepared for educational purposes only. As a rule, the dose-effect relationships are published in literature in tables where one

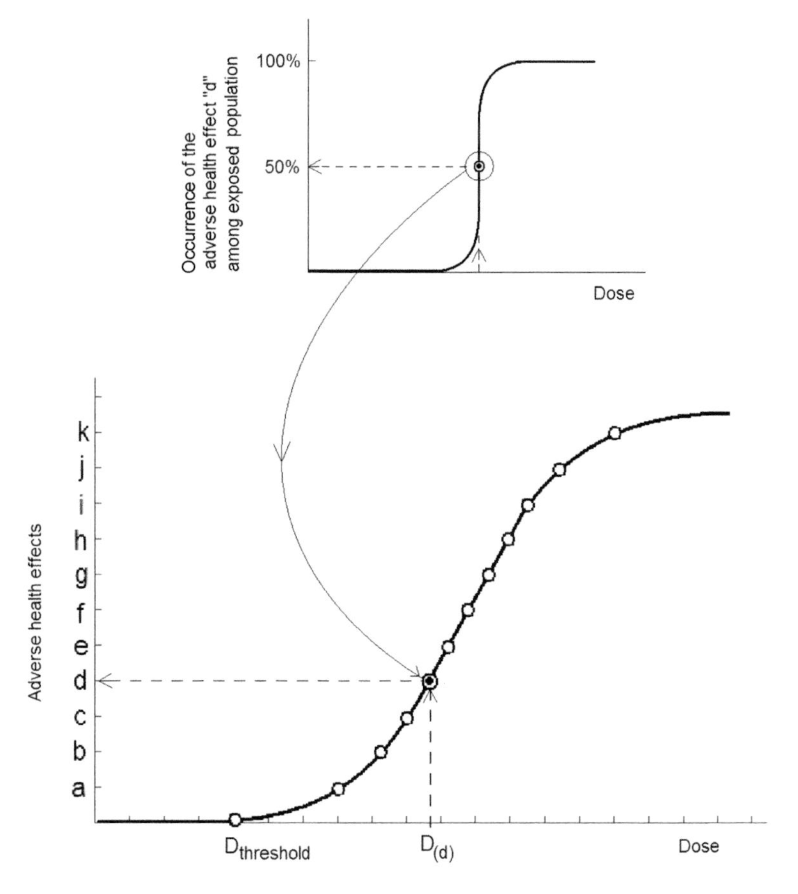

Fig. 11-1. Scheme of the preparation of the dose-effect relationship for a pollutant producing a threshold effect.

column contains the ranges of doses and the second, the description of the adverse health effects caused by these doses.

The dose-effect relationship can be obtained from toxicological studies (conducted on animals), epidemiological studies, and from limited inhalation studies carried out on voluntaries. The book covers all these kinds of studies.

It is important to note that all of the points on the dose-effect curve have been found during the time-consuming dose-response studies (see the top of Fig. 11-1). The aim of these dose-response investigations is to find how many subjects receiving the increasing dose of the specific pollutant will have one precisely defined adverse health effect. As it can be seen the value of dose for which the studied health effect (for example effect *d*) appears in 50% of the exposed subjects is next used as one point of the dose-effect curve. Continuing such response studies for other effects (*a*, *b*, c,.., *e*, *f*, ...), the full curve of the dose-effect relationship can be prepared.

Figure 11-1 illustrates our previous conclusion that the health effects of the exposure to a toxic substance depend on the received dose of this substance but the dose must exceed the so called "threshold dose", usually more or less precisely determined. There are some misclassifications of the toxic effects of chemicals which are usually classified as threshold and nonthreshold effects, as well as according to the effects' duration and route of exposure, reversibility, and toxic endpoints. To clarify these terms, some types of adverse health effects are precisely defined below, according to Rotenberg (1992).

Nonthreshold and threshold effects

A chemical or insult is said to produce a nonthreshold effect if any exposure could result in the effect; that is no safe dose or exposure can be established, and any exposure is associated with some risk. This does not imply, however, that any exposure *will* result in the effect. Only toxic effects that interact with genetic material are currently considered nonthreshold effects (carcinogenesis and mutagenesis).

A chemical or insult is said to produce a threshold effect if the effect occurs only at a certain level of exposure. Below the threshold exposure, the effect does not occur. It is important to determine if the toxic effects of a chemical exposure (or other insult) demonstrate a threshold, because risk is estimated differently for threshold and nonthreshold effects.

It is important to know that the existence or absence of a threshold cannot be proven scientifically. The long-time experience of toxicologists, epidemiologists, physicians, and environmental health experts has led to the following conclusions (Rotenberg 1992):

- For data showing no apparent threshold, it could be argued that data at lower doses must be analyzed to show no effect at such doses.
- For data showing an apparent threshold, it could be argued that a large number of animals (or humans) in the low dose group must be examined to show an effect at that dose.

Duration of exposure

Acute effect is a term describing the initial effects of exposure to very high doses or the effects that result from a single exposure, not the severity of the effects.

Chronic effect is a term referring either to lifetime exposure or to the longest period of exposure time needed to produce an effect.

Subchronic effect is a term referring to all lengths of exposure between acute and chronic.

Reversibility of effects-refers to the inherent ability of the biological system to return to normal function upon recovery; it does not refer to the ultimate toxic effect of an unlimited exposure. Many, if not most, toxic effects in humans are fully reversible.

Reversible effects occur when no permanent change is made in the biological system.

Irreversible effects occur when the biologic system does not return to its former state after removal of the toxic insult; some permanent change has occurred as a result of the chemical interaction.

Cumulative effects occur when the chemical is sequestered (usually in a target organ) and does not affect the organ system until a critical body burden is present.

Toxic endpoint—final outcome of toxic effects

Because a chemical has many toxic effects, a safe dose or acceptable daily intake level must be based on the most sensitive toxic endpoint and the most sensitive human population at risk.

Examples of toxic endpoints (Rotenberg 1992):

(1) Teratogenesis
(2) Carcinogenesis
(3) Mutagenesis
(4) Reproductive effects
(5) Immune system effects
(6) Central nervous system (CNS) effects
(7) Irritation effects
(8) Organ system effects
(9) Mortality

Examples of tests or measurements used to confirm or identify toxic endpoints:

(1) Physical observation
(2) Pathologic test
(3) Neurological test
(4) Specific enzyme level measurements
(5) Body weight and organ measurements
(6) Analytic tests for specific chemicals (or metabolites)
(7) Behavioral tests

Although most researchers define carcinogens and allergens as the subgroups of toxic pollutants, our proposal is to clarify and simplify this terminology. In our opinion, only chemicals causing adverse health effects due to absorption of the dose exceeding the threshold dose should be treated as toxic pollutants. Therefore, for toxic pollutants an S-shaped curve with a threshold level can be obtained; assumed to fit all toxic effects, except those that are produced by direct reaction with genetic material. These groups of pollutants (carcinogens and/or mutagenic pollutants) have the dose-effect curves with no threshold level, and often are linear, especially for relatively small doses.

Concentration standards: dose-based approach

Concerning the toxic chemicals it should be noted that decision-makers very often do not need the precise prognosis of the adverse health effects due to exposure of a large or small population to the air pollutants. They are rather interested in receiving only the fundamental information if the exposed population will have any adverse health effects or not. In this case, the preparation of the appropriate prognosis is very easy because it only requires determining if the maximal absorbed dose will exceed the threshold dose or not. Exposed people are safe if:

$$D_{MAX} = D_{LIFE} \leq D_{THRESHOLD} \tag{11-5}$$

where: D_{LIFE} is the dose which can be absorbed during the whole lifetime. Assuming that $D_{THRESHOLD}$ is related with the lifetime exposure it can be written:

$$Ct_{LIFE}w \leq C_{THRESHOLD}t_{LIFE}w \tag{11-6}$$

Hence,

$$C \leq C_{THRESHOLD} \tag{11-7}$$

and

$$C_{THRESHOLD} = \frac{D_{THRESHOLD}}{t_{LIFE}w} \tag{11-8}$$

t_{LIFE} can be assumed, according to WHO guidelines, as 70 years (WHO 2000).

This analysis shows that in common practice the prognosis of the influence of polluted air on human health may be limited in comparison to the actual concentration of the air pollutant/pollutants with the threshold concentration, usually named as the standard concentration. If the actual concentration is lower than the standard, that means that no adverse health effects will appear because the maximal possible dose which can be absorbed (during a lifetime) will be smaller than the threshold dose.

Equation (11-8) illustrates the general idea of the standard (threshold) concentration. However, realistic determination of the standard is much more complicated. First, from the experimental studies it is possible to find the threshold dose as the lowest observed adverse effect level (so-called LOAEL). On the other hand, the threshold dose should be precisely defined as the maximal dose for which no adverse effect is yet observed (named: no observed adverse effect level—NOAEL). Therefore, instead of threshold dose $D_{THRESHOLD}$ in the equation (10-8), the so-called reference dose (*RfD*) should be put, which can be calculated as follows:

$$RfD = \frac{NOAEL}{UF} \tag{11-9}$$

where the uncertainty factor UF is determined on a case-by-case basis, depending principally on the quality of the data base. Generally, a factor of 1 to 10 is used to account for interspecies variation (Meek et al. 1994):

$$UF \in \{1,...,10\} \tag{11-10}$$

An additional factor of 1 to 100 is used to account for inadequacies of the data base, which include, but are not necessarily limited to lack of adequate data on developmental toxicity, use of a LOAEL versus a NOAEL, and inadequacies of the critical study (Meek et al. 1994). An additional uncertainty factor ranging between 1 and 5 may be incorporated where there is sufficient information to indicate a potential for interaction with other chemical substances commonly present in the environment (Meek et al. 1994).

For a number of toxic pollutants, the occurrence of human adverse health effects depends not only on the magnitude of the dose but also on the time of inhalation (typically expressed as the number of days) and on the average body weight of exposed people (typical units: kg). For example, defined reference dose for inhalation of some selected pollutants are the following (Rotenberg 1992): for chromium $RfD = 5.7 \cdot 10^{-7}$ mg (kg day)$^{-1}$, for toluene $RfD = 0.57$ mg (kg day)$^{-1}$, while for acetone $RfD = 3.0$ mg (kg day)$^{-1}$.

The guideline values (standards) obtained by carefully using the appropriate reference dose should represent concentration of air pollutants that would not pose any hazard to the human population. However, because of the lack of information (especially as epidemiological data on the adverse effects of air pollutants are still sparse), compliance with the established standards does not provide absolute safety (WHO 1987, 2004). Standards are often set using extrapolation from animal data or experimental control human health studies without a clear idea of the shape of the dose-effect relationship between air pollutants and specific health effects (Romieu 1997). The fact that guidelines are established for single chemicals restricts their utility to protect the health of the population since it is likely that mixtures of air pollutants will have synergistic effects. Some researchers (for example, Romieu 1997) summarizing the existing knowledge have concluded that air quality guidelines and standards do not provide sufficient information on subject exposure/dose and potential adverse health effects to be useful in quantitative risk assessment. We can agree with their opinion; however our practical experience indicates that the standards and guidelines are very useful for obtaining a qualitative estimation of the general prognosis of the potential health effects among the population exposed to specific air pollutants, toxic in nature.

Carcinogenic risk assessment

In case of carcinogenic substances, the most important effect is the appearance of cancer, therefore the health prognosis for the selected human population exposed to carcinogenic pollutants concerns the calculation of the probability of cancer (i.e., "carcinogenic risk") among the people who received the real dose. Hence, instead of the dose-effect relationship, the dose-risk relationship is, as a rule, analyzed for carcinogenic substances.

In opposition to toxic substances, carcinogenic substances have no threshold level what means that only lack of exposure to these substances generates zero-level risk. As was mentioned in Chapter 5, risk is a function of dose. Hence:

$$R = f(D) \tag{11-11}$$

It should be noted that the cancer risk classification for the so-called general population has yet not been fully completed. The only general agreement is that the risk equal to 10^{-6} or less is classified as acceptable while the risk level 10^{-3} or higher is unacceptable. The recommendation for people living in the areas where the environmentally-generated cancer risk ranges between 10^{-4} and 10^{-5} varies in literature, and is still the subject of emotional discussions among experts.

In most of the environmental exposures, the inhaled dose of carcinogenic pollutants is relatively low and may be approximated by the linear function

$$R = (UCR) \cdot D \tag{11-12}$$

where "*UCR*" is named "Unit Carcinogenic Risk" and has been determined for all known carcinogens.

As pointed out by Schwartz et al. (2011), typical risk assessment for chemical contaminants often makes implicit assumptions that simplify the risk assessment, but sometimes fail. These assumptions include risk independence, risk averaging, risk non-transferability, risk synchrony, and risk accumulation.

On the other hand, carcinogenic risk should be calculated for precisely assumed time of exposure and dose—changing with time because of a number of reasons, including the time change of the concentration levels. Therefore, calculating the cancer risk for a long time, especially for the lifetime period is difficult. However, during the last thirty years another so-called "lifetime carcinogenic risk" became very popular in public health analysis. This risk (R_{LIFE}) is defined as the risk due to the inhalation of the dose of the specific carcinogenic pollutant during all lifetime (assumed to be 70 years) assuming that the present concentration of the carcinogen in the studied area will be unchanged for these 70 years and the exposed population will stay at this area for 70 years. Although such a defined factor seems to be completely detached from reality, this lifetime risk can be used in comparison of various areas polluted by carcinogenic chemicals. Besides, it may be a useful tool for environmental health management. Using the previous equations, the lifetime risk can be expressed as follows:

$$R_{LIFE} = (UCR)D = (UCR)Ct_{LIFE}w \tag{11-13}$$

Because UCR, t_{LIFE} (equal to 70 years), and w are constant for the specific carcinogens, the above equation can be written:

$$R_{LIFE} = K_{LIFE} C \qquad (11\text{-}14)$$

where C is the actual concentration of the analyzed carcinogen.

In Chapter 5, the above equation was used to calculate the lifetime risk for exposure to airborne asbestos.

The appropriate values of the coefficient K_{LIFE} for various carcinogenic pollutants are published in literature (for example, WHO 1987, 2000). For BaP regarded as an index of general Polycyclic Aromatic Hydrocarbons (PAH) mixtures in the air $K_{LIFE} = 8.7 \cdot 10^{-5}$ ng(BaP) m^{-3}.

The dose-risk relationships published in literature which can be used to estimate the cancer risk due to the exposure to specific carcinogenic pollutants are obtained during different, mostly long-time studies. According to Rotenberg (1992), these kinds of studies can be briefly summarized as follows:

1. Human epidemiological studies.
2. Animal bioassay tests for animal carcinogens are typically two years in length, which is the life span of rats or mice used in the test. Exposure is by ingestion or inhalation, and all tissues are examined at the conclusion of the study.
3. Genotoxic tests. Short-term tests to measure the ability of a chemical to interact with or alter DNA. Positive genotoxic test results are not considered sufficient to classify a chemical as a known carcinogen; however, positive results are suggestive and, thus, are often used to select chemicals for animal bioassays.

a) The Ames test, the best known and most widely used of all short-term tests, measures the ability of a chemical to mutate a specific gene in *Salmonella typhimurium.* There are two strong points of the Ames test:

* It was found that about 90% of chemicals that test positive in the Ames test are also carcinogenic in animal bioassay tests.
* It is a short-term, low-cost test that is easy to perform.

However, it should be noted that:

- Negative test results are common for metals and some chlorinated organics.
- It is difficult to test volatile compounds.
- Mutagenesis is not directly equated with carcinogenesis.

b) Other point mutation tests use bacteria, yeast, or animal cell systems.

c) Chromosome effects are measured as changes in the structure or number of chromosomes or in the production of chromosome fragments. Many cell types are used and exposure to the chemical can occur *in vivo* or *in vitro.*

d) DNA damage is measured as either direct breaks in DNA or the inability of a cell system to repair such breaks. Both bacterial and mammalian cell systems are used for DNA damage tests.

e) Mammalian cell transformations are changes in cells exposed to a chemical that cause the cell to induce tumors in recipient animals. Transformation is often accompanied by characteristic morphologic changes in the transformed mammalian cell.

f) Evaluation for structural similarities to known carcinogens.

In the absence of animal or human data, structural similarities can be important in determining the likelihood that a chemical is a carcinogen.

As was mentioned earlier, the dose-risk studies need a long time for realization, some of these more than twenty years. It is related, among others, with the so-called latency time, i.e., the time between the first exposure and the appearing of cancer, typically being very long. Therefore, the knowledge about the carcinogenetic properties of the studied pollutant increases very slowly, even if there is a suspicion, for some reasons, that the pollutant can be a carcinogen. Hence, to improve the risk management some classifications for carcinogens were prepared. One example is presented in Table 11-1.

It should also be noted that risk assessment, primarily developed for the exposure to carcinogenic pollutants and, next—more general for chemical hazards, can be applied to all hazards, whether they are chemical, physical, biological, or psychosocial in nature (Goldstein 2009; Mekel et al. 2014).

As can be seen, a further question concerns exposure-effect relationships. Although knowledge of the shape of the relationship and the existence of a threshold is needed, it is very difficult to obtain with the existing methods.

Table 11-1. Environmental Protection Agency (EPA) weight of evidence classification for carcinogens (after Rotenberg 1992).

Group	Description	Comment
A	Human carcinogen	Sufficient evidence in epidemiologic studies to support causal association.
B1	Probable human carcinogen	Limited evidence of carcinogenicity in humans.
B2	Probable human carcinogen	Sufficient evidence of carcinogenicity in animals.
C	Possible human carcinogen	Limited evidence of carcinogenicity in animals.
D	Not classified	Inadequate evidence of carcinogenicity in animals.
E	No evidence of carcinogenesis for humans	No evidence of carcinogenicity in animals or humans.

New view-point on the old relationships

Particulate matter

Unfortunately, accumulation of existing knowledge sometimes makes the described dose-effect relationship unclear again. One such example is the exposure to airborne particulates. As is known, airborne particles impact human health both through changes in the quality of environment and, if the concentration of aerosol is significantly higher than the level of natural background level of pollution (though it is difficult to be exactly precise), directly as a result of inhaling polluted air. The epidemiological studies indicate that air pollution with particulate matter was related to deterioration of lung functioning (Dassen et al. 1986), increased frequency of occurrence of symptoms and respiratory diseases (Graham 1990; Ware et al. 1986; Dockery et al. 1989), the increase of hospitalization range (Bates et al. 1990; Pope 1991) and increase of mortality (Ostro 1993), even at concentration levels currently occurring in many urbanized areas (Dockery et al. 1996). On the basis of long-term studies carried out in various scientific centers around the world, it can be stated that inhaling fine particles involves the increase of the mortality rate of the general population and respiratory and cardiovascular diseases. As was mentioned in the previous chapters, the results of epidemiological studies indicate that the increase of concentration of PM10 by 10 μg m^{-3} causes the increase of upper respiratory tract morbidity, including asthma. Studies indicating such conclusions have been carried out in cities, where the main source of airborne particles was combustion of various fuels. However, adverse health effects are related not only with living in areas highly polluted with dust particles of industrial origin, where systematic studies indicate the increase of asthmatic symptoms and number of hospitalized people, but also with exposure to high concentration of dust of natural origin. Some interesting results, which confirm this argument, were obtained among others in Alaska, where there are only a few industrial sources of air pollution. For example, in Anchorage, dust particles suspended in the air mainly consist of rock material and volcanic ash, but even there the increase of concentration of suspended particles—PM10 by 10 μg m^{-3} caused the increase of physician office visits because of asthma by 3–6% and the increase of physician office visits because of other upper respiratory tract diseases by 1–3% (Gordian et al. 1996).

Not only long-term epidemiological studies, but also toxicological ones indicated a negative impact of dust inhalation on human health of people exposed to concentration levels, which according to current standards are considered safe. These results were also unexpectedly confirmed by observations made in the so called highly developed countries during the last thirty years, which indicate lack of correlation between a significant improvement of the quality of environment and the morbidity rate from diseases which can be causally related to exposure to dust.

Generally, analysis of the available data leads to the surprising conclusion that airborne particles seem to be the nonthreshold pollutant (without threshold dose). In the WHO Guidelines for Europe (WHO 2000), it is stated that effects on mortality, respiratory, cardiovascular hospital admissions, and other health variables have been observed at levels below 100 μg m^{-3} (expressed as a daily average PM10 concentration). For this reason, no guideline value for short-term average concentrations is recommended either. A similar conclusion can be drawn from the analysis of long-term effects. Although a number of studies showed that the prevalence of bronchitis symptoms in children, and of reduced lung function in children and adults, are associated with particulate matter exposure, these effects have been observed at annual average concentration levels below 20 and 30 μg m^{-3} for PM2.5 and PM10, respectively. For this reason, according to WHO (WHO 2000), no guideline value for long-term averaged concentration is recommended.

In this context, it is worth mentioning the interesting conclusions published by Davidson et al. (2005), who reviewing the existing data found that despite the variability in particulate matter characteristics, which are believed to influence human health risk, the observed relative health risk estimates per unit PM mass falls within a narrow range of values. They concluded that furthermore, no single chemical species appears to dominate health effects; rather the effects appear to be due to a combination of species.

Bioaerosols

Another serious problem is related with attempting to establish some standards for bacterial and fungal aerosols. During every day work related with both environmental and public health management, the answer to the question if the concentration levels of bioaerosols measured or calculated in the analyzed environment are safe or not for exposed people is crucial. It seems that the simplest method could be to compare the obtained concentrations of airborne bacteria or fungi or other bioaerosol particles with appropriate standards. This idea, adopted from industrial hygiene and also used intensively in environmental health management in the case of a number of air pollutants, is really very attractive. Therefore, for the last thirty years, many experts tried to prepare such standards for airborne bacteria and fungi. Unfortunately, these efforts were fruitless. In my opinion, establishing such precisely defined standards for airborne bacteria and fungi is impossible because of some fundamental reasons.

First problem: Adverse health effects due to inhalation of bacterial or fungal aerosol heavily depend on the bacterial/fungal genera and species. To solve this problem, it should be assumed that the potential concentration standard should be established for some typical micro-environments only; like buildings, urban ambient air, country air, etc. Besides, it should be assumed that the comparison of the obtained concentration with the "standard level" will be treated only as the first stage of the prognosis of the health effects. More precise analysis would require microbiological identification of the sampled microorganisms.

Second problem: It should be decided whether the standard concentration level will be related only with the exposure to the viable microorganisms (in this case, the concentration of the number of colony forming units present in one cubic meter of air—CFU m^{-3} can be applied), or with the exposure to the total airborne bacterial/fungal particles; alive and dead (in this case, the concentration is the number of all airborne bacteria/fungi present in one cubic meter of air—m^{-3}). Although the adverse health effects certainly depend on the dose, the prognosis of different effects requires different information related with the absorbed dose—see the Table 11-2.

Table 11-2. Information about the inhaled dose of bioaerosol, needed for the assessment of the adverse health effects (Henningson 1990).

Kind of adverse health effects	Needed information about the inhaled dose
Infectious	Number of viable/alive microorganisms
Allergic effects	Number of the total microorganisms (alive and dead)
Toxic effects	Number of viable microorganisms, number of the total microorganisms (alive and dead), "composition" of bioaerosol, products of reaction/metabolism (example, mycotoxin, endotoxin)

Third problem: Bacterial and fungal aerosols cause different kinds of health effects. For allergic effects there is no threshold dose at all (Dotterud et al. 1996). For toxic effects it is not possible to find one precisely defined threshold dose (Rylander 1996). Therefore it seems that it could be possible to establish appropriate standards for airborne bacteria, fungi, and viruses but limited to infections only. Unfortunately, Table 11-3 shows that the dose-effect relationships should be analyzed separately for at least two size ranges of particles.

The above analysis also indicates that the knowledge about the influence of bioaerosols on human health is still very limited and should be extended. At present, the question of significant risk is difficult to assess for any bioaerosol. Where infectious agents are involved, some information is available on bacterial levels in reservoirs that have been associated with outbreaks of disease. However, for the majority of microorganisms likely to be recovered during routine air and source sampling, very little is known about potential health effects.

Table 11-3. Dependence of infectious dose on the size of bioaerosol particles (Henningson 1990).

Microorganism (infected organism, Author)	Dose in number of inhaled particles (characteristics of particles)		
Legionella pneumophilia (guinea pigs, Henningson 1990)	1–10 (single cells)	or	> 1,000 (aggregates)
Influenza virus (human, Knight 1973)	3 (small droplets with diameter of 1.5 μm)	or	30 (large droplets)
Francisella tularensis (monkeys, Druett 1967)	1 (fine particles with diameter of 1 μm)		200 (20 μm aggregates)

Synergy and other problems

Closing the discussion on the available tools for prognosis of the adverse health effects caused by the exposure to air pollutants, another problem should also be mentioned. Time-series studies are commonly used to examine the association between short-term effects of air pollution and health. In such time-series, a core model is first constructed and the effects of an air pollutant are then assessed by adding a linear term for the pollutant to the core model. However, the effect estimate for the air pollutants could be biased if all potential confounders have not been adequately controlled, such as the effect of influenza epidemics which may be seasonally correlated with both air pollution and health outcomes. Thach et al. (2010) examined effects of influenza on associations between nitrogen dioxide, sulfur dioxide, PM10, and ozone and health outcomes including all natural causes of mortality, cardiorespiratory mortality, and hospitalization. They found that generally influenza does not confound the observed associations of air pollutants with all natural causes of mortality and cardiovascular hospitalization but for some pollutants and subgroups of cardiorespiratory mortality and respiratory hospitalization, there was evidence to suggest confounding by influenza.

Conclusions

In this chapter, the dose-effect and dose-risk relationships were analyzed and it was documented that they can be very useful for the quantitative description of health implications of the inhaled pollutants. However, there are some exceptions: exposure to airborne particles, exposure to mixtures of airborne pollutants, as well as exposure to bacterial and fungal aerosols. Adverse health effects caused by these pollutants are very difficult for quantitative prognosis by the use of these tools. Additionally, there are some specific health effects, for example allergic effects, which generally cannot be predicted using the dose-effect and dose-risk relationships.

From the review of the literature data it can be concluded that currently it is only possible to determine the allergens present in the air and, knowing the list of these pollutants, predict which adverse health effects can appear among the exposed subjects. The probability of appearance of these effects is significantly higher in the case of atopic people, but the intensity of symptoms is only weakly related, or even not related at all, to the inhaled dose and to the concentration level of the specific allergenic pollutant.

Acknowledgements

The author is grateful to Dr. Hans-Peter Witschi, Professor-Emeritus from the University of California Davis, for the valuable discussion about the role of Professor Fritz Haber in the development of the modern toxicology.

References

Angerer, J., U. Ewers and M. Wilhelm. 2007. Human biomonitoring: state of the art. Int. J. Hyg. Environ. Health 210: 201–228.

Bates, D.V., M. Baker-Anderson and R. Sizto. 1990. Asthma attack periodicity: a study of hospital emergency visits in Vancouver. Environ. Res. 51: 51–70.

Dassen, W., B. Brunekreef, H. de Groot, E. Schouten and K. Biersteker. 1986. Decline in children's pulmonary function during an air pollution episode. J. Air Poll. Control Assoc. 36: 1223–1227.

Davidson, C.I., R.F. Phalen and P.A. Solomon. 2005. Airborne particulate matter and human health: a review. Aerosol Sci. Technol. 39: 737–749.

Dockery, D.W., F.E. Speizer, D.O. Stram, J.H. Ware, J.D. Spengler and B.G. Ferris, Jr. 1989. Effects of inhalable particles on respiratory health of children. Am. Rev. Respir. Dis. 139: 587–594.

Dockery, D.W., J. Cunningham, A.I. Damokosh, L.M. Neas, J.D. Spengler, P. Koutrakis, J.H. Ware, M. Raizenne and F.E. Speizer. 1996. Health effects of acid aerosols on North American children: respiratory symptoms. Environ. Health Perspect. 104: 500–505.

Dotterud, L.K., L.H. Vorland and E.S. Falk. 1996. Mould allergy in school children in relation to airborne fungi and residential characteristics in homes and schools in northern Norway. Indoor Air 6: 71–76.

Druett, H.A. 1967. The inhalation and retention of particles in the human respiratory system. pp. 165–202. *In*: P.H. Gregory and J.L. Monteith [eds.]. Airborne Microbes, Symposium 17 of the Society for General Microbiology. Cambridge University Press, Cambridge, U.K.

Fugaš, M., B. Wilder, R. Pauković, J. Hršak and D. Steiner-Škreb. 1972. Concentration levels and particle size distribution of lead in the air of an urban and an industrial area as a basis for the calculation of population exposure. pp. 961–967. *In*: The Proceedings of the International Symposium: Environmental Health Aspects of Lead. Commission of European Communities, Amsterdam, The Netherlands.

Fugaš, M. 1976. Assessment of total exposure to an air pollutant. pp. 1–3. *In*: The Proceedings of the International Symposium, Environmental Sensing and Assessment, Vol. 2. Institute of Electrical and Electronic Engineers, New York, N.Y.

Fugaš, M. 1986. Assessment of true human exposure to air pollution. Environ. Int. 12: 363–367.

Goldstein, B.D. 2009. Toxicology and risk assessment in the analysis and management of environmental risk. pp. 931–939. *In*: R. Detels, R. Beaglehole, M.A. Lansang and M. Gulliford [eds.]. Oxford Textbook of Public Health. The Methods of Public Health, Vol. 2. Oxford University Press, Oxford, UK.

Gordian, M.E., H. Özkaynak, J. Xue, S.S. Morris and J.D. Spengler. 1996. Particulate air pollution and respiratory disease in Anchorage, Alaska. Environ. Health Perspect. 104: 290–297.

Graham, N.H.M. 1990. The epidemiology of acute respiratory infections in children and adults: a global perspective. Epidemiol. Rev. 12: 149–178.

Gravesen, S. 1979. Fungi as a cause of allergic diseases. Allergy 33: 268–272.

Haber, F. 1924. Zur Geschichte des Gaskriegs. pp. 76–92. *In*: Fünf Vorträge aus den Jahren 1920–1923. Springer, Berlin, Germany.

Heffernan, A.L., L.L. Aylward, L.-M.L. Toms, P.D. Sly, M. Macleod and J.F. Mueller. 2014. Pooled biological specimens for human biomonitoring of environmental chemicals: opportunities and limitations. J. Exposure and Environ. Epidem. 24: 225–232.

Henningson, E. 1990. Unpublished material, private communication, Swedish Defense, Research Establishment, Umea, Sweden.

Knight, V. 1973. Airborne transmission and pulmonary deposition of respiratory viruses. pp. 175–182. *In*: J.F. Hers and K.C. Winkles [eds.]. Airborne Transmission of Airborne Infection. VIth International Symposium on Aerobiology. Wiley, New York, N.Y., USA.

Kozielska, B., W. Rogula-Kozłowska and J.S. Pastuszka. 2013. Traffic emission effects on ambient air pollution by PM2.5-related PAH in Upper Silesia, Poland. Int. J. Environ. Pollut. 53: 245–264.

Kruize, H., O. Hänninen, O. Breugelmans, E. Lebret and M.J. Jantunen. 2003. Description and demonstration of the EXPOLIS simulation model: two examples of modeling population exposure to particulate matter. J. Exposure Analysis and Environ. Epidem. 13: 87–99.

Lioy, P.J. 2010. Exposure science: a view of the past and milestones for the future. Environ. Health Perspectives 118: 1081–1090.

Meek, M.E., R. Newhook, R.G. Liteplo and V.C. Armstrong. 1994. Hazard and risk assessment for inhaled pollutants. pp. 365–378. *In*: P.G. Jenkins, D. Kayser, H. Muhle, G. Rosner and E.M. Smith [eds.]. Respiratory Toxicology and Risk Assessment. Proc. Int. Sem. Wissenschaftliche Verlagsgeselschaft mbH, Stuttgart, Germany.

Mekel, O., P. Martin-Olmedo, B. Ádám and R. Fehr. 2014. Quantification of health risk. pp. 199–232. *In*: G. Guliš, O. Mekel, B. Ádám and L. Cori [eds.]. Assessment of Population Health Risks of Policies. Springer, New York, N.Y., USA.

Miller, F.J., P.M. Schlosser and D.B. Janszen. 2000. Haber's rule: a special case in a family curves relating concentration and duration of exposure to a fixed level of response for a given endpoint. Toxicology 149: 21–34.

Ostro, B. 1993. The association of air pollution and mortality: examining the case for inference. Arch. Environ. Health 48: 336–342.

Özkaynak, H., L.K. Baxter, K.L. Dionisio and J. Burke. 2013. Air pollution exposure prediction approaches used in air pollution epidemiology studies. J. Exposure Sci. Environ. Epidem. 23: 566–572.

Pastuszka, J.S., W. Rogula-Kozlowska and E. Zajusz-Zubek. 2010. Characterization of PM10 and PM2.5 and associated heavy metals at the crossroads and urban background site in Zabrze, Upper Silesia, Poland, during the smog episodes. Environ. Monitoring Asses. 168: 613–627.

Pope, C.A. III. 1991. Respiratory hospital admissions associated with PM-10 pollution in Utah, Salt Lake, and Cache Valleys. Arch. Environ. Health 46: 90–97.

Romieu, I. 1997. Scientific basis for establishing air quality standards; their role in quantitative risk assessment and in monitoring the effectiveness of intervention strategies. pp. 87–101. *In*: T. Fletcher and A.J. McMichael [eds.]. Health at the Crossroads: Transport Policy and Urban Health. John Wiley & Sons Ltd., Chichester, UK.

Rotenberg, S.L. 1992. Environmental health issues. pp. 285–324. *In*: B.J. Cassens [ed.]. Preventive Medicine and Public Health, 2nd ed. Harwal Publishing, Philadephia, Baltimor, USA.

Rylander, R. 1996. Airway responsiveness and chest symptoms after inhalation of endotoxin or $(1 \rightarrow 3)$-β-*D*-glucan. Indoor Built Environ. 5: 106–111.

Ryswyk, Van K., A. Wheele, L. Wallace, J. Kearney, H. You, R. Kulka and X. Xu. 2014. Impact of microenvironments and personal activities on personal PM2.5 exposures among asthmatic children. J. Exposure Sci. Environ. Epidem. 24: 260–268.

Schwartz, J., D. Bellinger and T. Glass. 2011. Expanding the scope of environmental risk assessment to better include differential vulnerability and susceptibility. Am. J. Public Health 101(Suppl. 1): S88–S93.

Sexton, K. and P.B. Ryan. 1988. Assessment of human exposure to air pollution: methods, measurements, and models. pp. 202–238. *In*: A.Y. Watson, R.R. Bates and D. Kennedy [eds.]. Air Pollution, the Automobile, and Public Health. National Academy Press, Washington, D.C., USA.

Sexton, K., L. Needham and J. Pirkle. 2004. Human biomonitoring of environmental chemicals. Am. Sci. 92: 38–45.

Thach, T.-Q., Ch.-M. Wong, K.-P. Chan, Y.-K. Chau, G.N. Thomas, Ch.-Q. Ou, L. Yang, J.S.M. Peris, T.-H. Lam and A.J. Hedley. 2010. Air pollutants and health outcomes: assessment of confounding by influenza. Atmos. Environ. 44: 1437–1442.

Ware, J.H., B.G. Ferris, Jr., D.W. Dockery, J.D. Spengler, D.O. Stram and F.E. Speizer. 1986. Effects of ambient sulfur oxides and suspended particulates on respiratory health of children. Am. Rev. Respir. Dis. 133: 834–842.

Warren, E. 1900. On the reaction of *Daphnia magna* (Straus) to certain changes in its environment. Q.J. Microsc. Sci. 43: 199–224.

WHO Air Quality Guidelines for Europe. 1987 and 2000 (second edition), Copenhagen, Denmark.

WHO Report: Health aspects of air pollution. Results from the WHO project Systematic review of health aspects of air pollution in Europe. E83080, 2004, Copenhagen, Denmark.

Witschi, H.P. 1999. Some notes on the history of Haber's law. Toxicol. Sci. 50: 164–168.

PART 3
Synergistic Effects

12

Analysis of the Known Synergistic Effects of the Exposure to Selected Air Pollutants

Jozef S. Pastuszka,[1,*] *Aino Nevalainen,*[2]
Martin Täubel[2] and *Anne Hyvärinen*[2]

New philosophy of the prognosis of adverse health effects due to the inhalation of air pollutants

(*Jozef S. Pastuszka*)

While trying to change the philosophy of the prediction of adverse health effects due to the inhalation of polluted air, it should be noted again that in a typical urban environment, the population is exposed to about 200 air pollutants or classes of air pollutants. Therefore, instead of investigating the unique effects of specific pollutants, it has been suggested that it might be more reasonable to assume that it is a mixture of pollutants that might be considered harmful to health (Dominici and Butnett 2003; Moolgavkar 2003; Stieb et al. 2002; Roberts and Martin 2006; Dionisio et al. 2013). Potential interaction among pollutants seems to be a fundamental problem indispensable for explaining the relations between the exposure of people to air pollutants and their

[1] Silesian University of Technology, Division of Energy and Environmental Engineering, Department of Air Protection, 22B Konarskiego St., 44-100 Gliwice, Poland.
[2] National Institute for Health and Welfare, Department of Health Protection, Neulaniementie 4, FI-70701 Kuopio, Finland.
* Corresponding author: Jozef.Pastuszka@polsl.pl

health condition. However, before any analysis of this problem, some key terms should be precisely defined. According to U.S. EPA Guidance (2000), and supplemented by Mauderly and Samet (2009), the following terms may be introduced:

Additivity: effect of the combination equals the sum of individual effects.

Synergism: effect of the combination is greater than the sum of individual effects.

Antagonism: effect of the combination is less than the sum of individual effects.

Inhibition: a component having no effect reduces the effect of another component.

Potentiation: a component having no effect increases the effect of another component.

Masking: two components have opposite, cancelling effects such that no effect is observed from the combination.

For the prognostic reasons, the development of new models to concurrently estimate the adverse health effects of multiple air pollutants has been identified by statisticians, epidemiologists, and policy makers as an important topic of research (Dominici and Burnett 2003). Another new model provides for identification and estimation of a mixture of pollutants associated with adverse health effects has been also introduced by Roberts and Martin (2006).

Delfino et al. (2010) analyzed the role of the oxidative potential of particle mixtures containing probably hundreds of correlated chemicals. According to their opinion, oxidative potential likely depends on particle composition and size distribution, especially ultrafine particle concentration, and on the transition metals and certain semi-volatile and volatile organic chemicals. This problem needs, of course, advanced future studies using biomarkers of oxidative stress. Delfino et al. (2010) suggest to measure systematic inflammatory and thrombotic mediators in the blood as well as, simultaneously, clinical outcomes to assess the impact of air pollutant-related oxidative stress.

Mauderly and Samet (2009) reviewed selected published literature to determine whether synergic effects of combinations of pollutants on health outcomes have actually been demonstrated. They concluded that the plausibility of synergism among environmental pollutants has been established, although comparisons are limited, and most involved exposure concentrations were higher than typical of environmental pollutants.

Mauderly and Samet (2009) noticed that in the literature the term “synergy” is often used loosely and sometimes applied to any effect of a combination of pollutants that is greater than the effect of one of the components alone. Meanwhile it should be remembered that synergy is strictly defined as occurring if the effect of the combined exposure is greater than the sum of the effects of these two or more individual components of the mixture (see, for example, Mauderly and Samet 2009; USEPA 2000).

While analysing the synergy, another phenomenon should also be mentioned. It sometimes appears during the comparison of the low level of exposure to some air pollutants and unexpected strong health effects. Such phenomenon, related with an exposure to carbon dioxide and nitrogen dioxide has been recently reviewed by Delfino et al. (2010). They indicated that although epidemiologic studies have shown associations of cardiovascular morbidity and mortality with ambient CO and NO_2 (Brook et al. 2010; Bhaskaran et al. 2009), concentrations of these gases have been considerably lower in relation to levels causing effects in experimental studies and models. Therefore, Delfino et al. (2010) concluded that it is possible that these gasses serve as surrogates for other causal components from fossil fuel combustion. A recent review also reports a lack of consistent findings in time series studies for the relation of ozone to daily hospital data for myocardial infarction (Bhaskaran et al. 2009; Delfino et al. 2010).

Synergy and other interactions

Synergy and other interactions between human carcinogens

(*Jozef S. Pastuszka*)

It was mentioned in Chapter 5 that there is a significant synergy phenomenon between the exposure to asbestos and tobacco smoke. Generally, calculation of risk due to inhalation of some, different carcinogenic pollutants, is not easy. Kaldor and L'Abbé reviewed in 1990 the definition of interaction and the theoretical basis for different types of interaction in cancer causation. Actually, their analysis seems to still be absolutely relevant. They concluded that when a mixture of pollutants contains several carcinogens which fall within the same class (according to their physical and chemical properties), and if it appears that individual pollutants from this class have an effect on cancer risk, the carcinogenicity of the mixture could reasonably be expected to be the simple sum of the components' effects. However, if the mixture consists of pollutants from different classes of carcinogens, a multiplicative effect would provide a more appropriate representation of the overall carcinogenicity of the mixture (Kaldor and L'Abbé 1990).

Synergistic effects of biological particles and other air pollutants

(*Aino Nevalainen, Martin Täubel* and *Anne Hyvärinen*)

The major part of research that links outdoor biological particles to health are allergological studies connected with pollen and fungal allergens. Otherwise, health effects of exposures to airborne biological agents have mainly been studied in either occupational settings or in indoor environments. Although the research on health effects of atmospheric particulate matter has increased enormously during the last few decades (e.g., Cesaroni et al. 2014), it is evident that biological particles have had no major role in this paradigm. This is interesting, as many health effects have been linked with exposures to both PM and biological particles.

Microbial interactions between fungi and bacteria are a well-known phenomenon in, e.g., biocontrol research (Whipps 2001; Woo et al. 2002). While these interactions are recognized to be extremely complex by microbial ecologists, the paradigm of studies that would reveal the possibly health-relevant interactions is yet to be established. Furthermore, there is a limited amount of knowledge about synergistic health effects linking biological agents to other—non-biological—pollutants. One can expect that in the future, with improved DNA-based methodology for detection and quantifying biological material, and for better understanding microbial physiology with approaches such as proteomics and lipidomics, this situation will change. As presented before, a causal relationship can only in a few cases be established between a health outcome and a defined biological agent. Usually there is a complex mixture of exposing agents that leads to health symptoms or disease with pathophysiological pathways that are not fully known. This applies, e.g., to indoor environments with damp and mold, where health effects are manifold while exposure levels often remain relatively low, as opposed to occupational settings with very intense exposures. Are such indoor situations examples of synergistic effects of various pollutants? This possibility is often mentioned in the literature but rather in a sense of frustrated discussion than in the form of vigorous efforts to experimentally study such effects.

Mycotoxins have been suggested to potentially cause or contribute to adverse health effects observed in occupants of moisture damaged indoor environments. Toxins are found in such environments but exposure levels are very low and difficult to differentiate from non-damaged indoor environments (Peitzsch et al. 2012). Human studies where exposure to airborne microbial toxins would have been linked to observed health effects are scarce (Kirjavainen et al. 2015; Cai et al. 2011; Zock et al. 2014) and clearly, more such studies are needed.

There is more data on toxins from experimental studies that have used *in vitro* approaches. Using spores and pure toxins of the fungus *Stachybotrys chartarum* and toxin-producing bacterium *Streptomyces californicus*, synergistic interactions were observed as shown with inflammatory markers and cytotoxicity (Huttunen et al. 2004). Kankkunen et al. (2009) concluded that human macrophages sense trichothecene mycotoxins as a danger signal, which activates a number of inflammatory reactions. The toxins gliotoxin and patulin acted synergistically in a so called MIXTOX model, while the combined effect with sterigmatocystin was antagonistic, with switch to synergism if the toxicity of the mixture was mainly caused by sterigmatocystin (Mueller et al. 2013). Neurotoxicity and inflammation caused by the mycotoxin roridin A and the potentiation of the effects by bacterial LPS was documented in a mouse model with intranasal instillation (Islam et al. 2007). As these examples demonstrate, microbial metabolites may have significant potential for adverse health effects both individually and as combinations with other agents. The role of microbial toxins in outdoor particulate exposures is totally unknown today; however, such toxins are also found in the outdoor air (Täubel et al. 2013).

Even if the details of the exposures and development of health effects are poorly known today, there is need for policies that aim to control the observed health effects of environmental stressors. One such stressor that can potentially be avoided is the exposure due to moisture and mold associated with indoor environments.

Recently, the WHO has stated in their guidelines on dampness and mould (WHO 2009) that while causative agents of adverse health effects in damp buildings have not been conclusively identified, excess level of various microbial agents, including mycotoxins, in the indoor environment need to be considered a potential health hazard. For this reason, it is of key importance to prevent or remove microbial growth in response to moisture problems in buildings, which implies excessive microbial proliferation and distribution of microbial spores and fragments, allergens, cell wall components such as bacterial endotoxin and fungal β-glucans, microbial volatile organic compounds (MVOCs), and mycotoxins.

Other interactions between airborne microorganisms and particulate matter

(*Jozef S. Pastuszka*)

It is interesting that particle exposures also increase the risk for human infection. Recently Ghio (2014) indicated that one of the mechanisms for particle-related infections includes an accumulation of iron by surface functional groups of particulate matter (PM). Besides, air pollution can modify the concentration levels of bioaerosols, as well as their number size-distributions. This applies in particular to the interaction of bacterial aerosol with dust suspended in the air. It is estimated that about 80% of the microorganisms present in the air can be attached to dust particles, and lots of data points to the strengthening of an unfavorable, combined impact on the health of people exposed simultaneously to biological and dust aerosols (Seedorf et al. 1998; Haas et al. 2013). The phenomenon of the attachment of fine particles, including biological particles, to a coarse solid particle is illustrated in Fig. 12-1. This aggregate is surrounded by the bacterial colony developed from one bacterial particle (precisely: one colony forming unit—CFU) collected together with the airborne dust particle.

Recently, Brągoszewska et al. (2013) showed that in the atmospheric air in Gliwice, Poland, concentration of coarse bacterial particles (having the aerodynamic diameter d_{ae} > 3.3 μm) was highly correlated with the concentration of coarse

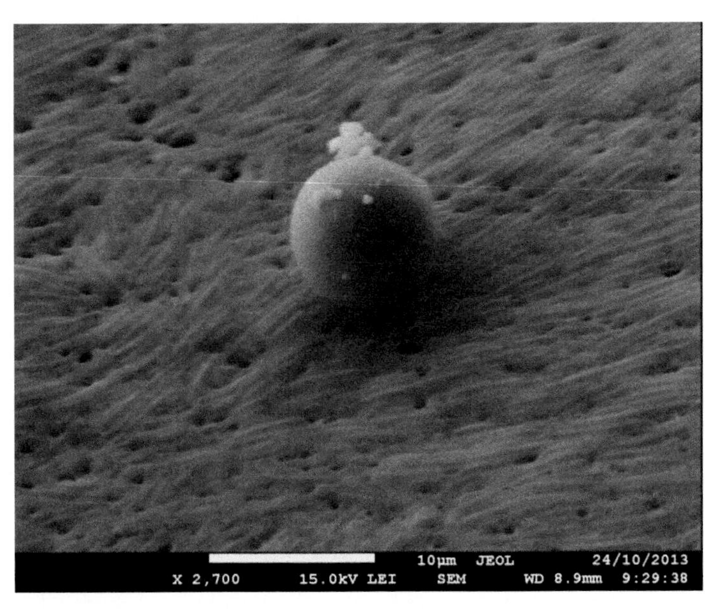

Fig. 12-1. Coarse solid particle collected together with airborne bacteria by using an Andersen 6-stage impactor. This particle is located in the center of a bacterial colony. Micrograph was taken by Ewa Talik, University of Silesia. With permission.

particles of atmospheric particulate matter (Fig. 12-2). Although the mentioned above phenomenon of the attachment of fine bacterial particles to a coarse solid particle can be responsible for this correlation, this problem needs future investigation. The equation of the regression line for the graph in Fig. 12-1 can be assumed as follows, $y = 0.0148 \cdot x + \text{const}$. At the significance level of 0.05, there is a relatively strong correlation between the concentration of coarse bacteria and dust equal to 0.79. It should be noted, however, that this relationship was a preliminary estimate only, and for the higher concentration levels of airborne dust—it probably will not be linear.

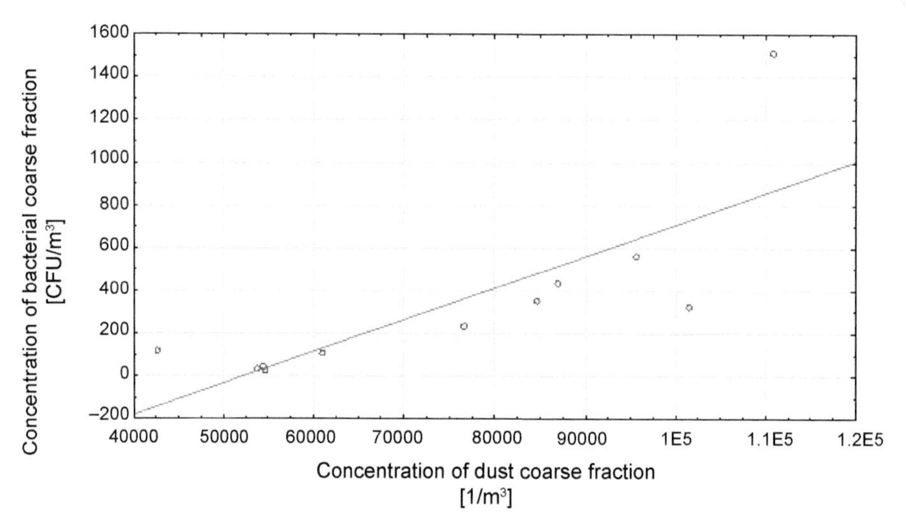

Fig. 12-2. Concentration of the coarse fraction of bacterial aerosol versus number concentration of coarse solid particles in ambient air in Gliwice (Brągoszewska et al. 2013).

Synergy and allergic symptoms

(*Jozef S. Pastuszka*)

Exposure to environmental substances triggering allergic symptoms of the respiratory tract

The number of registered allergy cases worldwide is increasing each year (Aas et al. 1997; IOM 2000). Although a relatively short time ago the problem of allergy occurred rather rarely, currently most of the population in many regions of the world suffers from it. Due to the fact that this drastic increase of morbidity occurred in a short period of time, genetic changes must be excluded as the reason of increasing cases of allergy, while environmental changes should be considered as one of the important causal factors. This hypothesis is confirmed, among others, by the large geographical variability of the common allergy based asthma symptoms. A map of these symptoms was created within the International Study of Asthma and Allergies (ISAAC). In the report of these studies (ISAAC 1998), the diversity of asthma symptoms in various countries was presented. In Western Europe, asthma occurs as much as ten times more frequently than in the eastern part of the continent, while the frequency of morbidity is higher in urbanized areas than in agricultural ones. The reason for this state of things is the subject of many speculations. It is assumed that the following factors may be important:

1) The so called "western lifestyle"
2) Infections during infancy
3) Type and range of immunization
4) Diet
5) Exposure to tobacco smoke (ETS)
6) Pets

Although all of the above mentioned factors seem to be important, and especially the exposure to tobacco smoke, or "passive smoking" (Wartenberg et al. 1994; Dayal et al. 1994), the greatest significance is attributed to the relation between the allergens and humidity at home (Bornehag et al. 2001).

The role of indoor air pollution (in rooms) in initiating and intensifying asthma symptoms has been widely discussed by the Institute of Medicine of National Academy of Sciences USA (IOM 2000). The review of this issue was concluded by the statement: "we still do not know whether or to what extent the reported increases in asthma can be attributed to indoor exposures". The authors of the report blame the existing literature data for the state of the art. Scientific articles usually do not discuss the issue of indoor environment (rooms) on an appropriate level and *vice verse*; scientific studies on the indoor exposure do not focus on asthma. As a result, the issues of asthma and the quality of the indoor environment are very poorly related. Because of this, the postulate of the above mentioned report is to initiate extensive interdisciplinary studies on asthma, which should involve not only clinicians, immunologists, and biologists but also engineers, architects, building materials specialists, and all the others who are responsible for the designing and functioning of the environment. This American point of view is confirmed in a number of other studies and reviews (Sundell and Kjellmam 1994; Sundell 1999; Andersson et al. 1997; Ahlbom et al. 1998; Bornehag et al. 2001; Wargocki et al. 2002).

Exposure to allergens is of, course, crucial to inducing allergy. The most significant sources of emission of allergens in the indoor environment are probably house dust, mites, and pets. It seems that less significant (in the scale of the issue), is exposure to formaldehyde and volatile organic compounds, although in certain cases they can be evidently related even to asthma (e.g., Wieslander et al. 1996). In a number of studies, the exposure to allergens such as airborne chemicals, microorganisms (Gravesen 1979) and their metabolites (Rylander 1996) and airborne dust (Koch 2000; Ezzati et al. 2000), or exposure to groups of select air pollutants (Bascom 1996; McConnel et al. 1999; Koenig 1999) are discussed.

Allergy to mites' deject is a common worldwide issue. Mites require moist environment and are therefore directly related to moisture and poor building ventilation. However, on the other hand, mites are scarce in some regions of the world where allergy cases are frequent. This situation occurs, for example, in northern Scandinavia (Sundell et al. 1995). The broadest studies on house dust mites in Poland were carried out by Solarz (1987). Horak and others (1996) studied the content of mites, bacteria, and microscopic fungi in settled dust in apartments in Upper Silesia. The authors determined that in houses with residents suffering from asthma, the concentration of fungi in dust settled on beds was five times higher than in houses without residents suffering from asthma. The lack of presence of visible mould in any of the studied houses is also important. These studies support the hypothesis put forward by Pastuszka, that the intensification of asthma symptoms (or attacks) in people allergic to fungi occurs as a consequence of inhaling fungal aerosol from close sources of emission (Pastuszka et al. 2000; Pastuszka 2001).

The hypothesis that micro-flora in apartments is dominated by *Micrococcaceae*, mainly by *Micrococcus* and *Staphylococcus*, put forward in the 1990s by Finnish and British researchers, was confirmed also in other countries, including Poland (Pastuszka et al. 2000). Generally, the obtained results indicated that the occurring concentration levels of bacterial and fungal aerosol in Polish apartments are comparable with the concentration levels registered in apartments in other parts of Europe. Figure 12-3 shows colonies of such Gram-positive cocci obtained from the air samples collected recently in the same area in Poland, while the example of Gram-positive rods is illustrated in Fig. 12-4.

It is interesting that Gaweł and others (1997) in their studies carried out in Rabka, Poland, did not find a correlation between the frequency of occurrence of specific antibodies IgE in children's blood sera and the concentration of pollen grains and spores of specific plant species in the atmosphere. The most frequent aeroallergens were

Fig. 12-3. Colonies of the bacteria (Gram-positive cocci) grown on the agar from the collected sample of airborne bacteria. Micrograph was made by Ewa Talik, University of Silesia. With permission.

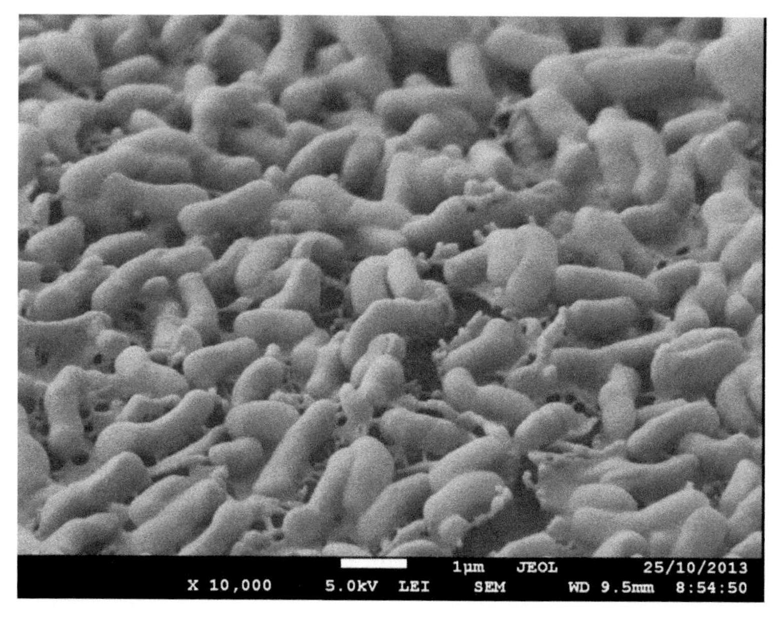

Fig. 12-4. Colonies of bacteria (Gram-positive rods) grown on the agar from the collected sample of airborne bacteria. Micrograph was made by Ewa Talik, University of Silesia. With permission.

spores, mainly *Cladosporium*, with relatively low frequency of increased concentration of specific antibodies IgE. Grass pollen, despite low year average concentration, resulted in the largest amount of increased concentrations of IgE.

In the study conducted in Germany (Nicolai and Mutius 1996), it has been determined that exposure of children aged 9–11 to high concentrations of sulfur dioxide and dust in the atmospheric air is related to the increased morbidity to bronchitis but is not correlated to asthma or hay fever.

Jędrychowski and Flak (1998) conducted epidemiological research on a population of over a thousand school children aged 9, attending schools in Cracow, Poland, with varied levels of atmospheric air pollution. The authors determined that only the occurrence of the symptom of chronic coughing was unrelated to the children's allergies but was related to atmospheric air pollution. Moreover, the risk of hay fever was related to the level of pollution and allergy, while the influence of moisture and mould was statistically significant in occurrence of hay fever, wheeze, and short breath.

The analysis of collected literature data indicates that the most significant environmental factors affecting the development of allergies in children are tobacco smoke and high humidity at home, often caused by poor ventilation. Moisture at home causes the increase of amount of a number of allergens, such as bioaerosols (Burge 1990), and especially microscopic fungi (e.g., Kemp et al. 2002), and mites, which can be causally related with occurrence or intensification of allergy. This conclusion directly supports the studies of Cuijpers and coworkers (1995).

Example of case-study

Preliminary studies on potential influence of a few airborne allergens indoors on children's asthma were carried out by Pastuszka and co-workers (2003a). They measured the concentration of TSP, PM5, fungi, bacteria as well as Gram-negative bacteria. They analyzed the selected air pollutants only and neglect other environmental factors promoting asthma, such as changes in the air temperature or different meteorological parameters (Hales et al. 1998). Aerosol particles were collected using the Casella samplers (TSP, PM5) and the Harvard impactors (PM10, PM2.5). Samples have been taken during three months, mostly at the flats (apartments) in 4–10 story buildings. Additionally, in some homes the airborne particles were fractionated in a 10-stage Andersen impactor. To determine the relative contribution of the indoor and outdoor environment to the total exposure to the respirable fraction of the aerosol, the selected participants of the two-day exposure experiment (15 people/voluntaries) filled out an activity diary and a questionnaire about their homes and their surroundings during the course of the study.

Subjects for indoor environmental testing were identified from a population of asthmatic children seen in the Clinic of Environmental Medicine at the Institute of Occupational Medicine and Environmental Health in Sosnowiec, Poland. The measurements of the concentration of TSP and PM5 as well as bacterial and fungal aerosol were carried out in two categories of homes in Sosnowiec: (A)—10 homes with asthmatic children, and (C)—15 control homes with healthy children. Airborne bacteria and fungi were collected using the 6-stage Andersen impactor. Microorganisms were collected on nutrient media (specific to either fungi or bacteria) in Petri dishes located on all impactor stages.

The results of this pilot cross-sectional study carried out in the homes with asthmatic children (A) and in the group of homes (C) are presented in Table 12-1. It can be seen that between the homes (A) and (C) the difference in the averaged concentration of airborne matter is for TSP only 14 $\mu g\ m^{-3}$ while for PM5, it is 28 $\mu g\ m^{-3}$. It is interesting to note that although the visual inspection has not detected the fungal contamination in any home, the averaged concentration of airborne fungi in homes (A) was significantly higher than in homes (C). The level of airborne bacteria was also elevated in homes with asthmatic children in comparison with the control flats. Especially, the observed tendency to elevating the concentration of Gram-negative bacteria, which can be the sources of endotoxin, seems to be very important. Unfortunately, this problem is very difficult to discuss because of the generally very low level of these bacteria indoors.

This study confirms that in allergy based asthma, the interaction of various separately occurring allergens, even in low concentrations, is significant. Besides, several different pathways should be proposed that contribute to the asthmatic response and which could be amplified by the particulate matter exposure. This result agrees with a well-known phenomenon that particulate matter (PM) and endotoxins are able to trigger inflammatory responses in the lung. Unfortunately, most studies have focused on the components separately and on the identification of chemical components associated with PM.

Table 12-1. Concentration levels of different indoor aerosols in homes with asthmatic and healthy children in Sosnowiec, Poland. Results of the cross-sectional study (Pastuszka et al. 2003a).

Aerosol	Concentration in homes with asthmatic children	Healthy children
Particulate matter (Geometric mean)		
TSP [μg m^{-3}]	179	165
PM5 [μg m^{-3}]	123	95
Fungi total [CFU m^{-3}]		
Geometric mean	230	129
Median	179	104
Respirable fungi [CFU m^{-3}]		
Geometric mean	177	96
Median	147	62
Bacteria total [CFU m^{-3}]		
Geometric mean	881	703
Median	944	526
Respirable bacteria [CFU m^{-3}]		
Geometric mean	445	331
Median	391	274
Gram-negative bacteria (total) [CFU m^{-3}]		
Geometric mean	31	29
Median	39	24
Number of flats studied	10	15

Analysis of other recent data

Since biological components may represent around 20% of airborne PM, and endotoxins may reach concentrations as high as 30 EU/mg, some studies have focused attention on the characterization of endotoxins present in PM and health effects (for example, Górny and Dutkiewicz 1998). Review of the available literature made recently by Degobbi et al. (2011) has suggested that the endotoxin adsorbed in PM is able to elicit immunological responses associated with increase in pro-inflammatory cytokine expression. On the other hand, the obtained results confirm the hypothesis that the mass might not be the best metric for describing the harmful fraction of airborne particles (Donaldson et al. 2000). The total mass is usually dominated by larger secondary particles such as sulfates and nitrates, which are generally considered to be of low toxicity, whereas the mass contribution of fungi, bacteria, or endotoxins is much less than the inhaled mass of other pollutants.

Besides, allergens from animals are frequent in an indoor environment (Custovic et al. 1997; Ahlbom et al. 1998). Currently a related issue is being discussed, whether exposure to allergens during infancy constitutes a risk factor, or on the contrary, it prevents an allergy in the subsequent period of life. The dominating opinion seems to be that early exposure to allergens is a significant risk factor of allergy (Ahlbom et al. 1998).

Meantime, Peden (2002) reviewed available literature on the relation between air pollutants and asthma. He stated that most asthma exacerbations that result in hospitalization are associated with viral upper respiratory tract infections. However, a more recently appreciated cause of asthma exacerbation is exposure to pollutants, including ozone and various components of particulate matter (PM), including transition metals, diesel exhaust, and biological compounds such as endotoxins.

Takano et al. (2002) experimentally documented that diesel exhaust particles (DEP) enhance neutrophilic lung inflammation related to bacterial endotoxins. The enhancement is mediated by the induction of proinflammatory molecules, likely through the expression of Toll-like receptors and the activation of p65-containing dimer(s) of NF-B, such as p65/p50.

Finally, it should be mentioned that finding detailed conclusions on the air pollutants-asthma relationship is often very difficult. Caution for the analysis of asthma data obtained in developing countries is especially needed. For example, Brożek et al. (2010) analyzed a 15-year trend in the prevalence of allergic disorders and respiratory symptoms in children living in an urban area of Upper Silesia, Poland. Three cross-sectional studies (1993, 2002, and 2007) in children aged 7–10 showed a statistically significant increase in the prevalence of all physician-diagnosed allergic disorders (1993–2002–2007): asthma (3.4%–4.7%–8.5%); allergic rhinitis (9.1%–13.7%–17.4%); atopic dermatitis (3.6%–8.4%–8.9%); allergic conjunctivitis (4.3%–11.8%–14.9%); allergy to pollen (5.9%–12.3%–17.3%); allergy to food (5.5%–11.0%–17.0%). A simultaneous decreasing trend in the prevalence of coughing correlated with a significant improvement of ambient air quality. However, this rising trends could result from both increasing incidence and improved diagnosis of allergic diseases. The quickly improving diagnosis of allergic symptoms is really a fact in a number of countries now and may be a reason (although the "technical" reason only) of the rising number of cases of environmentally generated asthma documented in the literature.

Other kinds of interactions

It is very interesting that some interactions can be found not only between the airborne allergens but also between allergens suspended in the air (aeroallergens) and present in food (food allergens). An example is studies of Kasznia-Kocot and Sąda-Cieślar (1993), who found the allergic background of the disease in 95.3% of children suffering from obturative bronchitis. Hypersensitivity to both respirable and food allergens was dominant. It was determined that, apart from atopy in the family, too early introduction to solid food and exposure to environmental allergens such as tobacco smoke (ETS—passive smoking), house dust, plumage, hay dust are favorable to the development of an allergy. These data confirm the leading role of allergens occurring in indoor air of flats in the induction and intensification of allergies of the respiratory system in children.

New tools in health care of the community exposed to air pollutants

(*Jozef S. Pastuszka*)

Aggregative risk index

Actually, there is well documented relation between human health, well-being, and air pollution levels. Therefore, some sanitary index which could be used to communicate information about the health risk as a consequence of air pollution in real-time and forecast of this health risk is needed. The first step in this direction was so called "air quality indices" but most of them are not accompanied by health advice on health protection. In the next step, the quality of outdoor air was classified according to the special index being an averaged concentration of some, arbitrary selected pollutants, and summarized together with the specific factors reflecting the generally assumed "toxicity" of these pollutants. The advanced formula of the air quality index is based on the exposure-response relationship (Cairncross et al. 2007; Pyta 2008) and defined in using the probability function for the occurrence of the specified health outcomes, on the assumption that there are no threshold concentrations which are safe for health (a risk based approach). Sicard et al. (2011, 2012) intensively worked on the new index associated with the corresponding daily risk increase, which was easy to understand and intuitive for the general public. They defined the Aggregative Risk Index (ARI), being the total attributable risk for simultaneous short-term exposure to several air pollutants, as follows (Sicard et al. 2011, 2012):

$$ARI = \sum_i (RR_i - 1) = \sum_i a_i C_i \qquad (12\text{-}1)$$

where: the *ARI* index reflects the contribution of individual pollutants to total risk,

C_i—is the corresponding time-averaged concentrations (in μg m^{-3})

a_i—is the coefficient proportional to the incremental risk values ($RR_i - 1$)

From the published RR_i values for each of *i* pollutants, it can be obtained the coefficients for the terms a_i in order to derive a numerical scale specific to each of the pollutants for the analyzed/studied area.

For each pollutant and pathology, the risk index coefficient a_i can be calculated using the following equation (Sicard et al. 2012):

$$a_I = \frac{4(RR_i - 1)}{10(1.120 - 1)} \qquad (12\text{-}2)$$

Table 12-2 shows, prepared by Sicard et al. (2012), the arbitrary index scale used to facilitate risk communication. The index values may extend beyond 10 for highly polluted areas. It can be seen that the index has four bands indicating "Low, Moderate, High and Very High" risk of increase. The choice of percentage change is arbitrary.

It is important to mention that the health response varies between individuals, therefore the prognosis of the adverse health level can be made only for the exposed

Table 12-2. Arbitrary generic scale for the Aggregate Risk Index, associated risk increase, health advice, and information on the short-term effects for information and communication in the European regions (Sicard et al. 2012).

Index	Information level	Excess relative risk	Health messages	
			At risk population	General population
0 1 2 3	Low	Low risk of increase: 0–11.9%	Enjoy your usual outdoor activities. Follow your doctor's advice about exercise.	Ideal conditions for outdoor activities.
4 5 6	Moderate	Moderate risk of increase: 12.0–20.9%	If you have heart or lung problems, consider reducing strenuous physical outdoors activities, or reschedule to times when the index is lower. Follow your doctor's advice about exercise.	No need to modify your usual outdoor activities unless you experience symptoms.
7 8 9	High	High risk of increase: 21.0–29.9%	Children, the elderly, and people with heart or lungs problems should reduce physical exertion outdoors, particularly if they experience symptoms or reschedule to periods when the index is lower. Follow your doctor's usual advise.	Anyone experiencing discomfort such as coughing or throat irritation should consider reducing or rescheduling strenuous outdoor activities to when the index is lower.
10	Very high	Very high risk of increase:	Health warnings of emergency conditions.	Health alert: everyone may experience more serious health effects.

population (i.e., for the set of the organisms), not for the one, determined person. Such a prediction for the individual subject is possible but needs cooperation with a physician who can modify the general prediction to the individual patient, knowing his/her specific health condition.

Synergy of small doses of air pollutants

Current preventive hygiene and preventive medicine programs led to significant improvement of health of the people of the subpopulation occupationally exposed to air pollutants, as well as the health of the population in general. However, these successes concerned only people absorbing very large doses and usually exposed to concentration levels of air pollutants which today would be rated as extremely high. Nowadays, the focus of environmental scientists, industrial hygienists, and experts on environmental health is on the health effects of exposure to much lower concentrations of air pollutants. Analysis of the results obtained so far from epidemiological and toxicological studies on people absorbing relatively small doses of air pollutants reveal a complicated picture. On the one hand, it turns out that exposure to relatively low pollutants' concentrations, oftentimes lower than the current standards, implies distinctive, measurable negative health effects. The example is exposure to airborne particulates. One intriguing aspect of the epidemiological data is that health effects of airborne particles are primarily seen in people with predisposing factors, including bronchitis, asthma, chronic obstructive pulmonary diseases, pneumonia, compromised immune systems, and age over 65 years old (after Takano et al. 2002).

As can be seen, prediction of health effects of the exposure of the general population to aerosol is very difficult, especially because the direct correlation of the observed health effects and the absorbed dose of pollutants prevent the estimation of a quantitative "dose-effect" relationship. These issues are especially important when assessing the quality of indoor environment, especially home environment (Owen et al. 1992), which is dominates the daily exposure. The increasing difficulties in the interpretation of a growing collection of measurement data (including the data obtained in the Upper Silesia region, Poland, e.g., Pastuszka 1997; Pastuszka et al. 1993, 1995, 2003b, 2010; Jabłońska et al. 2001; Ćwiklak et al. 2009; Kozielska et al. 2013) confronted with the research data concerning the state of the health of people exposed to specific air pollutants, especially indoors (e.g., Pastuszka 1999, 2001; Pastuszka et al. 2000, 2003a, 2005) indicate the necessity to verify the basic hypothesis relating air quality with health effects.

Another example has been described a long time ago by Bingham and Falk (1969), who found that a relatively nontoxic substance, together with a known carcinogen, can enormously increase the carcinogenic potency of the known carcinogen. They give the example of exposure of mice to benzo(a)pyrene and benzo(a)anthracene when n-dodecane is used to dilute these chemicals. The result is a 1,000-fold increase in potency although n-dodecane itself is considered relatively benign and certainly not carcinogenic. They also noticed that the effect is greatest or at least most apparent at low concentrations of the carcinogen (Bingham and Falk 1969).

It seems that at a relatively low level of exposure to "basic", until recently individually studied, air pollutants, the inhalation of the whole composition of co-occurring pollutants begins to play a significant role. Although, the phenomenon of synergism has already been studied for some pollutants, the synergism of relatively low doses of pollutants, especially of non-carcinogenic substances, has not been studied so far. Meanwhile, it seems, at the exposure to relatively low concentration levels of air pollutants, synergism of the inhaled pollutants is one of the basic phenomena and absolutely cannot be ignored. Wherefore this direction of research should be proposed as a new research strategy, especially in reference to the research of the indoor environment.

Concluding remarks about the mechanism of synergy

The mechanism of synergy between air pollutants is, generally speaking, still unknown. Our scientific experience and available literature indicate that probably no universal synergic mechanism exists. It seems that each case of the common influence of some, specific pollutants on human health needs an individual explanation. For example, there is a significant synergy phenomenon between the exposure to asbestos and tobacco smoke. It has been found that the cancer risk for smokers exposed to airborne asbestos is roughly ten times higher compared to non-smoking people exposed to asbestos of the same level (see Chapter 5). This synergy is probably due to carcinogenic compounds absorbed on the fibers. This hypothesis, if true, may be classified into the group of simple explanations related with very simple cases of synergy. Another example is synergy between gaseous pollutants and airborne bacteria when it is believed that the micro-damages of the respiratory tract made by toxic gases (SO_2, NO_2, or others) enable rapid penetration of inhaled live bacteria to the bloodstream.

Also much more complicated synergic phenomena are mentioned in this book. In our opinion, the synergy of airborne pollutants inhaled in very small doses, synergy of aeroallergens and food allergens, as well as, synergy between air pollutants and socio-economic factors seem to be especially difficult for a detailed, quantitative explanation and needs future interdisciplinary studies.

Acknowledgements

The author is grateful to Professor Ewa Talik, from the University of Silesia, August Chełkowski Institute of Physics in Katowice, for her electron micrographs of bioaerosol particles.

References

Aas, A., N. Aberg, C. Bachert, K. Bergemenn, S. Bonini, J. Bousqet, A. de Weck, I. Farkas, K. Heidenberg, St. Holgate, A.B. Kay, N. Kkjellman, K. Kontou-Fili, G. Marone, F.B. Michel, A. Pandit, M. Radermecker, P. Van Cauvenberge, U. Gent, D.Wang, U. Wahn and G. Walsch. 1997. European allergy white paper. Ghent University, Brussels, Belgium.

Ahlbom, A., A. Backman, J.V. Bakke, T. Foucard, S. Halken, M. Kjellman, L. Malm, S. Skerfving, J. Sundell and O. Zetterström. 1998. "Nordepets". Pets indoors—a risk factors or protection agains sensitisation/allergy. A Nordic interdisciplinary review of the scientific literature concerning the relationship between exposure to pets at home, sensitisation and the development of allergy. Indoor Air 8: 219–235.

Andersson, K., J.V. Bakke, O. Bjørseth, C.-G. Bornehag, G. Clausen, J.K. Hongslo, M. Kjellman, S. Kjærsgaard, F. Levy, L. Mølhave, S. Skerfving and J. Sundell. 1997. TVOC and health in non-industrial indoor environments. Indoor Air 7: 78–91.

Bascom, R. 1996. Environmental factors and respiratory hypersensitivity: the Americas. Toxicology Letters 86: 115–130.

Bhaskaran, K., S. Hajat, A. Haines, E. Herrett, P. Wilkinson and L. Smeeth. 2009. Effects of air pollution on the incidence of myocardial infarction. Heart 95: 1746–1759.

Bingham, E. and H.L. Falk. 1969. Environmental carcinogens: the modifying effect of carcinogens on the threshold dose response. Arch. Environ. Health 19: 779–783.

Bornehag, C.-G., G. Blomquist, F. Gyntelberg, B. Järvholm, P. Malmberg, L. Nordvall, A. Nielsen, G. Pershagen and J. Sundell. 2001. Nordic interdisciplinary review of the scientific evidence on associations between exposure to "dampness" in buildings and health effects (NORDDAMP). Indoor Air 11: 72–86.

Brągoszewska, E., A. Kowal and J.S. Pastuszka. 2013. Bacterial aerosol occurring in atmospheric air in Gliwice, Upper Silesia, Poland. Architecture, Civil Engineering, Environment 4: 61–66.

Brook, R.D., S. Rajagopalan, C.A. Pope 3rd, J.R. Brook, A. Bhatnagar, A.V. Diez-Roux, F. Holguin, Y. Hong, R.V. Luepker, M.A. Mittleman, A. Peters, D. Siscovick, S.C. Smith, Jr., L. Whitsel and J.D. Kaufman. 2010. Particulate matter air pollution and cardiovascular disease: an update to the scientific statement from the American Heart Association. Circulation 121: 2331–2378.

Brożek, G.M., J.E. Zejda, M. Kowalska, M. Gębuś, K. Kępa and M. Igielski. 2010. Opposite trends of allergic disorders and respiratory symptoms in children over a period of large-scale ambient air pollution decline. Polish J. Environ. Stud. 19: 1133–1138.

Burge, H. 1990. Bioaerosols: prevalence and health effects in the indoor environment. J. Allergy Clin. Immunol. 86: 687–701.

Cai, G.H., J.H. Hashim, Z. Hashim, F. Ali, E. Bloom, L. Larsson, E. Lampa and D. Norback. 2011. Fungal DNA, allergens, mycotoxins and associations with asthmatic symptoms among pupils in schools from Johor Bahru, Malaysia, Pediatr. Allergy Immunol. 22: 290–297.

Cairncross, E.K., J. John and M. Zunckel. 2007. A novel air pollution index based on the relative risk of daily mortality associated with short-term exposure to common air pollutants. Atmos. Environ. 41: 8442–8454.

Cesaroni, G., F. Forastiere, M. Stafoggia, Z.J. Andersen, C. Badaloni, R. Beelen, B. Caracciolo, U. de Faire, R. Erbel, K.T. Eriksen, L. Fratiglioni, C. Galassi, R. Hampel, M. Heier, F. Hennig, A. Hilding, B. Hoffmann, D. Houthuijs, K.H. Jöckel, M. Korek, T. Lanki, K. Leander, P.K. Magnusson, E. Migliore, C.G. Ostenson, K. Overvad, N.L. Pedersen, J. Penell, G. Pershagen, A. Pyko, O. Raaschou-Nielsen, A. Ranzi, F. Ricceri, C. Sacerdote, V. Salomaa, W. Swart, A.W. Turunen, P. Vineis, G. Weinmayr, K. Wolf, K. de Hoogh, G. Hoek, B. Brunekreef and A. Peters. 2014. Long term exposure to ambient air pollution and incidence of acute coronary events: prospective cohort study and meta-analysis in 11 European cohorts from the ESCAPE project. BMJ Jan 21; 348: f7412. doi: 10.1136/bmj.f7412.

Cuijpers, C.E.J., G.M.H. Swaen, G. Wesseling, F. Sturmans and E.F.M. Wouters. 1995. Adverse effects of the indoor environment on respiratory health in primary school children. Environ. Res. 68: 11–23.

Custovic, A., R. Green, A. Fletcher, A. Smith, C.A. Pickering, M.D. Chapman and A. Woodcock. 1997. Aerodynamic properties of the major dog allergen Can f 1: distribution in homes, concentration, and particle size of allergen in the air. Am. J. Resp. Crit. Care Med. 155: 94–98.

Ćwiklak, K., J.S. Pastuszka and W. Rogula-Kozłowska. 2009. The influence of traffic on particulate-matter polycyclic aromatic hydrocarbons (PAHs) in the urban atmosphere of Zabrze, Poland. Polish J. of Environ. Stud. 4: 579–585.

Dayal, H.H., S. Khuder, R. Sharrar and N. Trieff. 1994. Passive smoking in obstructive respiratory diseases in an industrialized urban population. Environ. Res. 65: 161–171.

Degobbi, C., P. Hilario, N. Saldiva and Ch. Rogers. 2011. Endotoxin as modifier of particulate matter toxicity: a review of the literature. Aerobiologia 27: 97–105.

Delfino, R.J., N. Staimer and N.D. Vaziri. 2010. Air pollution and circulating biomarkers of oxidative stress. Air Qual. Atmos. Health. DOI 10.1007/s11869-010-0095-2.

Dionisio, K.L., V. Isakov, L.K. Baxter, J.A. Sarnat, S.E. Sarnat, J. Burke, A. Rosenbaum, S.E. Graham, R. Cook, J. Mulholland and H. Özkaynak. 2013. Development and evaluation of alternative approaches for exposure assessment of multiple air pollutants in Atlanta, Georgia. J. Exposure Sci. and Environ. Epidem. 23: 581–592.

Dominici, F. and R.T. Butnett. 2003. Risk model for particulate air pollution. J. Toxicol. Environ. Health, Part A 66: 1883–1889.

Donaldson, K., M.I. Gilmour and W. MacNee. 2000. Asthma and PM10. Respiratory Res. 1: 12–15.

Ezzati, M., H. Saleh and D.M. Kammen. 2000. The contributions of emissions and spatial microenvironments to exposure to indoor air pollution from biomass combustion in Kenya. Environ. Health Pespect. 108: 833–839.

Gaweł, J., A. Halota, K. Sordyl, J. Radliński, R. Kurzawa, K. Pisiewicz and Z. Doniec. 1997. Częstość występowania swoistych przeciwciał IgE w zależności od stężenia aeroalergenów w Rabce w latach 1991–1995 (in Polish). Alergia, Astma, Immunologia 2: 43–46.

Ghio, A.J. 2014. Particle exposures and infections. Infection 42: 459–467.

Górny, R.L. and J. Dutkiewicz. 1998. Evaluation of microorganisms and endotoxin levels of indoor air in living rooms occupied by cigarette smokers and non-smokers in Sosnowiec, Upper Silesia, Poland. Aerobiologia 14: 235–239.

Gravesen, S. 1979. Fungi as a cause of allergic diseases. Allergy 33: 268–272.

Haas, D., H. Galler, J. Luxner, G. Zarfel, W. Buzina, H. Friedl, E. Marth, J. Habib and F.F. Reinthaler. 2013. The concentrations of culturable microorganisms in relation to particulate matter in urban air. Atmos. Environ. 65: 215–222.

Hales, S., S. Lewis, T. Slater, J. Crane and N. Pearce. 1998. Prevalence of adult asthma symptoms in relation to climate in New Zealand. Environ. Health Perspect. 106: 607–610.

Horak, B., J. Dutkiewicz and K. Solarz. 1996. Microflora and acarofauna of bed dust from homes in Upper Silesia, Poland. Ann. Allergy, Asthma and Immunol. 76: 41–50.

Huttunen, K., J. Pelkonen, K.F. Nielsen, U. Nuutinen, J. Jussila and M.R. Hirvonen. 2004. Synergistic interaction in simultaneous exposure to *Streptomyces californicus* and *Stachybotrys chartarum*. Environ. Health Perspect. 112: 659–665.

IOM. The National Institute of Medicine. 2000. Clearing the Air. Asthma and Indoor Air Exposures. The National Academy Press, Washington, DC, USA.

ISAAC. 1998. Worldwide variations in the prevalence of asthma symptoms: the international study of asthma and allergies in childhoods. European Resp. J. 12: 315–335.

Islam, Z., C.J. Amuzie, J.R. Harkema and J.J. Pestka. 2007. Neurotoxicity and inflammation in the nasal airways of mice exposed to the macrocyclic trichothecene mycotoxin roridine a: kinetics and potentiation by bacterial lipopolysaccharide coexposure. Toxicol. Sci. 98: 526–541.

Jabłońska, M., F.J.M. Rietmeijer and J. Janeczek. 2001. Fine-grained barite in coal fly ash from Upper Silesia Industrial Region. Environ. Geology 40: 941–948.

Jędrychowski, W. and E. Flak. 1998. Separate and combined effects of the outdoor and indoor air quality on chronic respiratory symptoms adjusted for allergy among preadolescent children. Int. J. Occup. Med. and Environ. Health 11: 19–35.

Kaldor, J.M. and K.A. Abbé. 1990. Interaction between human carcinogens. pp. 35–43. *In*: H. Vainio, M. Sorsa and A.J. McMichael [eds.]. Complex Mixtures and Cancer Risk. International Agency for Research on Cancer, Lyon, France.

Kankkunen, P., J. Rintahaka, A. Aalto, M. Leino, M.L. Majuri, H. Alenius, H. Wolff and S. Matikainen. 2009. Trichothecene mycotoxins activate inflammatory response in human macrophages. J. Immunol. 182: 6418–6425.

Kasznia-Kocot, J. and M. Sąda-Cieślar. 1993. Udział alergenów środowiskowych w indukcji wczesnej alergii u dzieci (in Polish). Materiały Krajowej Konferencji "Zdrowy dom", Politechnika Warszawska, Wydział Architektury, Warszawa, pp. 196–2001.

Kemp, P.C., H.G. Neumeister-Kemp, F. Murray and G. Lysek. 2002. Airborne fungi in non-problem buildings in a southern hemisphere Mediterranean climate: preliminary study of natural and mechanical ventilation. Indoor Built Environ. 11: 44–53.

Kirjavainen, P.V., M. Täubel, A.M. Karvonen, M. Sulyok, P. Tiittanen, R. Krska, A. Hyvärinen and J. Pekkanen. 2015. Microbial secondary metabolites in homes in association with moisture damage and asthma. Indoor Air doi: 10.1111/ina.12213.

Koch, M. 2000. Airborne fine particulates in the environment: a review of health effect studies, monitoring data and emission inventories. Interim report IR-00-004, International Institute for Applied Systems Analysis, Luxemburg.

Koenig, J.Q. 1999. Air pollution and asthma. J. Allergy Clin. Immunol. 104: 717–722.

Kozielska, B., W. Rogula- Kozłowska and J.S. Pastuszka. 2013. Traffic emission effects on ambient air pollution by PM2.5-related PAH in Upper Silesia, Poland. Int. J. Environ. Poll. 53: 245–264.

Mauderly, J.L. and J.M. Samet. 2009. Is there evidence for synergy among air pollutants in causing health effects? Environ. Health Perspect. 117: 1–6.

McConell, R., K. Berhanne, F. Gilliland, S.V. London, H. Vora, W.J. Gauderman, H.G. Margolis, F. Lurmann, D.C. Thomas and J.M. Peters. 1999. Air pollution and bronchitic symptoms in Southern California children with asthma. Environ. Health Perspect. 107: 757–760.

Moolgavkar, S.H. 2003. Air pollution and daily mortality in two US counties: season-specific analysis and exposure-response relationships. Inhalation Toxicol. 15: 877–907.

Mueller, A., U. Schlink, G. Wichmann, M. Bauer, C. Graebsch, G. Schüürmann and O. Herbarth. 2013. Individual and combined effects of mycotoxins from typical indoor moulds. Toxicol. *in Vitro* 27: 1970–1978.

Nicolai, T. and v.E. Mutius. 1996. Respiratory hypersensitivity and environmental factors: East and West Germany. Toxicology Letters 86: 105–113.

Owen, M.K., D.S. Ensor and L.E. Sparks. 1992. Airborne particle sizes and sources found in indoor air. Atmos. Environ. 26 A: 2149–2162.

Pastuszka, J., S. Hławiczka and K. Willeke. 1993. Particulate pollution levels in Katowice, a highly industrialized Polish city. Atmos. Environ. 27B: 59–65.

Pastuszka, J.S. 1997. Study of PM-10 and PM-2.5 concentrations in southern Poland. J. Aerosol Sci. 28(Suppl.1): 227–228.

Pastuszka, J.S. 1999. Occupational hygiene in non-industrial indoor places (in Polish). pp. 369–398. *In*: J. Indulski, E. Wiecek and H. Mikolajczyk [eds.]. Handbook of Occupational Hygiene, Vol. 2. Editorial Department of the Nofer Institute of Occupational Medicine, Łódź, Poland.

Pastuszka, J.S. 2001. Exposure of the General Population Living in Upper Silesia Industrial Zone to the Particulate, Fibrous and Biological (Bacteria and Fungi) Aerosols, Wroclaw Technical University, Wroclaw, Poland.

Pastuszka, J.S. 2005. Exposure to the airborne particles in the indoor environment in Upper Silesia, Poland. pp. 61–62. *In*: Z. Trzeciakiewicz [ed.]. Proc. Int. Conf.: Energy Efficient Technologies in Indoor Environment. Silesian University of Technology, Gliwice, Poland.

Pastuszka, J.S. and K. Okada. 1995. Features of atmospheric aerosol particles in Katowice, Poland. Sci. Total Environ. 175: 179–188.

Pastuszka, J.S., A. Kabała and K.T. Paw U. 1999. A study of fibrous aerosols in the home environment in Sosnowiec, Poland. Sci. Total Environ. 229: 131–136.

Pastuszka, J.S., K.T. Paw U, D. Lis, A. Wlazło and K. Ulfig. 2000. Bacterial and fungal aerosol in indoor environment in Upper Silesia, Poland. Atmos. Environ. 34: 3833–3842.

Pastuszka, J.S., J.E. Zejda, A. Wlazło, D. Lis and I. Maliszewska. 2003a. Exposure to PM in Upper Silesia and results of the pilot study on the airborne particles, bacteria and fungi in homes with asthmatic children. Proc. 5th Int. Technion Symp. "Particulate Matter and Health", 23–25 February 2003, Vienna. Austrian Academy of Sciences, Vienna, Austria, pp. 95–99.

Pastuszka, J.S., A. Wawroś, E. Talik and K.T. Paw U. 2003b. Optical and chemical characteristics of the atmospheric aerosol in four towns in southern Poland. Sci. Total Environ. 309: 237–251.

Pastuszka, J.S., E. Marchwińska-Wyrwał and A. Wlazło. 2005a. Bacterial aerosol in the Silesian hospitals: preliminary results. Polish J. of Environ. Stud. 14: 883–890.

Pastuszka, J.S., W. Rogula-Kozlowska and E. Zajusz-Zubek. 2010. Characterization of PM10 and PM2.5 and associated heavy metals at the crossroads and urban background site in Zabrze, Upper Silesia, Poland, during the smog episodes. Environ. Monitoring and Assess. 168: 613–627.

Peden, D.B. 2002. Pollutants and asthma: role of air toxics. Environ. Health Perspect. 110(Suppl. 4): 565–568.

Peitzsch, M., M. Sulyok, M. Täubel, V. Vishwanath, E. Krop, A. Borrás-Santos, A. Hyvärinen, A. Nevalainen, R. Krska and L. Larsson. 2012. Microbial secondary metabolites in school buildings inspected for moisture damage in Finland, The Netherlands and Spain. J. Environ. Monit. 14: 2044–2053.

Pyta, H. 2008. Classification of air quality based on factors of relative risk of mortality increase. Environ. Prot. Engin. 34: 111–117.

Roberts, S. and M.A. Martin. 2006. Investigating the mixture of air pollutants associated with adverse health outcomes. Atmos. Environ. 40: 984–991.

Rylander, R. 1996. Airway responsiveness and chest symptoms after inhalation of endotoxin or (1 → 3)-β-*D*-glucan. Indoor Built Environ. 5: 106–111.

Seedorf, J., J. Hartung, M. Schröder, K.H. Linkert, J.H.M. Metz, P.W.G. Groot Koerkamp and G.H. Uenk. 1998. Concentrations and emissions of airborne endotoxins and microorganisms in livestock buildings in Northern Europe. J. Agricul. Engin. Res. 70: 97–109.

Sicard, P., O. Lesne, N. Alexandre, A. Mangin and R. Collomp. 2011. Air quality trends and potential health effects—development of an aggregate risk index. Atmos. Environ. 45: 1145–1153.

Sicard, P., Ch. Talbot, O. Lesne, A. Mangin, N. Alexandre and R. Collomp. 2012. The aggregate risk index: an intuitive tool providing the health risk of air pollution to health care community and public. Atmos. Environ. 46: 11–16.

Solarz, K. 1987. Fauna alergogennych roztoczy (Acari) kurzu domowego w wybranych środowiskach Górnego Śląska (in Polish). Ph.D. Thesis, Silesian University of Medicine, Katowice, Poland.

Stieb, D.M., S. Judek and R.T. Burnett. 2002. Meta-analysis of time-series studies of air pollution and mortality: effects of gases and particles and the influence of cause of death, age, and season. J. Air and Waste Manag. Assoc. 52: 470–484.

Sundell, J. 1999. Indoor Air Environment and Health. National Institute of Health, Stockholm, Sweden.

Sundell, J. and M. Kjellman. 1994. The air we breathe indoors. The significance of the indoor environment for allergy and hypersensitivity. Summary of scientific knowledge. National Institute of Public Health, Stockholm, Sweden.

Sundell, J., M. Wickman, G. Pershagen and S.L. Nordvall. 1995. Ventilation in homes infested by house-dust mites. Allergy 50: 106–112.

Takano, H., R. Yanagisawa, T. Ichinose, K. Sadakane, S. Yoshino, T. Yoshikawa and M. Morita. 2002. Diesel exhaust particles enhance lung injury related to bacterial endotoxin through expression of proinflammatory cytokines, chemokine's, and intercellular adhesion molecule-1. Am. J. Resp. Crit. Care Med. 165: 1329–1335.

Täubel, M., M. Sulyok, V. Vishwanath, E. Bloom, M. Turunen, K. Järvi, E. Kauhanen, R. Krska, A. Hyvärinen, L. Larsson and A. Nevalainen. 2011. Co-occurrence of toxic bacterial and fungal secondary metabolites in moisture-damaged indoor environments. Indoor Air 21: 368–375.

Täubel, M., M. Sulyok, M. Peitzsch, M. Happo, V. Vishwanath, J. Jacobs, A. Borras, P. Jalava, E. Krop, H. Leppänen, J.P. Zock, J. Pekkanen, D. Heederik, L. Larsson, R.O. Salonen, R. Krska and A. Hyvärinen. 2013. Mikrobitoksiinit koulurakennuksissa ja niiden potentiaaliset lähteet (Microbial toxins and their potential sources in school buildings). FISIAQ Seminar Report 31/2013, 41–45 (in Finnish).

U.S. EPA. 2000. Supplementary Guidance for Conduction Health Risk Assessment of Complex of Chemical Mixtures. EPA 630/R-00/002. U.S. Environmental Protection Agency, Office of Research and Development, Washington D.C., USA.

Wargocki, P., J. Sundell, W. Bischof, G. Brundrett, P.O. Fanger, F. Gyntelberg, S.O. Hanssen, P. Harrison, A. Pickering, O. Seppanen and P. Wouters. 2002. Ventilation and health in non-industrial environments: report from a European Multidisciplinary Scientific Consensus Meeting (EUROVENT). Indoor Air 12: 113–128.

Wartenber, D., R. Erlich and D. Lilienfeld. 1994. Environmental tobacco smoke and childhood asthma: comparing exposure metrics using probability plots. Environmental Research 64: 122–135.

Whipps, J.M. 2001. Microbial interactions and biocontrol in the rhizosphere. J. Exp. Botany 52: 487–511.

WHO. 2009. WHO Guidelines for Indoor Air Quality: Dampness and Mould. World Health Organization, Copenhagen, Denmark.

Wieslander, G., D. Norbäck, E. Björnsson, C. Janson and G. Boman. 1996. Asthma and the indoor environment: the significance of emission of formaldehyde and volatile organic compounds from newly painted indoor surfaces. Int. Arch. Occup. Environ. Health 69: 115–124.

Woo, S., V. Fogliano, F. Scala and M. Lorito. 2002. Synergism between fungal enzymes and bacterial antibiotics may enhance biocontrol. Antonie van Leeuwenhoek Int. J. Gen. Mol. Microbiol. 81: 353–356.

Zock, J.P., A. Borras-Santos, J. Jacobs, M. Täubel, A. Espinosa, U. Haverinen-Shaughnessy, E. Krop, K. Huttunen, M.-R. Hirvonen, M. Peitzsch, M. Sulyok, J. Pekkanen, D. Heederik and A. Hyvärinen. 2014. Dampness and microbial secondary metabolites in schools and respiratory symptoms in teachers, ERS Conference 2014, Abstract no. 2161.

13

Socio-Economic Determinants and Population Health in the Context of Environmental Risk Exposure

Joanna Kobza

Introduction

Despite the fact that overall health has improved over the past few decades in most high developed countries, including the European Union, socio-economic inequalities in health persist both between Member States and between different regions within the same Member State. These inequalities could even be increasing in the near future (Marmot 2010). In most of the countries, clear differences are being observed in morbidity and mortality rate due to disorders dependent on socio-economic factors, their criterions being education, income (individual, family, material status), or employment status (Mackenbach et al. 2003). Environmental exposures also significantly contribute to health inequalities and associated health disparities and diseases. Evidence shows that air pollution at current levels in European cities is responsible for a significant burden of deaths, hospital admissions, and exacerbation of symptom, especially for cardiorespiratory disease in urban environments (WHO 2013b). Poor health is often made worse by the interaction between individuals and their physical environment. There are two major mechanisms that may act independently or together, through which environmental exposure may contribute to health inequalities; among the general population, disadvantaged and "higher-risk" groups are recognized as being more often exposed to sources of air pollution (differential exposure) and may also be more susceptible (differential susceptibility)

Silesian University of Medicine, School of Public Health, 18 Piekarska St., 41-902 Bytom Poland.
E-mail: koga1@poczta.onet.pl

to the resultant health effects (Deguen and Zmirou-Navier 2010). WHO assessment shows that inequalities in environmental exposure can reach extreme levels, with disadvantaged populations groups often being at least five times more exposed than advantaged groups (WHO 2013a). Over the past few decades, the socio-economic and environmental risk factors related to such phenomena like climate change, globalization, urbanization, economic, demographic, and health changes have significant influence on the poorest and most vulnerable groups.

There are population target groups who suffer from respiratory diseases caused by biological and chemical agents in the air because they live in places that have higher levels of air pollutants. WHO indicates that 300 million people in the world suffer from asthma (WHO 2008a). One of the most vulnerable groups are children, they suffer from multiple and cumulative exposure and are more susceptible to a variety of environmental toxicants and often lack environmental and other resources (for example, access to quality health care) to counterbalance environmental threats and to limit health consequences (Bolte et al. 2009). Exploring the risk factors through the entire lifespan that influence cognition in midlife to late life, the greatest impact comes from early life exposure and socioeconomic factors like: education, nutrition, social, and family environment in early life, which may have a long term influence on cognition; a concern developing countries especially (Lynch and Kaplan 2000). The issue of exposure and health inequalities in relation to environmental risk factors requires a broad and cross-sectorial engagement in the nearest future.

Socio-economic inequalities in health

Inequalities in population health are strictly connected with social and economic differences in societies. Income conditions, social status, education, quality of work, unemployment, social exclusion, and poverty are the most representative elements of the socioeconomic determinants of health. As was noticed above, highly developed countries reflect tremendous differences in mortality and morbidity nowadays between people living in the poorest and in the richest neighborhoods; there is also a big difference in DALY and HALE (WHO 2011). People from poor areas die sooner and also spend more of their lives with a disability. The gap in life expectancy between high and low income countries is about 20 years and in a disability-free life, expectancy is 15 years (WHO 2012). High AIDS mortality rates among African migrants compared to European citizens are related with socio-environmental and economic differences. Gender is also a very important factor in terms of differences in the risk of death' mortality in men is much higher than in women. Urban or rural place—region of residence has great significance concerning risk of death caused by respiratory and digestive system diseases. Tremendous differences in health status are observed between groups of people with various educations. Among less educated people, a higher death rate was observed. Another research showed that children with low SES, but who engaged in shift and persist strategies which enabled them to adapt to stressors, displayed less asthma impairments and inflammations (Chen et al. 2011). The same concerns adolescents; there are correlations between low SES and immune responses in asthma, what is connected with stress experience. Low socioeconomic status (SES) is related, for example, with higher morbidity and mortality rate due to coronary heart disease, since it was noticed that low social standing activates an autonomous, neuroendocrine, and immunological answer, which plays a significant role in pathogenesis of sclerosis. Also, an incorrect profile of health behavior with increased coronary disease morbidity is found in lower socio-economic groups. The relation between socio-economic factors and morbidity and mortality rate due to diseases of the circulatory system, especially coronary disease, was proven by numerous studies: prospective, cohort, epidemiological, and clinical. Differences in occurrence frequency of classical risk factors in particular groups were also noted. Groups with a higher socio-economic status are characterized by higher pro-health awareness, which manifests itself, for example, in increased physical activity and the use of a low-calorie diet. In turn, many studies show significantly more frequent cases of depression, social isolation, or chronic occupational stress in people with low socio-economic status (De Backer et al. 2003).

Socio-economic factors do not exist independently from each other and are strictly connected one with another, interweaving in a cause-effect relationship, and demonstrate a tendency to accumulate, so it is difficult to deal with each of them separately. One should take into account interactions within one group, as well as correlations with other health, social, economic, and demographic determinants. That's why it is necessary to monitor not only epidemiological indicators, such as morbidity and mortality rate, but also diseases risk factors arising from living conditions of the individuals. We must take into consideration wider context for prevention of diseases, as there is a significant amount of evidence showing that their influence is independent from classical risk factors.

The study from Nordic countries included people from Denmark, Finland, Norway, and Sweden. They highlighted significantly better health among people with a higher income. This relationship was strongest in Norway and Finland and weakest in Denmark. The relatively high health inequalities between income groups in Norway and Finland were visible in the youngest age groups. An important conclusion was the statement that health policy programs may be particularly important five to ten years prior to retirement and in early adulthood (Huijts et al. 2010).

The Commission on Social Determinants of Health stated that social inequalities in health are due to the inequalities in the conditions of daily life and the fundamental factors are inequities in power, money, and resources (WHO 2008b). Social position depends on education, occupation (employment and working conditions), income, standard of living (housing and neighborhood conditions), gender, ethnicity, and race. The benefits of reducing health inequalities are not only social but mainly also economic. It was estimated that inequality in illness causes the productivity losses at the level of Ł31–33 billion per year as well as higher welfare payments and lost taxes of Ł20–32 billion per year (Marmot 2010).

Socio-economic determinants

Socioeconomic factors are being currently recognized as the main determinants of population and individual health (Dahlgren and Whitehead 1991). In this group of factors we identify: income, level of education, living conditions, as well as place of work, local community, social support, and the sense of safety. Influence of these factors on health, especially on mental health, is observed in periods of political, economic and social changes.

Several international research studies proved the impact of social capital on health, and a particularly strong association was found between life expectancy and GNP per capita (Lindstrom and Lindstrom 2006). The proportion of health expenditure financed by the government and income (per capita) are both significantly and positively associated with better health outcomes (Kennelly et al. 2003). A strong, statistically significant and consistent relation between national income inequality and population health at the country level was also found; change in inequality is significantly related to change in life expectancy (Babones 2008). The experiences of Eastern and Central Europe Region, in the last decades, should also serve as an important example of how the profound socio-economic and political changes among society can influence the health of a population (Weidner and Cain 2003).

The WHO Commission on Social Determinants of Health recommended the application of four main indicators of social economic position (SEP) to assess their influence on population health:

1. Housing
2. Income
3. Education
4. Occupation

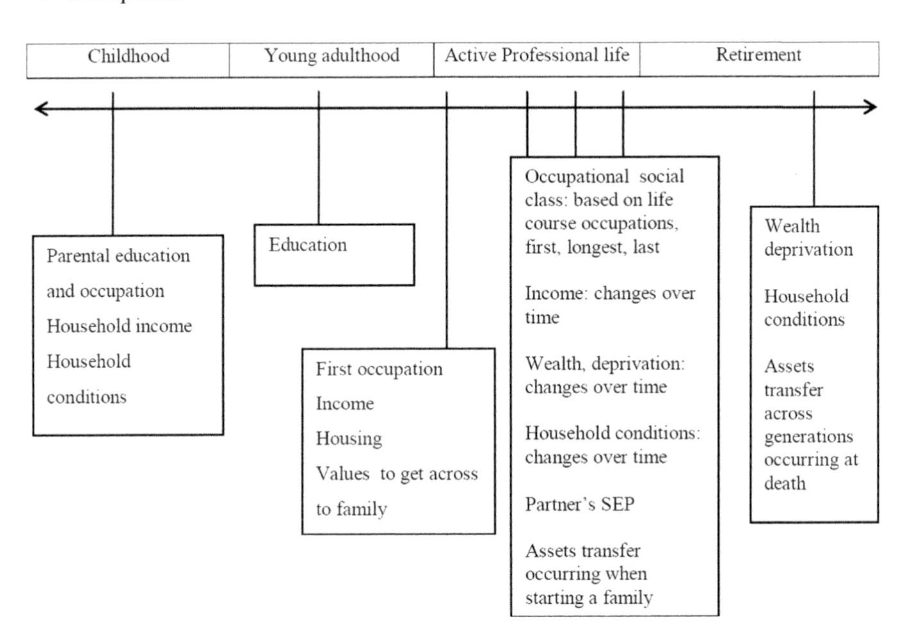

Fig. 13-1. Life course socioeconomic trajectory (Galobardes et al. 2006).

Housing

Housing is a material resources indicator and also reflects socio-economic position; it take into consideration:

- the status of the indicated person meaning ownership status; owner-occupied or renter of the apartment/house; being a renter meaning poorer housing conditions,
- basic equipment of the apartment/house (central heating, hot water, sanitary equipment),
- standard and conditions of living area (air quality, density, equipment, middle class house, green spaces).

Housing can be also defined as an indicator of cumulative prosperity and wealth. The main advantage of this indicator is that its' rather easy accessibility to the researchers and they are easy to compare between indicated societies, but geographical and cultural context must be also taken into consideration and in this cases it is more difficult to do comparative analysis.

Housing and health

The environment of residence influencing health and housing conditions are strictly associated with some diseases. Over the past decades, the meaning of the association between housing and health has expanded beyond infectious diseases or sanitation and includes a vast list of agents associated with mortality and morbidity like: asbestos, lead, radon, and other chemicals and toxics, pet and insect allergens, fungi, tobacco smoke, combustion byproducts (carbon monoxide, nitrogen dioxide, fine particular matter, cooking, gas stoves, smoking) and others. The indoor concentrations of airborne pollutants is related to a home's physical structure, products usage characteristics, attributes of furnishing, air infiltration pathways, the type of ventilation and heating, residents' behavior, and outdoor pollutant concentrations. Low income inhabitants, especially in multifamily buildings, have increased concentrations of NO_2, smaller unit size, increased source strength, and inadequate ventilation (Zota et al. 2005).

It seems to be necessary to connect urban planning and public health action to create healthier environments. Fortunately, the improvements in health gains are a very important expected effect of several housing infrastructure programs nowadays in developed countries. Pleasant environments (better residencies) are significantly associated with more amounts of physical activity, which is correlated with better health (Sigmundova et al. 2011).

Income

Income is the key measure of material resources, individual or household income; it reflects, in a way, the strictest material resources. We can take into consideration total achieved income or income in different defined categories. Very often household income is analyzed. It is also possible to apply relative indicators (poverty level, minimum level, the average monthly gross wage, and salary). Limited financial resources may have negative influence on human behaviors.

The calculation of minimum income for healthy living includes: the income needed for adequate housing, nutrition, medical care, hygiene, transport, social interaction, and physical activity. The European Union average is the best in Germany, Netherlands, France, and the Nordic countries (Eurostat 2012).

Poverty

Poverty is a lack of sufficient material resources of living, hardship, and deficiency. A position below certain variables in time is the break even point, or a threshold of meeting needs in relation to a family unit or social group. Poverty is a social condition characterized by lack of sufficient material means to meet the living expenses of an individual or family. People are suffering from poverty, when their income drops significantly below that which is available on average in a society, even if it is sufficient to survive. Reasons for emergence of poverty: health factors, including long-term illness, physical or mental handicap, disability, alcoholism, drug abuse. Family factors besides orphanhood, single parent or families with many children, helplessness in caring and upbringing matters, loneliness, or else difficulty or inability to adapt after leaving a prison. Poverty belongs also to labor market factors, including the category of unemployment.

Poverty levels

1. Minimum existence (biological minimum)—basket of goods essential to maintain life functions of a human being and his psychophysical efficiency. It includes only these needs, satisfaction of which cannot be postponed in time, and consumption below that level causes biological devastation and life threat. The minimum existence basket includes housing needs and foodstuffs. The total cost of acquiring these goods constitutes the value of the basket, which makes up the threshold of extreme poverty. The amount equal to the value of the minimum existence basket sets up the threshold of extreme poverty (absolute).
2. Social minimum—a rate indicating household living expenses based on a "basket of goods" needed to satisfy living and consumption needs on a low level, not only to sustain life, but for having and upbringing children, and to maintain minimum social links. The first group of elements includes expenses such as housing, food, clothing, shoes, healthcare, and hygiene; and the second: communication (e.g., commuting to work), expenses of education and upbringing of children, on family and social contacts, and modest participation in culture.

Poverty has both an individual and a social character. It relates to financial status and also to a mental state of an individual. Poverty becomes a social issue, when it grows to such a scale and intensity that it does not only constitute an attribute of living conditions of whole population groups, but at the same time, it has a destructive impact on the whole social and economic growth of a country.

Income and health

Income is a more seldom used socio-economic factor in studies evaluating health status of the population; however, it is a significant measure, because it enables access to goods and services, including education quality, housing environment, and health care protection against illnesses. This coefficient is not as stable as education, because during a life span it can change. A Finnish study confirmed that income level is strictly related to health behaviors (Laaksonen et al. 2003).

Influence of poverty on health has an indirect character, manifesting itself as an access barrier to health care, purchase of medicines, and access to certain medical procedures such as visits to a specialist or dental treatment. The investigations based on a sample of 11 countries of European Union provided evidence that income inequality is negatively associated with self-rated health status for both men and women, especially when measured at the national level (Hildebrand and Kerm 2009). A Canadian study, which examined the differences in health adjusted life expectancy across income groups, confirmed that generally worse health related quality of life of lower income population groups, and that those in higher income decidedly had more years of health adjusted life expectancy (McIntosh et al. 2009). The inequality of income distribution is reflected by the Gini coefficient which ranges from perfect equality (0) to absolute inequality (1). The Gini index for most highly developed countries is about 0.3, while for middle-income countries it is about 0.5–0.6.

Education level

Education level can partly define the socio-economic position. We take the number of years and education level in cohort groups into consideration most often. Education level determines future work, occupational position, and achieved income of individuals; it has tremendous impact on human behavior, and it contributes to the increase of self-esteem. It is very difficult to measure and compare the "quality" of education but it is easy data to access.

Education and health

Educational level is the most important socio-economic indicator which reflects human health behavior and influences more healthy behavior choices, for example those concerning diet, smoking, health activity, etc. Education is the most common socio-economic gauge, because it is characterized by ease and objectivity of measurement. Serious divergences between education of a population of a given country and its wealth, and what goes with it, that is, social marginalization, manifests itself, among other things, in the poor health status of these people. This fact can be explained by how usually lower income people in this social group have inadequate access to health care due to poverty, or in many cases, lower social awareness linked to a lack of knowledge about the discipline of widely understood health education.

Occupation

Occupation is significant factor which determines an individual's position in society and is strictly connected with intellect and the level of achieved income. The parents' occupation reflects the socio-economic position of their children. Occupation of one or both parents influence the level of expenses, so is strictly connected with the standard of living of the whole family.

Occupational status and education determine people's access to material resources and define their structural position in socioeconomic hierarchy. Income and housing reflect material resources but occupational status is strictly related with educational level, and is directly connected with working conditions. Universal WHO macroeconomic indicators are unemployment rate and work rate.

Occupation and health

Occupation is a commonly used variable in health research searching the links to social status and class, as well as to exposure to environmental hazards. Having good employment (employment stability and working conditions) is defined as a health protective factor and the opposite, that is, unemployment contributes to poor health (mental and physical) for each occupational class. We observed higher standardized mortality rate for the unemployed rather than the employed (Marmot 2010). The quality of work is also important; insecure and poor quality of work as well as low paid jobs have biologically negative health related effects. Working conditions influence health, but some research shows that the associations between health behaviors are not significantly associated with working conditions (Lallukka et al. 2008). Occupation can be connected with stable, negative for health, environmental exposure.

Unemployment

From the economic point of view, unemployment is a characteristic structural element, typical for the free market economy in all democratic, highly developed countries. A person is recognized as unemployed under following conditions: staying without work, excluding persons working part-time or working full-time; or looking for a job—the problem is verification of commitment a given person has to a job search and expresses readiness to start work. It is generally accepted that unemployment is the condition when there is lack of paid employment among people being in the age defined as the so called "working age", that is, who are able and ready to start work. One can divide unemployment into categories, e.g., seasonal, voluntary, compulsory, frictional, total, partial, unemployment in the place of work, repeated unemployment, excluding unemployment—applies to people with minimal probability of becoming employed due to low qualifications, unemployment related to particular occupational categories, etc. Traditionally, the most often indicated categories of unemployment are: frictional, structural, cyclical (depends on trade cycle), also balanced and unbalanced, and long- and short term as well. In practice, there are serious problems with a precise definition of each unemployment type, mainly due to correlation of various unemployment types, for example, structural and cyclical, which is voiced in contemporary concepts. The basic factor increasing the risk of unemployment is lack of qualifications—a share of people with education level below secondary is clearly higher in relation to the whole population. Unemployment very often does not have a homogeneous dimension. It has its economic dimension, including among others financial, political, social—especially sociological and physiological, as well as health consequences. Unemployment is generally recognized as a negative condition, having many unwanted consequences; however, one should also notice its positive effects especially in the case when it does not exceed socially acceptable limits. Economic and social outcomes of unemployment should be explored through individual costs of unemployment incurred by an individual, and macroeconomic, related to the society in general. Individual unemployment costs are directly incurred by the unemployed person and his/her family; they can be broken into costs associated with lack of income from work, costs of a psychosocial nature, as well as costs of a health nature. Deprival of income from work leads to a strong decrease in living standards and the related necessity to resign from many basic needs. These are not only needs related to well-being, but also basic needs related to education and culture. Families of the unemployed people cannot afford education of their children and their intellectual development; consequently, in the future, they have a much lower stance on the employment market (inheritance of unemployment). Negative effects of unemployment are most severe in single-parent families and families with many children. As regards the individual costs, which are borne by the unemployed, one can include decrease of vocational competences due to long-term inactivity, and related it with decrease in chances of finding work in the future. The unemployed person and his/her family also bear costs of psycho-sociological nature. These include the strain related with the degradation of their social stance leading to weakening of social ties, isolation from the community, and an increase in conflicts with the neighborhood and within the family (loss of respect in the family, upbringing problems, divorces), and also destabilization of identity. Unemployment leads to poverty, and therefore loss of social safety and a sense of social justice.

Outside the negative impact on the economic potential of the society, that is, getting used to a significant level to impoverishment and reduction of consumption demand, unemployment has also a negative impact on many areas of life of an individual and in a negative way disturbs an individual's mental, physical, and social well-being. The institutional solution of the effects of unemployment is a vital element reflecting the health status of society.

Unemployment and health

Unemployment causes a multiplicity of health risks; such as long term illness (Bartley 2004), mental illness (Thomas et al. 2005), cardiovascular diseases (Gallo et al. 2006); it is also related with overall mortality and with suicide (Voss et al. 2004). Unemployed persons have a much higher use of medications (Jin et al. 1995). Unemployment has short term and long-term effects on health. The immediate negative impact (Stuckler et al. 2009) is that the negative effects are proportional to the duration of unemployment (Maier et al. 2006). Health effects of unemployment are difficult to evaluate, because it is generally an indirect influence and is hard to measure. Some symptoms of health deterioration already show up 3-4 years before a job loss. Those mental and physical complaints escalate in this period, but in the face of risk of job loss these people give up sick leave and resign from treatment. Unemployment is a stressful experience, especially in the face of events which one cannot control. In this situation, one has to deal with an especially destructive, uncontrolled type of stress, in which the individual has no control over the stressor, which in the long term results in numerous mental disorders, and in consequently, depression. The feeling of alienation, social isolation, and low control of what is happening increases the risk of cardiovascular disorder, psychical illnesses, risk of alcoholism, or suicide. Individual factors of unemployment risk include not only a low education level, poor qualifications, place of residence or profession, but also improper lifestyle, worse quality of human relations, as well as some character traits. Lack of job devastates mentally, physically and socially; on the other hand, bad psychical health, and lack of qualifications contribute to job loss. Lack of any financial security also constitutes a strong stressor, negatively affecting the subjective perception of a health condition. It releases a stress reaction with all its consequences of mental nature like fear, depression, and frustration, as well as physical, in the form of many dangerous somatic disorders: CVD, digestive system and immunological system diseases. Long term stress—distress has a destructive character and can stimulate emersion of many organic diseases, can lead to a general exhaustion of the body, and in extreme cases even to death. It can cause an occurrence of a new disease, and also an intensification of an already existing one. Chronic stress related to a difficult financial situation has an adverse influence on quality of life, especially when the period of being without work is excessively prolonged. The role of chronic stress in pathogenesis of circulatory system diseases is mainly related to the functioning of stress hormones, and as a result there are changes in the functioning of this system, such as: increase in blood pressure, contraction of blood vessels in abdominal cavity organs, flow increase through the heart and brain blood vessels. The influence of stress reaction on the circulatory system is multi-directional. A stress reaction can directly cause high blood pressure and arteriosclerosis and influence indirectly through being overweight, indulging in so-called "stress eating", an inclination to addictions, and depression. Poverty related to unemployment is linked with a lack of money for everyday expenses, and so with a limited access to health care, which manifests itself in the deterioration of the health status of not only people affected by unemployment, but also their families. Hence, among children of the unemployed, one can see more often, e.g., caries or obesity caused by incorrect eating habits and hygiene standards. In families with a difficult financial situation, non-vaccination of children is also a significant problem, which predisposes them to suffering from a number of dangerous diseases of bacterial or viral character. Quite often unemployment is a multi-generation condition. The reduction of living standards of unemployed parents is a cause of reduction of the children's education level, which in the future can be a cause of unemployment among the young, so called "unemployment inheritance". Within people without stable incomes, that is, both adults and children from poor families, alarming occurrences having a character of social pathology have spread, such as alcoholism, smoke addiction, or drug abuse, with all of their destructive consequences for the individual, family, or the whole of society. Relative differences in incomes are related to a health status and death rate on all hierarchy levels of the society. The risk of falling ill increases if differences in income deepen. Differences in the health status of each social group are shown in all age groups, whereas in the economically weakest, a larger number of premature births and low birth mass of infants, more cases of heart diseases, strokes and some tumors in older people have been recorded. In groups of the lowest socio-economic status, the most elements of risk are accumulated, such as, e.g., smoking tobacco, lack of physical activity, overweight, high blood pressure, bad diet, and others.

There are three ways in which unemployment affects morbidity and mortality:

- it causes financial problems and results in the decrease of the living standard (Maier et al. 2006), which can reduce self-esteem and social integration, and provoke anxiety, distress, and finally depression (Voss et al. 2004); several psychosocial stressors influence health and affect partners and children in these groups,
- unemployment influences life style, that is, increase of smoking, alcohol, drinking, etc., and causes a decrease of physical activity (Maier et al. 2006),
- unemployment rate seems to be one of the major determinant of suicide (Qi et al. 2009).

The comparative study of the relationship between unemployment and self-reported health in selected 23 European countries, which represented different welfare state regimes, Scandinavian, Anglo-Saxon, Bismarckian, Southern and Eastern, confirmed the negative health effect of unemployment, particularly strongly for women and especially represented in the Anglo-Saxon and Scandinavian welfare state regimes (Bambra and Eikemo 2009; Bambra 2010).

Socio-economic disparities in environmental exposure and its health consequences

Environmental health risk factors exposure are strictly connected with the socio-economic situation of individuals. People with better socio-economic positions have better health, thanks to more healthy behaviors, better access to health care services, and live in safer neighborhoods. There is interaction between infant mortality and socio-environmental vulnerability (Lara-Valencia et al. 2012). Some studies have investigated various impacts of air pollution on more susceptible groups of population based on area of residence, educational attainment, race, sex, and age (Amy et al. 2009).

Most of related risk factors strongly associated with socio-economic and environmental inequity could be prevented in large part. Tobacco use is one of them. It is one of the main risk factors for a number of associated chronic diseases, including cancers, CVD, and lung diseases. Tobacco use is the main cause of morbidity and mortality in highly developed countries (WHO 2012).

Many international researches indicate that socio-economic and socio-environmental factors have a strong association with the prevalence of childhood allergic diseases like asthma, allergic rhinitis, and eczema (Li et al. 2011). Asthma is one of the most common chronic diseases especially affecting young people in the United States and Europe; globally almost 10 million youth under eighteen have received diagnosis and 6.8 million have active asthma (Basch 2010). Asthma morbidity varies with socio-economic position (Kozyrskyj et al. 2010). The research of children asthma in Chicago highlighted that annual average asthma prevalence is approximately 45% higher for Black children as compared to White children; the same concerned asthma attacks, asthma mortality rates in Chicago are among the highest in US, and the highest rates concern low-income populations and minorities (African Americans) (Curtis et al. 2012). Inequities in access to continuous health care was indicated as also one possible reason which contributed to higher death rates from asthma in this city. The prevalence rates of asthma were also higher than 98% in schools with African Americans than in other schools. The same study reported the strong positive association between community vitality, and economic potential and low asthma prevalence. As was mentioned above, childhood asthma prevalence is higher in neighborhoods with a higher black population ratio, which indicates that social capital and community vitality as well as socio-economic determinants strongly contribute to asthma variation, and that there is also an association between community violence and asthma risk. One of the crucial consequences is negative impact on academic achievement among youth.

Another facet of socio-economic and environmental inequity is the increase of ambulatory physician consultations given the large population at risk. The excess of ambulatory care associated with air pollution constitute a significant societal cost (Burra et al. 2009).

London research proved that traffic air pollution concentrations are generally higher in low-SEP areas, but several exceptions indicate that traffic related air pollution may show different relations to different SEP dimensions (Goodman et al. 2011). Exposure to air pollution can irritate the nose, eyes, and throat, may decrease resistance to respiratory infections, exacerbate lung diseases, such as asthma, bronchitis, and chronic obstruction pulmonary disease and is the reason of premature death in individuals with cardiac and pulmonary diseases. The higher the exposure, the more severe are the adverse health effects. The research in Czech Republic, the country in transition, where 39 cities were included, indicated that exposure to air pollution cause adverse health effects in many Czech regions; inhabitants with an unfavorable socioeconomic position mainly live in smaller towns with higher concentration levels of combustion related air pollution, typical for residential heating (SO_2, PM10) and large cities with residents with higher SEP are exposed to higher levels of traffic related air pollution (NO_2) (Branis and Linhartova 2012).

The literature review provide the evidences of socio-economic inequalities in indoor and outdoor environmental exposures and exposure related health risk factors (Adamkiewicz et al. 2011). There is inverse association between main indices of SEP and numerous environmental risk factors like ambient air pollutants, toxins, water quality, noise, wastes, housing quality, working environments, residential crowding, educational facilities (Evans and Kantorowitz 2002).

In most European countries, Asia and the United States the research showed strong association between multiyear exposure to higher levels of particulate pollution and higher mortality rates (Ou et al. 2008). The lower educational level was also connected with bad health outcomes, including higher mortality rates. Several studies of effect modification of pollution exposure by education showed a strong gradient of increasing relations between PM and mortality with decreasing educational status in these parts of world, but with different patterns in Latin America (O'Neil et al. 2008).

Air pollution is a modifiable risk factor identified as being particularly dangerous for people from a lower socioeconomic class, living close to main roads, worse house conditions, and permanently exposed to elevated concentrations. Several research reports confirm the cumulative risk assessment of the socio-economic situation and respiratory and cancer air pollutants exposure hazards (Evans and Kantorowitz 2002). Lower socioeconomic status is related to significant environmental disparities, such as poor hygiene and nutritional conditions, poor residential environment, and environmental inequities. The quality of indoor air is as important as outdoor air, as generally people spend more time indoors.

Indoor particulate exposure can be affected by various factors: the building materials, such as furnishings, equipment, carpet, paint, plastics, activities of building occupants (smoking, cooking, cleaning), outdoor contamination levels, the season, temperature, humidity, type of ventilation, and others. What is often highlighted is that indoor PM10 concentrations are determined more by the socio-economic situation than human behavioral activities. Households with cleaning, cooking, or smoking activity had higher indoor PM10 concentrations than households with no such activity (Byun et al. 2010). Many epidemiological researches provide the proof that the PM10 concentrations, for example, in children's rooms and living rooms, as well as indoor and outdoor concentrations, depend on parental education, place of residence, and average monthly household expenses and that a lower socio-economic position is significantly related to higher indoor PM10 concentrations (Byun et al. 2010).

Indoor PM10 concentrations are directly associated with parental education, neighborhood, number of children, and type of housing and the neighborhood is often defined as the strongest socioeconomic factor affecting indoor PM10 concentrations. The same association concerns outdoor exposure; groups with a low SES tend to experience more exposure to air pollutants and toxicants, especially due to the proximity of their homes to various pollution sources, including high traffic roads, waste disposal sites, and industrial facilities (Havard et al. 2009). It is often concluded that neighborhood socioeconomic deprivation increases mortality risks associated with air pollution (Wong et al. 2008; Mohai et al. 2009).

Globalization has resulted in changes in industries acknowledged for the pollution in poor areas, where the environmental regulations are not so rigorous and costs of production are cheaper. Environmental health risk inequalities have become greater between countries over past few decades.

The research findings from Strasbourg confirm the association between the socio-economic deprivation areas and NO_2 levels, and the midlevel deprivation areas were the most exposed (Havard et al. 2009). The same results were obtained in Hong-Kong; PM10 and NO_2 were associated with greater risk of mortality in people living in public rental housing than in private housing (Ou et al. 2008). This research provides strong evidence of the associations between an individual's SES and short term effects of air pollution on mortality; the effects of such a pollutants like PM10, NO_2, SO_2, and O_3 on an individual's health were significantly stronger in lower socio-economic groups than in the higher SES groups.

Population exposure to multiple air pollutants, which include carcinogens presents significant challenges for environmental public health. As was mentioned above, identification of the most susceptible socio-economic groups at the greatest health risk from air pollution exposure is very important for exact estimates of air pollution impact. American research highlighted that environmental justice studies have mostly used communities as the main unit of analysis, while community-based reports make sense only when the location of polluting sources are taken into consideration. They are much less informative about studies of national air quality rules which affect multiple pollutants and sources (Post et al. 2011).

Environmental tobacco smoke exposure

Key health behaviors are connected with social and environmental determinants where smoking is one of the examples. The WHO indicates the group of tobacco related communicable diseases like: lower respiratory infections, tuberculosis, and non-communicable causes: malignant neoplasm, especially of trachea, bronchus, and lung, other neoplasm, CVD; mainly ischemic heart disease, cerebrovascular disease, and other, respiratory diseases: chronic obstructive pulmonary disease and other respiratory diseases (WHO 2012).

Smoking is the main public health problem which reflects socioeconomic inequalities. Less educated people have limited knowledge concerning long-term health consequences of smoking, which is associated with psychosocial stress connected with bad material conditions and low social position. Socio-environmental determinants influence initiation of tobacco use. On the other hand active involvement of family, religious leaders, community, local policy makers, and health professionals support and sufficient knowledge in tobacco cessation are key activities in creating tobacco free norms (Jarvis and Wardle 1999). People with lower education and income, for example manual and non-manual workers, smoke more often than semiprofessionals, managers, and professionals and the indicators in men and in women are similar (Stronks et al. 1997). In England, there are differences in tobacco consumption strictly connected with social classes. The prevalence of cigarette smoking among people in routine and manual households is 26% (28% for men and 24% for women) and the prevalence in managerial and professional households is 15% (16% for men and 14% for women) (Marmot 2010). The same situation is found in other countries; many researches confirm the increase of smoking from the higher to lower social groups (Laaksonen et al. 2005). Income is inversely related to smoking but this economic explanation seems not to be sufficient as regards limited financial sources and smoking. Psychological background must also be taken into consideration, which defines smoking as an answer to stress caused by bad material well-being conditions. People who live in worse socio-economic conditions are more often exposed to stress. Socioeconomic disadvantages are related with smoking but the relationship is more complex because we must take into consideration material, structural, and perceived aspects of socioeconomic differences (Laaksonen et al. 2005). Findings from a Korean study provided decision makers the information that any health policy to promote smoking cessation must take into account not only the SES of individuals but also their residential areas (Okhee et al. 2012).

Children especially are more susceptible than the general population. Several epidemiological studies suggest that the parental socio-economic situation influence directly the tobacco exposure of children, and the risk of exposure to ETS is higher among children whose parents have low education or income levels (Bolte and Fromme 2008; Yi et al. 2012). It has also been proved that most environmental tobacco smoke exposure of children occurs at home (Byun et al. 2010).

The social consequences for smoking are family poverty because the tobacco expenditures consume a part of household income which could be spent on other goods or services (for example healthy nutrition, physical and cultural activity, education) and increase the potential cost of medical treatment of the tobacco related diseases.

The negative impact of tobacco smoking includes higher rates of cardiovascular death, especially stroke and ischemic heart disease. In Europe, 25% of all deaths among men aged 30 years and over is attributed to tobacco. WHO indicated that globally 12% of all deaths (5 million) adults who are more than 30 years old is attributed to tobacco; 5% of all deaths from communicable diseases (within which 7% of all death are due to tuberculosis, 12% due to lower respiratory infections) and 14% of deaths due to non-communicable diseases are tobacco related (aged 30 and more); 10% of all death from CVD is attributable to tobacco exposure; 22% of cancer deaths and 36% of all deaths from respiratory system diseases. Globally, deaths from ischemic heart disease in 38% people are attributable to tobacco, 71% of all lung cancer deaths and 42% of all chronic obstructive pulmonary disease deaths are due to tobacco (WHO 2012). The regions with the highest proportion of deaths related to tobacco exposure are Europe and the Americas. Globally, tobacco product use and rates of tobacco-related mortality are higher in men than in women, but the use of tobacco has accelerated among women over the last decades.

It seems to be necessary to protect non-smokers; globally, one-third of adults are regularly exposed to tobacco smoke and they must be enabled to breathe air free of tobacco smoke. It is estimated that "second hand" tobacco smoke causes about 600,000 premature deaths in the world per year (WHO 2011) and 1 billion people worldwide are projected to die in the 21st century (WHO 2012).

This problem especially includes tobacco exposure in the work place; many researches confirm that both workers and customers are exposed to harmful levels of a carcinogens and toxin, and policies that prohibit smoking in public places significantly reduce exposure and improve worker and patron health assessment of exposure to second-hand smoke in indoor workplaces, i.e., the level of indoor air pollution was 96% lower in smoke-free sites compared to non-smoke-free sites (Koong et al. 2009).

More and more countries have introduced legislation restricting tobacco consumption in public areas because the evidence provided positive proof concerning business and public acceptance. A Scottish study confirmed that smoke-free legislation implemented in March 2006 has reduced exposure to second hand smoke among children in this country (Akhtar et al. 2010).

Ireland became the first nation in Europe which implemented smoke-free worksite regulations in March 2004 that included bars and restaurants. Norway implemented its policy in June 2004, then Italy and several Central European countries, New Zealand, Sweden, Scotland, and the United Kingdom have also passed the same regulations. Through provincial or state regulations, large parts of Canada, Australia, and the US have also implemented strong clean indoor air regulations. But till now, only about 11% of the global population is protected by national legislation.

Another element to increase people's knowledge about possible dangers of smoking are public comprehensive campaigns, whose key element are warnings about health risks and consequences to discourage adolescents and young adults from smoking. Many countries also incorporate special anti-tobacco programs into primary care services.

Another tool to limit smoking is the increase in tobacco products prices due to the increase in taxes. In highly developed countries, a 10% increase of tobacco products prices caused a 4% decrease of consumption (WHO 2013c). For governments and communities, the use of tobacco products and exposure to second hand smoke constitutes a significant social and economic burden; it means higher expenditure on health care of tobacco related diseases, and also contributes to the decrease of national productivity because of premature deaths and disabilities. The main WHO act aiming to protect people from the health, social, environmental, and economic consequences of tobacco exposure and consumption is the WHO Framework Convention on Tobacco Control. The treaty obliged countries to implement indicated evidence-based tobacco control measures.

Health outcomes in the, socio-environmental context

Non-communicable diseases are the main reason of morbidity, mortality and disability in high-developed countries; among them are: ischemic heart disease, cerebrovascular diseases, depression diseases, chronic obstructive pulmonary disease, and lung cancer. There is a significant correlation between socio-environmental determinants and the incidence of cancer and CVD. Mortality and morbidity from CVD and cancer are higher among people with poor mental health and a bad socio-economic position (Eurostat 2012). The association between low income status and a lack of motivation considering health prevention services which resulted in disease progression is well documented and it is an important factor in combating chronic diseases (Berkman 2011). The racial and ethnic inequalities contribute to the burden of cancer, and they caused, for instance, the new congressional legislation and the establishment of National Center for Minority Health and Health Disparities in the US.

Cancers and socio-environmental determinants

Cancer is a multi-factorial and complex disease and the investigations of cancer disparities involve several individual factors. Despite the special role played in carcinogenesis by genetic factors, cancers are influenced by environment and life style factors that extremely often reflect the affected individual's socioeconomic situation.

Socio-economic differences among population groups in health risk exposure, level of co-morbidity, the access to screening, the stage of diagnosis, and optimal and modern treatment, are the key explanations for the inequalities in cancer survival.

Lung cancer is the leading cause of death in Europe and in the United States, and approximately 60% of patients with lung cancer die within 1 year of diagnosis (WHO 2012). Lung cancer is extremely influenced by life style factors, such as smoking, environmental, and occupational exposure combined with socio-economic status, age, sex, and personal characteristics.

The nationwide cohort of lung cancer patients investigations exploring the associations between socio-economic position, stage at diagnosis, and length of period between referral period and diagnosis, confirmed that patients representing vulnerable groups (income, education) need special attention (Dalton et al. 2011). The incidence rate of lung cancer is strictly related with the socio-economic situation of individuals due to different smoking habits. The Danish studies showed the difference by SES in short time survival after lung cancer; 1-year survival was only 28% in men with low education and 34% in men with a higher education diploma. It was also noted the associations between socio-economic status and tumor progression (defined as an advanced stage lung cancer) and the length of the period between referral and diagnosis. Longer than recommended time between referral and diagnosis were noted for low income patients; both for patients with a diagnosis of early stage cancer and for patients with advanced stage lung cancer. Although the Danish health-care system is an example of an extremely friendly for citizens, tax-funded welfare system, which is based on free access to general practice, outpatient and hospital care. The other investigations confirmed that education status is connected with lung cancer incidence; low level of education is related with higher lung cancer incidence (Menvielle et al. 2009).

Several international studies illustrate the existing inequalities in cancer outcomes concerning incidence, morbidity, mortality, and survival, relating to an individual's socio-economic position. Cancer incidence is associated with socio-economic patterns and varies for specific cancers, being generally constant for a stage across cancers, but late-stage diagnoses are correlated with a lower socioeconomic position and there were observed constant gradients in incidence rates for such cancers as female breast, cervix, lung, prostate, and melanoma by family income and educational level (Clegg et al. 2009). Educational gradients in lung cancer was noted for both sexes, but especially for men, where men with a high school education had a cancer rate ratio significantly lower than men with less than a high school diploma. For breast and prostate cancers also, a positive correlation between higher education level and higher cancer incidence was noted, the same concerned colorectal and cervical cancers, risk for melanoma of skin cancer. British national studies reported strong associations between lung and cervical cancer and socioeconomic factors while the incidence of those cancers was the highest for the most vulnerable groups. The opposite relation concerned malignant melanoma and breast cancer, where the incidence of those both cancers was the highest in among the least vulnerable population (Shack et al. 2008). A multicenter 10-year retrospective clinical epidemiological study in China also confirmed that women in a low socio-economic position are diagnosed at later breast cancer stages than those in a high socio-economic situation (Wang et al. 2012).

Low socio-economic status is a prognostic factor for poor survival in patients with early and advanced stage of non-small cell lung cancer; patients with low education attainment and those living in high poverty neighborhoods have higher mortality (Erhunmwunsee et al. 2012). The other studies confirmed that the racial minorities in US have poorer prognostic of survival than Caucasians in this type of cancer because a large percentage of the African-American and Hispanic races are in low socio-economic position (Ou et al. 2008). A considerable positive relation between income and lung cancer incidence was also proved. Men with lower incomes represented reduced risk of prostate cancer compared to those with higher income and the same was observed for the melanoma of skin. Women with low incomes had higher cervical cancer incidence respectively than those with higher incomes.

Ethnic and racial disparities in incidence rates were observed for all cancers. Most findings indicate significantly increased rates of lung cancer related to unemployment (Clegg et al. 2009). People from a lower income group have a statistically higher risk of being diagnosed with a late stage prostate and breast cancer. The investigations among women in 26 states in US provided the evidence about the education disparities in the aspect of decline in cervical cancer mortality; the decline in mortality over the fifteen years was twice higher for those with highest education levels comparing to lower education and the risk of late stage diagnosis increased for uninsured women significantly versus privately insure. The main hypothesis of this investigation was that during one year even ¾ of cervical cancer deaths in the US may have been averted by eliminating socio-economic inequalities (Simard et al. 2012). A woman survival after a diagnosis of breast cancer is associated with socio-economic conditions (Dasgupta et al. 2011).

Cervical and breast cancer have both high morbidity and mortality rates in highly developed countries. The burden of these diseases could be reduced if more cases would be detected and treated early. Screening (mammography and pap smear) are known as effective preventive procedures in the early identification of pre-cancer or cancer before the symptoms are recognizable, and is associated with reduction in cancer mortality. The Italian investigations showed significant socio-economic inequalities in access to female cancer screening among Italian women (Damiani et al. 2012). Both, the determinants' occupation and education attainment were strongly related to the uptake of screening, and what is also important in this example is that the coverage of most expenditures for those services are financed from public sources by the Italian National Health System. This investigation also reported that among women who attended screening, those from the lower social class and lower education attainment declared more preferences to attend organized screenings than use this service of their own initiative. These results provide the information to create a more effective public policy to ensure more equal access to preventive care and for tackling socio-economic inequalities.

Colorectal cancer is the third most common cancer worldwide. Low socioeconomic position is among the key determinants for the development of colorectal cancer and investigations mostly define three main lifestyle health risk factors; obesity connected with unhealthful dietary patterns, smoking, and physical inactivity, which account for up to 70% of colorectal cancers. In highly developed countries, behavioral risk factors are much more common in a low socio-economic status population. Obesity affects molecular pathways involved in the carcinogenesis process. Smoking, diet, and physical inactivity together contribute in about almost 38% individuals to the excessive risk of colorectal cancer related with low educational attainment, and in 27%, to an increase incidence of colorectal cancer attributable to living in poor areas. The socio-economic disparities account for from 9% to 22% in colorectal cancer incidence (Doubeni et al. 2012a). A more in-depth prospective study of the same authors reported that the association between socio-economic position and colorectal cancer is statistically much more significant in the rectum and left colon cancers and much weaker in right colon cancers (Doubeni et al. 2012b). Among highly developed countries, we observed tremendous change in the social gradient in colorectal cancer mortality in the US. Till the late 70s, we noted a consistent difference in colorectal cancer mortality people living in neighborhoods of higher socio-economic status were at greater risk than people living in lower SES areas and from the beginning of 80s, this gradient began to reverse as people living in higher socio-economic position neighborhoods experienced higher reduction in cancer mortality than those in lower SES areas and what must be noted is that low-SES people were slower to experience decrease in colorectal cancer mortality than those with middle- and high-SES position after the introduction of health education programs and screening policies (Saldana-Ruiz et al. 2013). These result provide the evidence that development and better knowledge about prevention and treatment of disease became a significant factor in modifying mortality rates and are not distributed equally among the population in this country.

Similar results presenting existing social disparities and cancer mortality are represented in most highly developed countries, no matter their welfare health-care system. A ten year study in Barcelona identified the cancer types reflecting the greatest social inequalities over past years and provided the information that the highest educational disparities in women corresponded to cervix, uteri, liver, and colon cancer and men with a lower education level had higher mortality by lung cancer, larynx, esophagus, mouth, and pharynx cancer (Puigpinos et al. 2009). Spain has a National Health System that guarantees equal access to health services.

CVD and socio-environmental determinants

Cardiovascular diseases affects 4.1 million people and kills 170,000 every year and is responsible for a fifth of all hospital admissions (WHO 2009). There is strong evidence for the relation between heart diseases and socio-economic and environmental factors, among those unemployment is a key one (Weber and Lehnert 1997). The retrospective study of Medicare beneficiaries, patients hospitalized with acute myocardial infarction, heart failure, and pneumonia in the United States showed an existing association between exposure to income inequalities and increased risk of hospital readmission (Lindeneuer et al. 2013). A Dutch 12-year prospective population-based investigation reported that hostility and depressive symptoms are psychological risk factors for ischemic heart disease, and those symptoms are more common in deprived socio-economic groups (Klabbers et al. 2009). Another research showed that low income women represented higher risk for CVD, reflected by three risk factors: higher BMI, hypertension, hypercholesterolemia (Ahluwalia et al.

2009). High neuroticism is a risk factor for CVD mortality in women with low SEP (Hagger-Johnson et al. 2012). There is also strong evidence that a lower socio-economic position is significantly associated with hypertension prevalence (Kaplan et al. 2010). In recent years, the contribution of work related social and environmental conditions to the development and disparities in CVD has been reported. The investigations done in USA and Canada reported a strong inverse correlation between household income and the hypertension prevalence rate in the US and no such evidence was found in Canada. This is important information about the complex interactions between such determinants like social inequalities, primary prevention, and access to health care limitations.

Ambient air pollution increases morbidity and mortality in the general population. Air pollution is a complex mixture of particulate matter, gaseous, volatile and semi-volatile, and its composition varies. The composition in a single location depends on the time of the day, day of the week, meteorological condition, traffic density, and industrial activity. The contribution of exposure to air pollutants to CVD varies in different regions of world because of differences of the air pollutants' components and sources. Meteorological conditions and demographic and socio-economic patterns may also differ in developing countries from those in Europe and North America. The relation between ambient air pollution exposure and respiratory system diseases has been commonly acknowledged. Although most studies in recent years have reported the relations between long term exposure to particulate matter air pollution and CVD mortality, however, the development of CVD diseases resulting from contact with air pollutants is inconsistent and the pathways are poorly described. The mechanisms of these relations are very complex and likely to be multi-factorial. Some epidemiological studies have recently demonstrated the fundamental physiological mechanisms of acute and chronic influence of pollution exposure on the human cardiovascular system (Langrish et al. 2012). One of the mechanisms is vasoconstriction of the brachial artery succeeding the exposure to ambient particles in combination with ozone or to diluted diesel exhaust. In both cases, the exposure to particulate matter is related to increasing arterial vasoconstriction and arterial stiffness (Brook et al. 2010). The final response to airborne PM exposure is an increase in mean arterial blood pressure (two hours after exposure). Another research confirmed that exposure to particulate air pollution increases the risk of ST-segment depression in elderly patients with coronary heart disease (Chuang et al. 2008; Lanki et al. 2008).

Epidemiological evidence demonstrated that short term exposure to air pollution is related to an increase in hospital admissions concerning myocardial infarction, heart failure, and arrhythmias (Brook et al. 2010), and venous thromboembolic disease acute cardiovascular events, correlated with arterial thrombosis. The Brazilian research conducted in São Paulo state confirmed the association between environmental pollutants exposure and hospital admissions where exposure to particulate matter was strictly related to hospitalization for cardiovascular diseases three days after exposure (Nascimento 2011).

Another pathway is the endothelial mechanism. Increased coronary atherosclerosis following exposure to particulate air pollution may be by inflammatory responses connected with oxidative stress.

Three main mechanisms may be involved in the increase in cardiovascular diseases as a result of air pollution exposure: systemic inflammation and oxidative stress, short term exposure can increase blood pressure and can be the reason of arterial vasoconstriction, and finally those effects that influence the autonomic nervous system, which is connected with changes in heart rate variability. At the same time, blood became more prothrombic. The final result is the occlusion of the coronary artery and acute myocardial infarction.

In a retrospective cohort study in China, it was examined if exposure to ambient air pollution increases the risk for CVD and the relations between compound air pollutants PM10, sulfur dioxide, nitrogen dioxide, and mortality. The analysis was stratified by education and income. The findings confirmed that long-term exposure to ambient air pollution is related to CVD and cerebrovascular diseases mortality and that socio-economic conditions were also linked to them (Zhang et al. 2011). Another research in Spain reported that socio-economic deprivation was related to greater CVD mortality (IHD and stroke) and physical environment attributes are also associated with CVD mortality (Domingues-Berjon et al. 2010).

As was mentioned above, air pollution is indicated as a risk factor for CVD morbidity and mortality and could be responsible for about 3 million premature deaths globaly each year (WHO 2009). It is estimated that urban air pollution contributes to 5% of all cardiopulmonary deaths, which means around 100,000 premature deaths in Europe and 1.3 million deaths global each year (Hunter et al. 2012). Knowing that mechanisms of cardiovascular events are the results of air pollution exposure is important in helping to inform environmental health policy. Socioeconomic and physical environmental attributes must be taken into consideration in planning health policies improving cardiovascular health indicators of the general population.

Conclusions

The environmental justice and health justice in the context of socio-economic inequity have become more and more critical components in public health policy debates. The persistence of socio-environmental health disparities, even in highly developed countries, remains a significant public health problem. Recently, interest has been more concentrated on how socioeconomic factors and air pollution together contribute to population health inequalities. The epidemiological studies confirm that SES is an important factor in making environmental policy decisions. The lower the SES is, the higher are the PM10 concentrations and it is also highlighted that socioeconomic factors have an even higher effect on PM10 concentrations than human indoor activities. The reduction of health risk related to air pollution for a socially disadvantaged population should be a high priority in both environmental and public health policies.

The main purpose of health systems should be the reduction of inequalities in access to health services to provide the citizens health care based on health needs and not depending on socio-economic factors. Some solutions provide the proofs that health care can be determined basically by health needs, even among people living in areas with the worse socio-environmental indicators. The conceptual model of the determinants of health inequalities of the WHO Commission on Social Determinants of Health describes the influence of the main dimensions that affect inequalities in health, the majority of them being beyond the health sector (WHO 2008b). Following this model and considering the factors involved in the impact on equity in health and well-being, it is necessary to reinforce the need for considering the concept of "health in all policies". Aiming for economic growth in highly developed countries, it is important to remind public health professionals that environmental sustainability should also be a very serious and necessary societal aim. Exposure to environmental factors is for a large part beyond an individual's control and requires the action of public authorities at the international (EC, WHO), national (governments), regional, and local level to provide a more equitable distribution of health risk. The reduction of health risks attributable to air pollution for socio-economic disadvantaged populations should be a high priority in public health and environmental policies. Especially when implementing air pollution legal regulations and developing new air quality guidelines.

The negative relationships between social status and environmental health risk exposure are consistent across highly developed countries in Europe, North America, and Asia, but vary by welfare state regime, indicating that levels of social protection may have a moderating impact. Policy makers' attention therefore needs to be paid to more effective and multi-sector legislation and interventions, and especially the extent to which the welfare state is able to support the needs of SES deprived groups.

There were many programs as well in the US as in Europe against CVD and cancer. The Europe Against Cancer Program from 1985 is one of the examples. Its main objective was to achieve 15% decrease in cancer mortality by the year 2000. It concentrated the fight on the development of new technologies for prevention, diagnosis, and treatment. These new technologies should be applied equally and the reduction of socio-economic disparities in the burden of cancer make this task challenging, especially now, in the time of economic crisis and limited social policy, a few years after the admission of new member states with poorer socio-economic situations into the European Union. Policy aimed at decreasing socio-environmental disparities in health is a common demand and their implementation is necessary for governments to ensure sustainable development.

References

Adamkiewicz, G., A.R. Zota, M.P. Fabian, T. Chahine, R. Julien, J.D. Spengler and J.I. Levy. 2011. Moving environmental justice indoors: understanding structural influences on residential exposure patterns in low-income, communities. Am. J. Public Health 101: 238–245.

Ahluwalia, I.B., I. Tessaro, S. Rye and L. Parker. 2009. Self-reported and clinical measurement of three chronic disease risks among low-income women in West Virginia. J. Womens Health 8(11): 1857–62.

Akhtar, P.C., S.J. Haw, K.A. Levin, D.B. Currie, R. Zachary and C.E. Currie. 2010. Socioeconomic differences in second-hand smoke exposure among children in Scotland after introduction of the smoke-free legislation. J. Epidemiol. Community Health 64: 341–346.

Amy, L., S.S. Mudhasakul and W. Sriwatanapongse. 2009. The social distribution of neighborhood-scale air pollution and monitoring protection. J. Air & Waste Management Assoc. 59: 591–602.

Babones, S.J. 2008. Income inequality and population health: correlation and causality. Social Science & Medicine 66: 1614–1626.

Bambra, C.T. and A. Eikemo. 2009. Welfare state regimes, unemployment and health: a comparative study of the relationship between unemployment and self-reported health in 23 European countries. J. Epidemiol. Community Health 63: 92–98.

Bambra, C.T. 2010. Yesterday once more? Unemployment and health in the 21st century. J. Epidemiol. Community Health 64: 213–215.
Bartley, M. 2004. Health Inequality: An Introduction to Theories, Concepts and Methods. Polity Press, Cambridge, UK.
Basch, Ch. E. 2010. Healthier Students are Better Learners: A Missing Link in School Reforms to Close the Achievement Gap. A Research Initiative of the Campaign for Educational Equity. Teachers College, Columbia University, New York, USA.
Berkman, L.F. 2011. Unintended consequences of social and economic policies for population health: towards a more intentional approach. European Journal of Public Health 21(5): 547–549.
Bolte, G. and H. Fromme. 2008. Socioeconomic determinants of children's environmental tobacco smoke exposure and family's home smoking policy. Eur. J. Public Health 19(1): 52–58.
Bolte, G., G. Tamburlini and M. Kohlhuber. 2009. Environmental inequalities among children in Europe—evaluation of scientific evidence and policy implications. Eur. J. Public Health 20(1): 14–20.
Branis, M. and M. Linhartova. 2012. Association between unemployment, income, education level, population size and air pollution in Czech cities: evidence for environmental in equality? A pilot national scale analysis Health & Place 18: 1110–1114.
Brook, R.D., S. Rajagopalan and C.A. Pope. 2010. Particulate matter air pollution and cardiovascular disease. An update to the scientific statement from the American Heart Association. Circulation 121: 2331–2378.
Burra, T.A., R. Moineddin, M.M. Agha and R.H. Glazier. 2009. Social disadvantage, air pollution, and asthma physician visits in Toronto, Canada. Environ. Res. 109: 567–574.
Byun, H., H. Bae, D. Kim, H. Shin and Ch. Yoon. 2010. Effects of socioeconomic factors and human activities on children's PM10 exposure in inner-city households in Korea. Int. Arch. Occup. Environ. Health 83: 867–878.
Chen, E., R.C. Strunk, A. Trethewey, H.M.C. Schreier, N. Maharaj and G.E. Miller. 2011. Resilience in low socioeconomic status children with asthma: adaptations to stress. J. Allergy Clin. Immunol. 128: 970–976.
Chuang, K.J., B.A. Coull, A. Zanobetti, H. Suh, J. Schwartz, P.H. Stone, A. Litonujua, F.E. Speizer and D.R. Gold. 2008. Particulate air pollution as a risk factor for ST-segment depression in patients with coronary artery disease. Circulation 118: 1314–1320.
Clegg, L.X., M.E. Reichman, B.A. Miller, B.F. Hankey, G.K. Singh, Y.D. Lin, M.T. Goodman, Ch. F. Lynch, S.M. Schwartz, V.W. Ch. L. Bernstein, S.L. Gomez, J.J. Graff, Ch C. Lin, N.J. Johnson and B.K. Edwards. 2009. Impact of socioeconomic status on cancer incidence and stage at diagnosis: selected findings from the surveillance, epidemiology, and end results: National Longitudinal Mortality Study. Cancer Causes Control. 20(4): 417–435.
Curtis, L.M., M.S. Wolf, K.B. Weiss and L.C. Grammer. 2012. The impact of health literacy and socioeconomic status on asthma disparities. Asthma 49(2): 178–183.
Dahlgren, G. and M. Whitehead. 1991. Policies and Strategies to Promote Social Equity in Health. Institute for Future Studies, Stockholm, Sweden.
Dalton, S.O., B.L. Frederiksen, E. Jacobsen, M. Steding-Jessen, K. Østerlind, J. Schuz, M. Osler and C. Johansen. 2011. Socioeconomic position, stage of lung cancer and time between referral and diagnosis in Denmark, 2001–2008. British Journal of Cancer 105: 1042–1048.
Damiani, G., B. Federico, D. Basso, A. Ronconi, C.B.N.A. Bianchi, G.M. Anzellotti, G. Nasi, F. Sassi and W. Ricciardi. 2012. Socioeconomic disparities in the uptake of breast and cervical cancer screening in Italy: a cross sectional study. BMC Public Health 12: 99.
Dasgupta, P., P.D. Baade, J.F. Aitken and G. Turrell. 2012. Multilevel determinants of breast cancer survival: association with geographic remoteness and area-level socioeconomic disadvantage. Breast Cancer Res. Treat. 132: 701–710.
De Backer, G., E. Ambrosioni, K. Borch-Johnsen, C. Brotons, R. Cifkova, J. Dallongeville, S. Ebrahim, O. Faergeman, I. Graham, G. Mancia, C. Manger, K. Orth-Gomér, J. Perk, K. Pyörälä, J.L. Rodicio, S. Sans, V. Sansoy, U. Sechtem, S. Silber, T. Thomsen and D. Wood. 2003. Third joint task force of European and other societies on cardiovascular disease prevention in clinical practice. Eur. Heart J. 24(17): 1601–10.
Deguen, S. and D. Zmirou-Navier. 2010. Social inequalities resulting from health risks related to ambient air quality—a European review. Eur. J. Public Health 20(1): 27–35.

Domınguez-Berjon, M.F., A. Gandarillas, J. Segura del Pozo, B. Zorrilla, M.J. Soto, L. Lopez, I. Duque, M. Marta and I. Abad. 2010. Census tract socioeconomic and physical environment and cardiovascular mortality in the region of Madrid (Spain). J. Epidemiol. Community Health 64: 1086–1093.

Doubeni, Ch. A., A.O. Laiyemo, J.M. Major, M. Schootman, M. Lian, Y. Park, B.I. Graubard, A.R. Hollenbeck and R. Sinha. 2012a. Socioeconomic status and the risk of colorectal cancer. Cancer 118: 3636–44.

Doubeni, Ch. A., J.M. Major, A.O. Laiyemo, M. Schootman, A.G. Zauber, A.R. Hollenbeck, R. Sinha and J. Allison. 2012b. Contribution of behavioral risk factors and obesity to socioeconomic differences in colorectal cancer incidence. J. Natl. Cancer Inst. 104: 1353–1362.

Erhunmwunsee, L., M.B.M. Joshi, D.H. Conlon and D.H. Harpole. 2012. Neighborhood-level socioeconomic determinants impact outcomes in nonsmall cell lung cancer patients in the southeastern United States. Cancer 118: 5117–5123.

Eurostat. 2012. Mortality Rate in Europe. Luxembourg, http://www.epp.eurostat.ec.europa.eu/portal/page/portal/eurostat/home (last accessed 1 October 2013).

Evans, G.W. and E. Kantrowitz. 2002. Socioeconomic status and health: the potential role of environmental risk exposure. An. Rev. Public Health 23: 303–331.

Gallo, W., H. Teng, T. Falba, S. Kasl, H. Krumholz and E. Bradley. 2006. The impact of late career job loss on myocardial infarction and stroke: a 10 year follow up using the health and retirement survey. Occupational Environment Medicine 63: 683–687.

Galobardes, B., D. Shaw, A. Lawlor, J.W. Lynch and G.D. Smith. 2006. Indicators of socio-economic position. J. Epidemiol. Comm. Health 60: 7–12.

Goodman, A., P. Wilkinson, M. Stafford and C. Tonne. 2011. Characterizing socio-economic inequalities in exposure to air pollution: A comparison of socio-economic markers and scales of measurement. Health & Place 17: 767–774.

Hagger-Johnson, G., B. Roberts, D. Boniface, S. Sabia, G.D. Batty, A. Elbaz, A. Singh-Manoux and I.J. Deary. 2012. Neuroticism and cardiovascular disease mortality: socioeconomic status modifies the risk in women (UK Health and Lifestyle Survey). Psychosom. Med. 74(6): 596–603.

Havard, S., S. Deguen, D. Zmirou-Navier, Ch. Schillinger and D. Barda. 2009. Traffic-related air pollution and socioeconomic status. a spatial autocorrelation study to assess environmental equity on a small-area scale. Epidemiology 20: 223–230.

Hildebrand, V. and P. Van Kerm. 2009. Income inequality and self-related health status: evidence from the European Community Household Panel. Demography 46(4): 805–25.

Huijts, T., T.A. Eikemo and V. Skalická. 2010. Income-related health inequalities in the Nordic countries: examining the role of education, occupational class, and age. Soc. Sci. Med. 71(11): 1964–72.

Hunter, A.L., N.L. Mills and D.E. Newby. 2012. Combustion-derived air pollution and cardiovascular disease. British Journal of Hospital Medicine 73(9): 492–497.

Jarvis, M. and J. Wardle. 1999. Social patterning of individual health behaviours: the case of cigarette smoking. pp. 240–255. *In*: M. Marmot and R. Wilkinson [eds.]. Social Determinants of Health. Oxford University Press, New York, USA.

Jin, R., C.P. Shah and T.J. Svoboda. 1997. The impact of unemployment on health: A review of the evidence. J. Public Health Policy 18(3): 275–301.

Kan, H., S.J. London, G. Chen, Y. Zhang, G. Song, N. Zhao, L. Jiang and B.C. Season. 2008. Sex, age, and education as modifiers of the effects of outdoor air pollution on daily mortality in Shanghai, China: The Public Health and Air Pollution in Asia (PAPA) Study. Environ. Health Perspect. 116: 1183–1188.

Kaplan, M.S., N. Huguet, D.H. Feeny and B.H. McFarland. 2010. Self-reported hypertension prevalence and income among older adults in Canada and the United States. Soc. Sci. Med. 70(6): 844–9.

Kennelly, B., E. O'Shea and E. Garvey. 2003. Social capital, life expectancy and mortality: a cross national examination. Social Science & Medicine 56: 2367–2377.

Klabbers, G., H. Bosma, F.J. Van Lenthe, G.I. Kempen, J.T. Van Eijk and J.P. Mackenbach. 2009. The relative contributions of hostility and depressive symptoms to the income gradient in hospital-based incidence of ischaemic heart disease: 12-year follow-up findings from the GLOBE study. Soc. Sci. Med. 69(8): 1272–80.

Koong, H.N., D. Khoo, Ch. Higbee, M. Travers, A. Hyland and K.M.C. Dresler. 2009. Global air monitoring study: a multi-country comparison of levels of indoor air pollution in different workplaces. Ann. Acad. Med. Singapore 38: 202–6.

Kozyrskyj, A.L., G.E. Kendall, P. Jacoby, P.D. Sly and S.R. Zubrick. 2010. Association between socioeconomic status and the development of asthma: analyses of income trajectories. Am. J. Public Health 100: 540–546.

Laaksonen, M., R. Prättälä, V. Helasoja, A. Uutela and E. Lahelma. 2003. Income and health behaviours. Evidence from monitoring surveys among Finnish adults. J. Epidemiol. Commun. Health 57: 711–717.

Laaksonen, M., O. Rahkonen, S. Lahelma and E. Karvonen. 2005. Socioeconomic status and smoking. Analysing inequalities with multiple indicators. Eur. J. Public Health 15(3): 262–269.

Lallukka, T., E. Lahelma, O. Rahkonen, E. Roos, E. Laaksonen, P. Martikainen, J. Head, E. Brunner, A. Mosdol, M. Marmot, M. Sekine, A. Nasermoaddeli and S. Kagamimori. 2008. Associations of job strain and working overtime with adverse health behaviours and obesity: evidence from the Whitehall II study, Helsinki health study, and the Japanese civil servants study. Social Science and Medicine 66: 1681–1698.

Langrish, J.P., J. Bosson, J. Unosson, A. Muala, D.E. Newby, N.L. Mills, A. Blomberg and T. Sandstrom. 2012. Cardiovascular effects of particulate air pollution exposure: time course and underlying mechanisms. J. Intern. Med. 272: 224–239.

Lanki, T., G. Hoek, K.L. Timonen, A. Peters, P. Tittanen, E. Vaninen and J. Pekkanen. 2008. Variation in fine particle exposure is associated with transiently increased risk of ST segment depression. Occup. Environ. Med. 65: 782–786.

Lara-Valencia, F., G. Álvarez-Hernández, S.D. Harlow, C. Denman and H. García-Pérez. 2012. Neighborhood socio-environmental vulnerability and infant mortality in Hermosillo, Sonora. Salud. Publica Mex. 54(4): 367–74.

Li, F., Y. Zhou, S. Li, F. Jiang, X. Jin, C. Yan, Y. Tian, Y. Zhang, S. Tong and X. Shen. 2010. Prevalence and risk factors of childhood allergic diseases in eight metropolitan cities in China: a multicenter study. BMC Public Health 6;11: 437.

Lindenauer, P.K., T. Lagu, M.B. Rothberg, J. Avrunin, P.S. Pekow, Y. Wang and H.M. Krumholz. 2013. Income inequality and 30 day outcomes after acute myocardial infarction, heart failure, and pneumonia: retrospective cohort study. BMJ 14: 346: 21.

Lindström, Ch. and M. Lindström. 2006. Social capital, GNP per capita, relative income, and health; an ecological study of 23 countries. International Journal of Health Services 36: 679–696.

Lynch, J. and G. Kaplan. 2000. Socioeconomic position. pp. 13–35. *In*: L.F. Berkman and I. Kawachi [eds.]. Social Epidemiology. Oxford University Press, New York, USA.

Mackenbach, J.P., V. Bos, O. Adersen, M. Cardano, G. Costa, S. Harding, A. Reid, O. Hemstrom, T. Valkonen and A.E. Kunst. 2003. Widening socioeconomic inequalities in mortality in six Western European countries. International Journal of Epidemiology 32: 830–837.

Maier, R., A. Egger, A. Barth, R. Winker, W. Osterode, M. Kundi, C. Wolf and H. Ruediger. 2006. Effects of short- and long-term unemployment on physical work capacity and on serum cortisol. International Archives of Occupational and Environmental Health 79(3): 193–8.

Marmot, R. 2010. Fair Society, Healthy Lives. The Marmot Review strategic review of health inequalities in England, http://www.instituteofhealthequity.org/projects/fair-society-healthy-lives-the-marmot-review (10.10.2013).

McIntosh, C.N., P. Finès, R. Wilkins and M.C. Wolfson. 2009. Income disparities in Health adjusted life expectancy for Canadian adults, 1991 to 2001. Health Rep. 20(4): 55–64.

Menvielle, G., H. Boshuizen, A.E. Kunst, S.O. Dalton, P. Vineis, M.M. Bergmann, S. Hermann, P. Ferrari, O. Raaschou-Nielsen, A. Tjñneland, R. Kaaks, J. Linseisen, M. Kosti, A. Trichopoulou, V. Dilis, D. Palli, V. Krogh, S. Panico, R. Tumino, F.L. Büchner, C.H. van Gils, P.H.M. Peeters, T. Braaten, I.T. Gram, E. Lund, L. Rodriguez, A. Agudo, M.J. Sánchez, M.J. Tormo, E. Ardanaz, J. Manjer, E. Wirfält, G. Hallmans, T. Rasmuson, S. Bingham, K.T. Khaw, N. Allen, T. Key, P. Boffetta, E.J. Duell, N. Slimani, V. Gallo, E. Riboli and H.B. Bueno-de-Mesquita. 2009. The role of smoking and diet in explaining educational inequalities in lung cancer incidence. J. Natl. Cancer Inst. 101: 321–330.

Mohai, P., P.M. Lantz, J. Morenoff, J.S. House and R.P. Mero. 2009. Racial and socioeconomic disparities in residential proximity to polluting industrial facilities: evidence from the Americans' changing lives study. Am. J. Public Health 99: 649–656.

Nascimento, L.F.C. 2011. Air pollution and cardiovascular hospital admissions in a medium-sized city in São Paulo State, Brazil. Braz. J. Med. Biol. Res. 44(7): 720–724.

Okhee, Y., H.J. Kwon, D. Kim, H. Kim, M. Ha, S.J. Hong, Y.C. Hong, J.H. Leem, J. Sakong, C.G. Lee, S.Y. Kim and D. Kang. 2012. Association between environmental tobacco smoke exposure of children

and parental socioeconomic status: a cross-sectional study in Korea. Nicotine and Tobacco Research 14(5): 607–615.
O'Neill, M.S., M.L. Bell, N. Ranjit, L.A. Cifuentes, D. Loomis, N. Gouveia and V.H. Borja-Aburtof. 2008. Air pollution and mortality in Latin America. The role of education. Epidemiology 19: 810–819.
Ou, I., J.A. Zell, A. Ziogas and H.A. Culver. 2008. Low socioeconomic status is a poor prognostic factor for survival in stage i nonsmall cell lung cancer and is independent of surgical treatment, race and marital status. Cancer 112(9): 2011–2020.
Ou, Ch. Q., A.J. Hedley, R.Y. Chung, T.Q. Thach, Y.K. Chau, K.P. Chan, L. Yang, S.Y. Ho, Ch. M. Wong and T.H. Lam. 2008. Socioeconomic disparities in air pollution-associated mortality. Environ. Res. 107: 237–244.
Post, E.S., A. Belova and J. Huang. 2011. Distributional benefit analysis of a national air quality rule. Int. J. Environ. Res. Public Health 8: 1872–1892.
Puigpinós, R., C. Borrell, J.L.F. Antunes, E. Azlor, M.I. Pasarín, G. Serral, M. Pons-Vigués, M. Rodríguez-Sanz1 and E. Fernández. 2009. Trends in socioeconomic inequalities in cancer mortality in Barcelona: 1992–2003. BMC Public Health 9: 35.
Qi, X., S. Tong and W. Hu. 2009. Preliminary spatiotemporal analysis of the association between socio-environmental factors and suicide. Environ. Health 8: 46–58.
Saldana-Ruiz, N.S., S.A.P. Clouston, M.S. Rubin, C.G. Colen and B.G. Link. 2013. Fundamental causes of colorectal cancer mortality in the United States: Understanding the importance of socioeconomic status in creating inequality in mortality. Am. J. Public Health 103(1): 99–104.
Shack, L., C. Jordan, C.S. Thomson, V. Mak and H. Møller. 2008. Variation in incidence of breast, lung and cervical cancer and malignant melanoma of skin by socioeconomic group in England. BMC Cancer 8: 271.
Sigmundová, D., W.E. Ansari and E. Sigmund. 2011. Neighbourhood environment correlates of physical activity: a study of eight Czech regional towns. Int. J. Environ. Res. Public Health 8(2): 341–357.
Simard, E.P., S. Fedewa, J. Ma, R. Siegel and A. Jemal. 2012. Widening socioeconomic disparities in cervical cancer mortality among women in 26 States, 1993–2007. Cancer 118: 5110–6.
Stronks, K., H.D. van de Mheen, C.W. Looman and J.P. Mackenbach. 1997. Cultural, material, and psychosocial correlates of the socio-economic gradient in smoking behavior among adults. Prev. Med. 26: 754–66.
Stuckler, D., S. Basu, M. Suhrcke, M. Coutts and M. McKee. 2009. The public health effect of economic crisis and alternative policy responses in Europe: an empirical analysis. The Lancet 374(9686): 315–323.
Thomas, C., M. Benzeval and S. Stansfeld. 2005. Employment transitions and mental health: an analysis from the British household panel survey. J. Epidem. Com. Health 59: 243–249.
Voss, M., L. Nylén, B. Floderus, F. Diderichsen and P.D. Terry. 2004. Unemployment and early cause-specific mortality: a study based on the Swedish Twin Registry. American Journal of Public Health 94(12): 2155–2161.
Wang, Q., J. Li, S. Zheng, J.Y. Li, Yi Pang, R. Huang, B.N. Zhang, B. Zhang, H.J. Yang, X.-M. Xie, Z.H. Tang, H. Li, J.J. He, J.H. Fan and Y.L. Qiao. 2012. Breast cancer stage at diagnosis and area-based socioeconomic status: a multicenter 10-year retrospective clinical epidemiological study in China. BMC Cancer 12: 122.
Weber, A. and G. Lehnert. 1997. Unemployment and cardiovascular diseases: a causal relationship? Int. Arch. Occup. Environ. Health 70(3): 153–60.
Weidner, G. and V.S. Cain. 2003. The gender gap in heart disease: lessons from Eastern Europe. Am. J. Public Health 93: 768–770.
WHO. 2008a. Action plan of the Global Alliance Against Chronic Respiratory Diseases 2008–2013. Geneva, Switzerland.
WHO. 2008b. Commission on Social Determinants of Health. SDH Final Report: Closing the gap in a generation: health equity through action on the social determinants of health. Geneva, Switzerland.
WHO. 2009. Global Health Risks. Mortality and Burden of Disease Attributable to Selected Major Risks. Geneva, Switzerland.
WHO. 2011. WHO Report on the Global Tobacco Epidemic. Geneva, Switzerland.
WHO. 2012. WHO Global Report: Mortality Attributable to Tobacco. Geneva, Switzerland.
WHO. 2013a. Review of Evidence on Health Aspects of Air Pollution—REVIHAAP project. WHO Regional Office for Europe. Copenhagen, Denmark.

WHO. 2013b. Health risks of air pollution in Europe—HRAPIE project. Recommendations for concentration–response functions for cost–benefit analysis of particulate matter, ozone and nitrogen dioxide. Copenhagen, Denmark.

WHO. 2013c. WHO report on the global tobacco epidemic. Geneva, Switzerland.

Wong, Ch. M., Ch. Q. Ou, K.P. Chan, Y.K. Chau, T.Q. Thach, L. Yang, Y.N. Roger, G. Chung, N. Thomas, J. Sriyal, M. Peiris, T.W. Wong, A.J. Hedley and T.H. Lam. 2008. The effects of air pollution on mortality in socially deprived urban areas in Hong Kong, China. Environ. Health Perspect. 116: 1189–1194.

Yi, O., H.J. Kwon, D. Kim, H. Kim, M. Ha, S.J. Hong, Y. Ch. Hong, J.H. Leem, J. Sakong, M.Ch. G. Lee, S.Y. Kim and D. Kang. 2012. Association between environmental tobacco smoke exposure of children and parental socioeconomic status: a cross-sectional study in Korea. Nicotine and Tobacco Research 14(5): 607–615.

Young, G.S., M.A. Fox, M. Trush, N. Kanarek, T.A. Glass and F.C. Curriero. 2012. Differential exposure to hazardous air pollution in the United States: a multilevel analysis of urbanization and neighborhood socioeconomic deprivation. Int. J. Environ. Res. Public Health 9: 2204–2225.

Zhang, P., G. Dong, B. Sun, L. Zhang, X. Chen, N. Ma, F. Yu, H. Guo, H. Huang, Y.L. Lee, N. Tang and J. Chen. 2011. Long-term exposure to ambient air pollution and mortality due to cardiovascular disease and cerebrovascular disease in Shenyang, China. Plos ONE 6(6): 20827.

Zota, A., G. Adamkiewicz, J.I. Levy and J.D. Spengler. 2005. Ventilation in public housing: implications for indoor nitrogen dioxide concentrations. Indoor Air 15(6): 393–401.

Summary: Documented Facts, Hypothesis, Speculations, and Final Conclusions*

Various chemicals as well as biological particles are emitted into the air, not only from natural but also from man-made (anthropogenic) sources. Knowledge of the nature, quantity, physicochemical (or biological) behavior, and effect of air pollutants on the exposed human population has greatly increased in recent decades. Nevertheless, more needs to be known. The reader can find in this book brief information about the key properties of air pollutants, their environmental distributions, and the nature of the health hazards related with exposure to these pollutants, as well as the attendant problems for science and society. Many of the recent research methods, especially in such matters as exposure assessment and the handling of interactive effects, were discussed in detail.

Well-documented facts and conclusions

Gaseous pollutants

In Chapter 1, the review of main gaseous air pollutants, made by Professor Lucyna Falkowska, indicates that gaseous pollutants of the atmosphere can be primary or secondary. The former are directly emitted from biological, geogenic, and anthropogenic sources, and include:

- carbon compounds—CO, CO_2, CH_4, VOC;
- nitrogen compounds—NO, N_2O, NH_3;
- sulphur compounds—H_2S, SO_2, $(CH_3)_2S$;
- mercury compounds—Hg^0, CH_3Hg.

*Prepared by Jozef S. Pastuszka by using the contents of all the chapters.

Secondary pollutants are not emitted directly from land and ocean sources, but are formed in the atmosphere out of primary pollutants—their precursors. Major secondary pollutants include:

- ozone—created during photochemical reactions involving NO_x, CO, VOC,
- NO_2 and HNO_3—formed from NO,
- both inorganic and organic aerosols which are formed as a result of gas-to-particle reactions, and which contain primary gaseous pollutants and water vapor.

The influence of air pollutants on human health is often very complicated. For example, human organisms can be indirectly affected by air pollutants deposited in plants, animals, and other environmental media, resulting from the chemicals entering the food chain or being present in drinking water, and thereby constituting additional sources of human exposure. However, the subject of this book was a discussion only about the direct consequences for health by the pulmonary deposition and absorption of inhaled chemicals.

Primary gaseous pollutants can cause adverse effects in the fauna and flora as well as in people. For instance, when leaves are exposed to 10 ppm NO, the process of photosynthesis is impaired. SO_2 damages both plants' respiratory pores and the protective layer of wax which covers needles. Sulphur compounds are responsible for the creation of reductive smog, which can be lethal to humans. Respiration is the main way of introducing noxious gases into the system and the inhalation of CO reduces the amount of oxygen in the bloodstream. High concentration of this gas can lead to headaches, dizziness, unconsciousness, and death. Generally, CO is a toxic gas which can cause fatal asphyxiation.

Concerning some examples of the secondary pollutants, the toxicity of NO_2 consists in causing oxygen deficiency in the body, which in turn results in reduced immunity to bacterial infections. This compound causes irritation of the eyes and the airways, resulting in breathing disorders. It also triggers allergies, including asthma (especially in children living in smog-ridden cities). Prolonged exposure to nitrogen dioxide may lead to biochemical, immunological, and morphological changes in animals.

Ozone influences human health in quite a different way. First of all, it is a greenhouse gas and absorbs UV radiation, particularly in the upper part of the troposphere where positive radiative forcing amounts to between 0.2 and 0.35 W m^{-2}. This can account for 10–25% of radiative forcing caused by carbon dioxide. Secondly, tropospheric ozone participates in the production of hydroxyl radicals. Being a potent oxidant, ozone contributes to the formation of another extremely powerful oxidant, which is predominant in reactions with methane and other gases.

Sulphur dioxide in the air is a pollutant which can be also used for testing of general air quality. In chemical transformations in the atmosphere, all reduced forms of sulphur of both natural and anthropogenic origin undergo oxidation to one gaseous compound—sulphur dioxide. Of course, the presence of sulphur dioxide in the air is also harmful to humans. It causes breathing difficulties, which particularly affect individuals suffering from chronic diseases of the respiratory system, such as asthma.

Common knowledge about the health implications of the exposure to volatile organic compounds (VOC) has been significantly increased during the last two decades but it should be noted that monitoring of VOC only from organic waste material, led to the detection of 155 volatile compounds. Among them, benzene, toluene, ethyl benzene and naphthalene have been identified, all of which are classified as major pollutants due to their harmful influence on human health. Benzene in particular possesses carcinogenic qualities.

Generally, health effects of the exposure to hydrocarbons are difficult to prognoses due to the huge number of these compounds which can be detected in the air. On the other hand, there are the single pollutants which are still intensively studied in the context of their toxicology. One such example is mercury. The health consequences of the exposure to metallic mercury fumes depend not only on the absorbed dose of this element but also on the age and the health condition of the exposed individuals. Accumulation in the brain tissue leads to the occurrence of fits, emotional instability, insomnia, memory loss, fatigue, headaches, polyneuropathy, cognitive dysfunction, psychomotoric disorders, hearing difficulties, and impaired vision, slower conduction of nerve impulses, concentration disorders, false electromyography (EMG) results and paraesthesia. Motoric disorders are usually reversible after terminating the exposal, but cognitive dysfunction (mostly memory deficits), can last longer or be permanent in the case of chronic occupational exposures. Acrodynia is a syndrome of symptoms, occurring mostly in children and is characterized by pain of the skin covering the limbs, the redness of limbs and the nose, excessive perspiration, and sometimes gastro-intestinal symptoms. It can be combined with anorexia and increased sensitivity to light. Inhalation of a high dose of mercury over a few days leads to the inflammation of the oral cavity, the intestines, the bronchial tubes, the bronchiole, and the lungs. Besides, it should be remembered that kidneys are particularly sensitive to the influence of metallic mercury.

Atmospheric aerosol: Emission sources, concentration levels, and routes of migration

In Chapter 2, Dr. Kikuo Okada presented the state of the art on the atmospheric aerosol, describing their sources, levels, and routes of transport in the environment. Since there are many aerosol sources, atmospheric aerosol particles composed of various materials are present in a wide size range.

Analyzing the chemical composition of fine particles with an aerodynamic diameter less than 1 μm and collected near the surface, Dr. Okada pointed out in Chapter 2 (using the review paper of Heintzenberg) that the existing data reflect a long residence time of submicrometer particles and the proximity to industrial regions. Consequently, no drastic changes in fine particle composition occur when moving away from the urbanized areas. In the "urban" and "non-urban continental" areas, 2/3 of the fine particle mass is composed of sulfates and carbon compounds and about 15% is nitrogen compounds. It should be noted that 1/4 to 1/6 of the carbonaceous material is elemental carbon (EC). It is this minor component that is responsible for most of the absorption of solar radiation. The data sets from remote regions show that even there sulfate compounds are a major component. About 1/5 of the fine particle mass is sulfur compounds. Organic carbon is the next most important component comprising about 11% of these particles. Elemental carbon, which was the most clearly identifiable anthropogenic component, comprises only about 0.3% corresponding to an average concentration of 10 ng m^{-3}.

Most of the existing studies of the chemical composition of aerosol particles in the urban atmosphere used the size-segregated bulk samples. These studies have supplied information on particle composition as a function of size. However, the mixing properties of aerosol particles cannot be evaluated by the analysis of the bulk samples alone because atmospheric aerosols consist of many particles with different compositions even in a narrow size range. This was the reason why some decades ago the mixing properties of aerosols have been studied by size-segregated samples using hygroscopic growth. Actually, single particle analysis is a useful method for evaluating mixed properties of aerosol particles. More explicit measurements of the mixing properties of individual particles with respect to the degree of internal and external mixtures of water-soluble and -insoluble material were carried out by dialysis with water. Okada measured the mixed state of individual particles with radii between 0.03 and 0.35 μm in the urban atmospheres of Yokkaichi and Nagoya in Japan using a dialysis method for the extraction of water-soluble material by electron microscopy and found that more than 80% of the aerosol particles were hygroscopic. The results also showed that 34% of the hygroscopic particles in the Aitken size range (0.03–0.1 μm radius) and 67% in the large size range (0.1–0.35 μm radius) were mixed particles. Differences in the mixed properties were found to be associated with the formation process of aerosol particles by gas-to-particle conversion.

Reading Chapter 2, it is easy to understand that aerosol particles influence the atmospheric processes such as cloud formation and atmospheric radiation and then modify the earth's climate, and that they also cause adverse health effects.

PAHs and heavy metals in ambient air

Despite their relatively small mass share in particulate matter (PM), trace elements and polycyclic aromatic hydrocarbons (PAH), being by weight a tiny fraction of the total organic matter content of PM, are the most intensely investigated PM components. The available information on this subject was reviewed by Dr. Wioletta Rogula-Kozłowska in Chapter 3.

The elemental analysis of PM characterizes partly the PM chemical composition; it usually suffices for understanding the temporal and spatial relationship between the source and the receptors of PM, and for the assessment of the efficiency of emission abatement methods. The hazard from PM-bound PAH and heavy metals to humans depends on the PAH or metal ambient concentrations, their mass distribution in respect to particle size, and PM physicochemical properties; the health condition and habitat of the population may enhance or suppress the PM toxicity.

All heavy metals in PM are considered toxic. They accumulate in body tissues (bones, kidneys, brain). Exposure to their salts or oxides can cause acute or chronic poisonings, tumors, diseases of the cardiovascular and nervous systems, and of kidneys. Some heavy metals can weaken the immune system in humans. Some heavy metals are not bioavailable in their elemental forms. Recent studies show that PM-bound heavy metals are bioavailable; the majority occurs in well water-soluble compounds.

Polycyclic aromatic hydrocarbons (PAHs) are other components of PM, extremely different from heavy metals and equally important. In the atmospheric air, they can occur in solid, vapor, and gas phases; the phases usually differ in their toxicity. The lightest, two- or three-ring PAHs can only be vaporous in the air, while four-ring PAHs can change their phase depending on weather conditions. Five- or more ring PAHs occur only in the PM-bound phase.

According to the International Agency for Research on Cancer (IARC), 48 PAHs can be human or animal carcinogens. They are the first air pollutants that have been identified as carcinogens. The strength of their carcinogenicity grows with their molecular weight. After entering the human body, PAHs are transported by blood to various organs, where they are metabolized. The metabolites, bound covalently to DNA or RNA, cause neoplasm and affect replications, transcriptions, and biosynthesis of proteins. All PAHs can accumulate in the body tissues.

For a long time, airborne BaP, or rather its concentration has been used to indicate the health hazard from PAHs. However, some PAHs being more carcinogenic than BaP have been identified recently, such as dibenzo(a,h)anthracene and dibenzo(a,l)pyrene, which are at least five times more carcinogenic than BaP, and which can probably represent the carcinogenicity of PAHs mixtures better. Unfortunately, they are not as well-studied as BaP, especially PM10-bound BaP, which is routinely monitored in many countries (the most common limit for its yearly concentrations is 1 ng m^{-3}). Therefore, before they replace BaP as the new hazard indicators, they will need to be intensively investigated first.

Adverse health effects caused by the exposure to airborne particles

To better understand the inhalation of airborne particles, the anatomy and physiology of the respiratory system as well as the deposition of particles in the airways were briefly discussed in Chapter 4 by Professor Renata Złotkowska, M.D., D.Sc. Next, the toxicological characteristics of airborne particles were summarized. However, the first remark is that the identification of specific toxicological effects of particulate matter is very difficult because particulate matter is a mixture of particles, different in terms of chemical and physical properties, origin, mass, number, and shape.

The toxicological properties of particles depend on their size and chemical composition. Particles exceeding the size of 10 μm are removed from the upper part of the respiratory system by the mucocilliary and other defense mechanisms, as described in detail in Chapter 4. Particles of sizes below 10 μm penetrate into the respiratory system. The chemical properties affect the toxicological potential of causing health effects of exposure to the specific substance and they depend on the source of pollution.

In the last part of Chapter 4, there is an interesting discussion about nanoparticles being ultrafine particles of the dimension of up to 100 nm. The main factor characterizing nanoparticles and affecting the potential toxicity is the high ratio of area size to mass, which is a result of reduced size. This special property of nanoparticles may increase the ability to catalyze the chemical reactions. Thus, due to their changed structure and small size, nanoparticles may have greater reactivity than other particles with identical chemical composition. Another factor affecting toxicity of nanoparticles is their chemical composition. Transition metals occurring in the form of nanoparticles induce the oxidative stress process and the free radicals which could cause cell damage. It is also very important that there are three pathways for the nanoparticles to enter the human body: inhalation, skin absorption, and gastrointestinal tract absorption.

Fibrous particles

Among different particles, those with aspect ratio (length/diameter) of at least three are called "fibers". However, in addition to isolated fibers, assemblages of spherical particles and fibers frequently occur in air samples. Groupings of fibers and spherical particles, referred to as "structures" (for example, "asbestos structures"), are defined as fibrous bundles, clusters, and matrices. Knowledge about the emission of fibers, their atmospheric and indoor transport, and penetration from outdoor into indoor environment, as well as, deposition and resuspension is still very poor. Characteristics of fibres and health implications of exposure to this kind of airborne particle is reviewed in Chapter 5. There are various diseases, including lung cancer, which can be caused by the inhalation of airborne fibrous dust clouds. Inhaled asbestos fibers have been especially shown to induce fibrosis, lung cancer, mesothelioma, and probably other kinds of intestinal cancer.

The fundamental property of fiber toxicity is that, in contrast to chemicals, fibers are believed to cause disease through a physical interaction, which means that the health effects depend not only on the type of fiber but also upon its diameter and length. It is also possible that the physical form of a fiber is even more important than its chemical composition.

World Health Organization officially recognized that asbestos is a proven human carcinogen. No safe level can be proposed for asbestos because a threshold is not known to exist. Exposure should therefore be kept as low as possible.

To reach a lung, the aerodynamic diameter of inhaled fiber should be less than 3.5 μm (such particles are respirable), which means that the fiber should be thin. On the other hand, such carcinogenic fibers must be long. The reason is that the specific cells —macrophages, living in the lung, biochemically destroy various particles, including fibers, present in the lung and only large/long particles can remain (partially) there for a period of time long enough to generate the cascade of processes leading to cancer. Typically, it is assumed that the most hazardous fibers are those longer than 5 μm and of diameter up to 3–4 μm.

In Chapter 5 it was documented that the asbestos-cement slabs, very popular in buildings due to their thermo-isolating properties, are susceptible to emitting fibrous particles generated by impact. Factors such as vibrations of the slabs caused by the turbulence/gust of wind can cause emission of fibers from the elevation of buildings made from asbestos-cement sheets. For this reason, inside buildings with asbestos-cement facades, the mean concentration of respirable fibers longer than 5 μm is higher than in reference dwellings by approximately three times.

Bioaerosols

A specific kind of aerosol is biological aerosol, which is a two-phase system consisting of airborne biological particles. These particles, suspended in the air, may be viruses, bacteria, fungi, or microscopic parts of living organisms. Bioaerosol's particles may be viable or non-viable. Sources, description, and occurrence of biological particles were reviewed by Professor Aino Nevalainen, Dr. Martin Täubel, and Dr. Anne Hyvärinen in Chapter 6 while in the next chapter they discussed the health effects of fungi, bacteria, and other bioparticles.

Most health effects associated with biological particles are various respiratory conditions and skin reactions. Their mechanisms vary from infections, irritation symptoms, and allergic diseases to toxic or immunotoxic reactions and other conditions with less evident pathophysiology.

Examples of common infectious agents are pathogenic bacteria, such as *streptococci, staphylococci, pneumococci,* and *clostridia.* In the case of infectious diseases, the causal agent is usually detectable and the route of infection is most often human to human contact, either directly or indirectly through surfaces or via an airborne route.

Allergies are conditions where the contact with the allergen causes an immunological reaction via the IgE-mediated pathway in the exposed individual, leading to symptoms of varying severity. A prerequisite to allergy is atopy, a genetic tendency towards such a reaction to allergens.

Bacterial endotoxins and glucans of fungi are normal constituents of bacterial and fungal cells and they have immunotoxic potential that probably contributes to the health effects associated with exposures to biological particles and dusts. Another type of toxic compounds linked with bacteria and fungi are microbial toxins that are produced as secondary metabolic products of these organisms.

In contrast to the common opinion that the relationships between the exposure to bioaeroasols and the adverse health effects are rather easy to explain, Prof. Nevalainen, Dr. Täubel, and Dr. Hyvärinen indicated that sometimes, the mechanism responsible for the occurrence of illnesses is still unclear. It is often the case that the epidemiological and clinical evidence of the association between biological agents and the adverse health effect is strong, but the causal link between the exposing agents and the disease is poorly known. Furthermore, the mechanistic pathways leading to a certain condition are not well known.

Role of environmental epidemiology in prognosis of adverse health effects

Epidemiological studies may show a relation between air pollutants and some adverse health effects. The main types of environmental studies are introduced in Chapter 9 by Dr. Kinga Polańska, M.D. and Professor Wojciech Hanke, M.D. In the next chapter, they discussed in detail the epidemiological methods used to receive the information about the influence of air pollution on human health. They also pointed the limitations which can be related to exposure and outcome assessment and control for confounding factors and co-exposures, concluding that animal studies can also be crucial and together with environmental epidemiology, can be useful for the identification of the biological mechanism of the impact of air pollution on health outcomes. In fact, inhalation exposure studies are necessary in order to identify the mechanisms triggered by air pollution and causing these health effects. The experimental nature of the inhalation exposure studies conducted on animals and sometimes on people offers a chance for controlled investigations. A short introduction to the inhalation exposure methodology is presented in Chapter 8.

Adverse health effects as a function of an inhaled dose

When trying to precisely predict the adverse health effect of the specific air pollutant inhaled by the exposed population, it should be clearly stated that the factor responsible for the kind of effect and its intensity is not the concentration (C) but the dose (D) being, as a rule, the mass or amount of absorbed pollutant. On the basis of a number of studies on adverse health effects caused by exposure to air pollutants, it was possible to prepare the dose-effect relationship for specific pollutants. Dose-effect assessment is the quantitative relationship between the magnitude of dose (or exposure) and the occurrence of human health effects. For a number of toxic pollutants, the occurrence of human adverse health effects depends not only on the magnitude of the dose but also on the time of inhalation and on the average body weight of the exposed people.

The health prognosis for the selected human population exposed to carcinogenic pollutants concerns the calculation of the probability of cancer (i.e., "cancer risk") among the people who received the real dose. Hence, instead of the dose-effect relationship the dose-risk relationship is, as a rule, analyzed for the carcinogenic substances. In Chapter 11, the dose-effect and/or dose-risk relationships as the tools for the quantitative description of the adverse health effects of the inhaled pollutants were discussed.

It should be noted that the dose-effect, as well as dose-risk relationship can be obtained from toxicological studies (inhalation experiments conducted on animals), epidemiological studies, and from limited inhalation studies carried out on voluntaries. The book covers all these kinds of studies.

Synergic effects

It was documented that the dose-effect and dose-risk relationships can be very useful for the quantitative description of health implications of the inhaled pollutants. However, there are some exceptions: exposure to airborne particles, exposure to mixtures of airborne pollutants, as well as exposure to bacterial and fungal aerosols. Adverse health effects caused by these pollutants are very difficult for quantitative prognosis by the use of these tools. Additionally, there are allergic effects which generally cannot be predicted using the dose-effect and dose-risk relationships. In Chapter 12, it was noted that in a typical urban environment, the population is exposed to about 200 air pollutants or classes of air pollutants. Therefore, instead of investigating the unique effects of specific pollutants, it has been suggested that it might be more reasonable to assume that it is a mixture of pollutants that might be considered harmful to health. Potential interaction among pollutants seems to be a fundamental problem indispensable for explaining the relations between the exposure of people to air pollutants and their health condition. The analysis of the known synergic effects of the exposure to the selected air pollutants, including the synergy during the absorption of small doses of pollutants, is presented in Chapter 12. It was also noted that some interactions can be found not only between the airborne allergens but also between allergens suspended in the air (aeroallergens) and present in food (food allergens).

Socio-economic factors

Although evidence shows that air pollution is responsible for a significant burden of deaths, hospital admissions, and exacerbation of symptoms, in most of the countries clear differences are being observed in morbidity and mortality rate also due to disorders dependent on socio-economic factors, their criterions being education, income (individual, family, material status), or employment status. Many international studies indicate that socio-economic and socio-environmental factors have a strong association with the prevalence of childhood allergic diseases like asthma, allergic rhinitis, and eczema. In Chapter 13, Dr. Joanna Kobza, M.D. discussed these problems and showed that poor health effects are often made worse by the interaction between individuals and their polluted environment. Inequalities in environmental exposure can reach extreme levels, with disadvantaged population groups often being at least five times more exposed than advantaged groups. Over the past few decades, the socio-economic and environmental risk factors related to such phenomena like climate change, globalization, urbanization, economic, demographic, and health changes had significant influence on the poorest and most vulnerable groups. Finally, it can be concluded that environmental health risk factors are strictly connected with the socio-economic situation of individuals. People with better socio-economic positions have better health, thanks to more healthy behaviors, better access to health care services, and because they live in safer neighborhoods.

Concluding remarks

Summarizing the facts presented in this book, it should be remembered that the prognosis of the adverse health effects resulting from the exposure to air pollutants strongly depends on both chemical and/or biological characteristics of the inhaled pollutants, including their physical properties, and on the absorbed dose. Therefore, precise prediction of these health effects needs not only skills in using the schematic procedures but also deep knowledge about the properties of the pollutants and about their possible interactions. One of the examples of such an interaction is the recently documented fact that exposure to numerous different particles is associated with an elevated risk for infection (bacterial, fungal, viral, and protozoa). Another example is the well-known fact that the cancer risk for smokers exposed to airborne asbestos is roughly ten times higher compared to non-smoking people exposed to asbestos of the same level. Although the synergic effect between exposure to asbestos and tobacco smoke has been well documented, the mechanism of this synergy is still not completely clear. Therefore, the explanation of this phenomenon, like many others synergic mechanisms, should be classified as hypothesis only, or even as speculation.

Hypotheses

Here are some hypotheses of the processes and phenomena described in this book.

1. Analysis of the available data leads to the surprising conclusion that airborne particles seem to be a nonthreshold pollutant (without threshold dose).
2. The mixing state and its associated hygroscopic growth may be important in assessing the deposition of particles in the human airway.
3. All ultrafine particles (nanometric particles) are toxic although it is still unclear why materials which are generally considered inert should suddenly assume toxic properties below a certain particle diameter.
4. The carcinogenetic mechanism of asbestos fibers is still poorly recognized. There are three basic hypotheses concerning:
   a) chemical properties of asbestos (for example, presence of iron or other transition metals on fibers, ability of asbestos fibers to generate free radicals)
   b) carcinogenic compounds absorbed on the fibers (such compounds could be, for example, PAHs)
   c) physical properties of the asbestos fibers (mainly the fiber geometry)
5. A relatively nontoxic substance, together with a known carcinogen, can enormously increase the carcinogenic potency of the known carcinogen.
6. One of the mechanisms for particle-related infections includes an accumulation of iron by surface functional groups of particulate matter.
7. The health effects of exposure to certain components of particulate matter could be modified by the physiological and socioeconomic factors.
8. Allergens occurring in indoor air of flats probably have a leading role in the induction and intensification of allergies of the respiratory system in children.

Unfortunately, the mechanism of synergy between air pollutants is up to now only partially explained. Our scientific experience and available literature indicate that probably no universal synergic mechanism exists. Perhaps an individual explanation is needed for each case in which a common influence of some specific pollutants occurs. Also, the "general mechanism" of synergy between air pollutants and other factors influencing human health is still undiscovered (if exist).

Speculations

Although the role of the speculations in the explanation of the various observations and phenomena is rather weak compared to classical conclusions formulated on the research data, and even hypotheses, speculations should be treated as the first step in the understanding of environmental problems and their health impacts. Unfortunately, very often our strong, well-documented knowledge is poor. On the other hand, we need to better understand why people become ill from polluted air. Therefore, speculations should be included in scientific books but they shouldn't be mixed with strong facts and hypotheses. Some examples of such speculations, contained in this book are summarized below.

Carcinogenic mechanism of asbestos

It is possible that in the case of asbestos, not only one but two or three carcinogenic mechanisms are responsible for the occurrence and development of cancer.

If the assumption is that airborne asbestos particles are only vehicles for other carcinogenic compounds absorbed on the fibers, or/and hypothesis indicating the key role of the physical properties of the asbestos fibers (mainly the fiber geometry) are true and sufficient to initiate the cascade of the processes leading to cancer, other, non-asbestos, respirable and long fibers should also be considered as potential carcinogens.

Role of short fibers in elevation of carcinogenic risk

Despite some papers suggesting that short asbestos fibers could also be carcinogenic, the common opinion about the carcinogenicity of long fibers is based on the fact that due to the activity of the macrophages only parts of long fibers can remain in the lung for a period of time long enough to generate the cascade of processes leading to cancer. However, if the amount of short fibers deposited in the lung quickly increases, the efficiency of biochemical destruction of fibers by macrophages (due to the phagocitosis process) certainly rapidly decreases. Therefore, it may be concluded that some of the short asbestos fibers deposited in the lung will not be totally destroyed and will remain there for a long time. If this picture is true, the cancer risk is additionally elevated because of the contribution of the inhaled short fibers.

Possible modification of carcinogenicity of particle-related PAHs

It is unknown so far if the carcinogenicity of PAHs, always a mixture, never a single compound in the air, may be ascribed to individual hydrocarbons or if it is due to the concerted effects of some number of PAHs. The carcinogenicity of PM-bound PAHs may be enhanced or suppressed by other PM components.

Main task for future studies

It should be noted that the synergy of airborne pollutants inhaled in very small doses, synergy of aeroallergens and food allergens, as well as, synergy between air pollutants and socio-economic factors seem to be especially difficult to analyse for a detailed, quantitative explanation and needs future interdisciplinary studies.

Index